**Variable Frequency Transformers for
Large Scale Power Systems Interconnection**

Variable Frequency Transformers for Large Scale Power Systems Interconnection

Theory and Applications

Gesong Chen
State Grid Corporation
China

Xiaoxin Zhou
Chinese Academy of Sciences
China

Rui Chen
University of Liverpool
United Kingdom

Registered Offices
John Wiley & Sons, Inc., 111 River Street, Hoboken, NJ 07030, USA
John Wiley & Sons Singapore Pte. Ltd, 1 Fusionopolis Walk, #07-01 Solaris South Tower, Singapore 138628

Editorial Office
The Atrium, Southern Gate, Chichester, West Sussex, PO19 8SQ, UK.

For details of our global editorial offices, customer services, and more information about Wiley products visit us at www.wiley.com.

Wiley also publishes its books in a variety of electronic formats and by print-on-demand. Some content that appears in standard print versions of this book may not be available in other formats.

Library of Congress Cataloging-in-Publication Data:

Names: Chen, Gesong, 1968- author. | Zhou, Xiaoxin, author. | Chen, Rui, 1996- author.
Title: Variable frequency transformers for large scale power systems interconnection : theory and applications/ by Gesong Chen, Xiaoxin Zhou, Rui Chen.
Description: Singapore; Hoboken, NJ : John Wiley & Sons, 2018. | Includes bibliographical references and index. |
Identifiers: LCCN 2018009379 (print) | LCCN 2018021193 (ebook) | ISBN 9781119129080 (pdf) | ISBN 9781119129073 (epub) | ISBN 9781119128977 (cloth)
Subjects: LCSH: Electric machinery–Alternating current. | Electric power systems. | Electric transformers.
Classification: LCC TK2744 (ebook) | LCC TK2744 .C44 2018 (print) | DDC 621.31/4–dc23
LC record available at https://lccn.loc.gov/2018009379

Cover design by Wiley
Cover image: © Bubball/iStockphoto

Set in 10/12pt WarnockPro by SPi Global, Chennai, India
Printed in Singapore by C.O.S. Printers Pte Ltd
10 9 8 7 6 5 4 3 2 1

Contents

About the Authors

Gesong Chen: PhD, Professor Senior Engineer, IEEE Senior Member, Deputy Director of General Office of Global Energy Interconnection Development and Cooperation Organization.

His specialization and research interests include VFT, EMTP, FACTs, HVDC, UHV, TCSC, and SSR.

He was awarded 2nd Prize in the National Award for Science and Technology Progress, 1st Prize and two 2nd Prizes in the China Electric Power Science and Technology Progress.

Xiaoxin Zhou: Professor, IEEE Fellow, Academic of the Chinese Academy of Sciences, Honorary President of the China Electric Power Research Institute (CEPRI).

His specialization and research interests include Power System Analysis, Models and Simulations, Power System Stability Analysis and Control, and Power System Security Analysis.

He was awarded one 1st Prize and two 2nd Prizes in the National Award for Science and Technology Progress, and the IEEE PES Nari Hingorani FACTS Award.

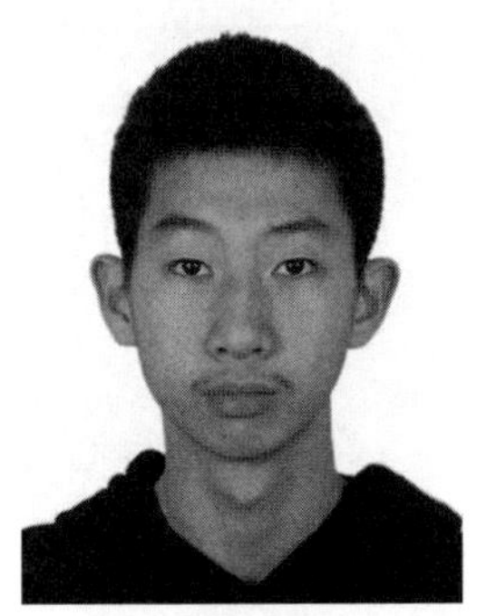

Rui Chen: EE student of Liverpool University.

His specialization and research interests include, Power System Analysis, Power and Electronics, VFT, Artificial Intelligence, and Robotics.

He was awarded 1st Prize in the National Robot Competition Award for students.

About the Authors

Preface to the English Version

In modern society, electricity has a profound impact on the development of human livelihood and production, as well as economic and social development. All equipment and devices, from the locomotive driving devices to mobile phones, from electric lights to factory machinery, are required to be powered by electricity. It's the responsibility of us as electrical engineers to ensure a safe and reliable power supply and make electricity available everywhere so that everyone can enjoy sustainable and affordable power supplies. To ensure power supply is not only a technical issue, but is also an economic one. In addition, it has a close relationship with resources and the environment, and has to rely on scientific and technological innovation. From Faraday's discovery of electromagnetic induction to Edison's invention of electric light, technological innovation has made the history of power industry and promoted the innovation and development of the world.

Power grid is the basic platform for ensuring power supply. With the growth in demand for electricity and advances in electric power technology, the world's power grids have evolved from initial single-machine and single-line systems to small town grids, urban grids, provincial grids, national grids, and continental grids. With increasing grid scale and strengthened grid functions, large-scale grid interconnection has become the trend. Innovation is the most important driving force for the development of power grids. Thanks to the development of high voltage and ultra-high voltage transmission technology, and the application of AC and DC transmission technology, power grids are more closely interconnected with each other and the power supply is becoming far more safe and reliable.

The variable frequency transformer (VFT) is a new type of grid interconnection device. By integrating technologies of power transformer, phase shifter, hydro generator, doubly fed generator, DC drive, and other technologies, and with functions such as smooth adjustment of transmission power, suppression of low frequency oscillation, and adjusting system frequency function, it can realize the interconnection of asynchronous power grids, supplying power to weak power systems and passive systems, and providing black-start power. VFT is an important achievement of power technology innovation in the new century. It provides a new option for the interconnection of large power grids, has been widely applied in North America, and has achieved good results.

As an engineer who has been long engaged in research on new power technologies, I specialize in power system simulation, overvoltage and insulation coordination, flexible

AC transmission technology, UHV power grids, smart grids, Global Energy Interconnection, new energy sources, and technical standards. I have undertaken major research projects such as China's first 500 kV compact line, the first thyristor controlled series compensation project, and the first UHV AC and UHV DC transmission project. VFT is the main research content of my doctoral dissertation. Learning from the VFT-related articles published by GE and based on years of research and practice in power systems, I have profoundly studied the work principle, mathematical models, simulation methods, and control systems of VFTs and solved problems of digital simulation analysis tools, operation control strategy, and large system low-frequency oscillation in the VFT. In addition, I have analyzed and verified the operational characteristics of the VFT in power systems and its good performance. Through in-depth study, I firmly believe that VFT will be a promising development and have prospective application in the power system as a new interconnection technology.

Funded by the electrical power scientific work publication program of the State Grid Corporation of China (hereinafter referred to as SGCC), the Chinese version of *Large Electric Power System Grid Connection New Technology – Theory and Application of Variable Frequency Transformers* was finally published and released by the China Electric Power Press in 2013, receiving positive responses from the industry. In 2014, Wiley organized experts to study the book and decided to publish the book in English in order to promote the application of this new technology to the world. The publication and release of this book as the world's first English monograph on the VFT reflects the forward-looking awareness and innovation spirit of Wiley. In order to enrich the research content of variable frequency transformer and respond to Wiley's requirements, the author has added 40% extra content to the book based on the new development of UHV power grid, smart grid, and Global Energy Interconnection. I would say that the English monograph is more than just a translation of the Chinese version, as it has richer content.

Through the joint efforts of all parties, the English version of *Variable Frequency Transformers for Large Scale Power Systems Interconnection: Theory and Applications* has finally been published and released. As the world's first monograph that systematically studies and introduces VFT, it has won the concern and support of a lot of experts and friends. Wiley, the China Electric Power Press, and Mr. R.J. Piwko from GE, Mr. Wang Qiankun from GEIDCO, Mr. He Wei from China Jibei Electric Power Company LTD. have given me a lot of support and help. I would like to take this opportunity to express my heartfelt thanks to all my peers and friends who have supported and shown concern for the publication and release of the book. I hope this book can play a positive role in spreading some basic knowledge about VFT and promoting the development of the world's power technology.

Despite repeated checks and modifications, it is inevitable that this book still has some inadequacies. We sincerely invite readers to criticize and correct them.

May 18, 2018   *Gesong Chen*

Preface

Electricity is an important substance in modern society. The rapid growth in electricity demand advances higher requirements for power grid development. The essence of ensuring sustainable electricity supply is scientific and technological innovation. By reviewing the world's power grid development in the past 100 years, we find it is a history of constant innovation, reform, and improvement. Research shows that the interconnection of large power grids can effectively improve the reliability of power supply, reduce spare capacity, advance economic operation of power system, and further achieve great benefits such as peak load shifting, hydro and thermal power mutual support, as well as inter-basin complementation, which is also the trend for the world's power grid development. The interconnection of traditional power grids includes synchronous AC and asynchronous DC. VFT is a new type of asynchronous grid interconnection device. Compared with back-to-back DC, it has obvious advantages such as less land occupation, lower transmission loss, and no harmonics. The world's first VFT, with a capacity of 105 MVA, was developed by GE and put into operation at the Langlois Substation, Quebec, Canada in October 2003. Years of practical operation proves the excellent performance of VFTs and effectively promotes their popularity and application. By 2010, a total of seven sets of VFTs had been put into commercial operation in North America.

Due to China's extremely uneven distribution of energy resources and energy demands, it is required that the power grid should be strengthened by interconnection to effectively improve the ability for optimal allocation of energy resources over large areas to ensure a safe, economical, good quality, green, and sustainable power supply. Since 1949, China's power grids have experienced rapid development from urban to provincial, regional, and inter-regional interconnections. By the end of 2011, with the integration of the Tibet power grid into the Northwest Power Grid through the Qinghai-Tibet DC interconnection project, all China's power grids were interconnected (except for those in Taiwan). Currently, SGCC is speeding up the construction of UHV AC/DC hybrid power grids. Up until 2013, China formed five regional synchronous power grids: North China-Central China, East China, Northeast China, Northwest China, and South China. Asynchronous interconnection of these power grids was achieved through HVDC and UHVDC.

The proposal and development of VFT technology is the progress of the world's power grid interconnection technology and provides a new choice for China's power grid interconnection. On the one hand, in asynchronous power grid interconnection, with its unique advantages, the VFT is more suitable for marginal interconnection between asynchronous networks. Particularly, it can help solve problems of weak or isolated power grids as well as wind-farm integration. On the other hand, with the higher voltage level of an AC power grid, an electromagnetic loop caused by the various voltage transmission lines' parallel operation has become increasingly prominent, which leads to large-scale power flow transfer and affects voltage stability. However, simply disconnecting low voltage AC circuits will bring about wasteful investment, power supply stability problems, decline in economic efficiency, and other problems. On the base of ensuring a power grid's safety, a VFT can achieve controllable interconnection and power exchange between adjacent power grids by utilizing existing AC power transmission lines to improve reliability and economic efficiency of the power supply.

Research into VFT technology is quite necessary and challenging. In April 2004, GE's expert Dr. Hamid Elahi came to the China Electric Power Research Institute (CEPRI) for a technical exchange and displayed the company's newly invented VFT technology for the first time. In order to learn this new technology, I encouraged and guided doctoral student Gesong Chen to take the VFT as a research direction in preparing his dissertation and conduct systematic research on it. In the following 6 years, with the support and help of relevant experts and peers, he overcame practical difficulties such as lack of data and no ready-made simulation tools. Through in-depth theoretical research and simulation, he gradually mastered the basic theory of VFT; derived VFT calculation models covering steady state, electromechanical transients, and electromagnetic transients of electric power systems; acquired the ability to use mature power system software packages such as PSASP, EMTPE, and PSCAD/EMTDC to carry out all-digital simulations of VFTs; set up VFT control systems applicable to different applications; and conducted plenty of simulation research by using simplified power systems, typical multi-machine systems, and actual large power grid systems, based on which, the dissertation entitled *Research on Modeling, Simulation and System Control of the Variable Frequency Transformer* was completed in 2010, and also constitutes the basic content of this book.

At present, China is speeding up the construction of information-based, automated, and interactive smart power grids with UHV as the backbone network and characterized by the coordinated development of power grids at all voltage levels. Various new technologies, devices, and materials will be more widely and effectively used in smart grids. As a new power grid interconnection technology, the VFT shows promising development and application prospects in China. The systematic introduction of the technical features, research methods, and simulation tools of the VFT is of great significance to promoting its development and application and advancing power grid technical levels in China. For this purpose, based on previous research as well as the basic trends in current energy reform, electric power development, and modern grid technology innovations, we have further enriched and improved VFT theories and application research, reinforced its theorization and readability, and revealed the general law of modern power grid development, thereby producing this book for your reference.

As China's first professional book focusing on systematic research of VFTs, this book has integrated research results accomplished over 8 years and received the support from experts and friends in many fields. In the research process, Mr. R.J. Piwko, Mr. Einar Larsen, and Dr. Hamid Elahi from GE; Professor Yuan Rongxiang from Wuhan University; Engineer Chen Ying from the Sichuan Electric Power Company Dispatching Center; Dr. Song Ruihua from CEPRI; and Dr. Wu Xuan, director of the research office, SGCC; as well as family members of the authors offered great support and help. The research results in the reference documents cited in this book also laid a good foundation for our VFT research and the drafting of this book.

In 2012, this book won the financial support of the electric power scientific works publication program of the SGCC. Through the joint efforts of all parties, *Large Electric Power System Grid Interconnection New Technology – Theory and Application of Variable Frequency Transformers* was finally published and released. I would like to take this opportunity to extend our heartfelt thanks to all peers and friends for their support and hope this book will play a positive role in spreading the basic knowledge about variable frequency transformers and promoting China's electrical power technology development.

Despite repeated checks and modifications, it is inevitable that this book still has some inadequacies. We sincerely invite readers to criticize and correct them.

June 18, 2013   *Xiaoxin Zhou*

1

Power Grid Development and Interconnection

1.1 Overview

The power grid is an important infrastructure needed for the development of national economies and electrical power transmission, and it is an important platform for the optimal allocation of energy resources. In particular, with the accelerated development of clean energy and the promotion and application of smart grid technology, the function of power grids has become increasingly prominent. The world's primary energy resources and energy consumption are generally reversely distributed where energy production is far away from power load centers. In order to meet the growing electricity demand needed for economic and social development, it has become more and more urgent to strengthen the interconnection of power grids, and achieve flexible control of large power grids and their capability to optimize the allocation of energy resources. For this purpose, the major countries and regions in the world are accelerating the development of large power grids, and some have realized cross-border power grid interconnection. Existing large interconnected power grids in the world include those in North America, Europe, and the Russia-Baltic Sea. Countries in the Gulf, South America, Southern Africa, and other regions are also developing cross-border interconnected power grids. With rapidly growing energy and electricity consumption, China is under more pressure from energy and electricity development. Therefore, China has made great effort to develop and deploy Ultra High Voltage (UHV) transmission technology for AC 1000 kV, DC $\pm$ 800 kV, and DC $\pm$ 1100 kV, focusing on constructing strong and smart power grids with UHV as the backbone network and characterized by coordinated development of power grids at all voltage levels. Grids of this kind are expected to form a new energy development pattern by connecting large energy bases and major load centers with power grids all over China closely interconnected with the "West-East and North-South power flow" [1–3]. Both for China and the rest of the world, there is an urgent requirement to accelerate the development and application of power grid interconnection technology. As a new power grid interconnection device and piece of flexible AC transmission system equipment, the variable frequency transformer (VFT) will play a positive role in the future power grid development in China and in the world.

This chapter introduces the prominent problems in current as well as future trends for world energy development. World energy development is divided into three important stages; namely, the first generation, the second generation, and the third generation of power grids. Based on China's power grid development, the structure, functions, and technical features of power grids in each stage have been compared comprehensively

Variable Frequency Transformers for Large Scale Power Systems Interconnection: Theory and Applications,
First Edition. Gesong Chen, Xiaoxin Zhou, and Rui Chen.

[4]. The trends for UHV grids, smart grids, and clean energy are discussed in detail. In addition, the important functions of large power grids interconnection and four different modes of grid interconnection method are analyzed and compared: these are synchronous AC, asynchronous DC, AC/DC parallel, and VFT asynchronous [5]. Finally, based on the development of the electrical power systems and the research results from VFTs, the main content of each chapter in this book will be briefly introduced.

1.2 Energy Reform and the Third Generation of Power Grids

1.2.1 Objectives of Energy Reform and the Mission of Power Grid Development

Energy is an important material base for economic and social development. Since 1965, the world's energy demand has been increasing rapidly. The world's primary energy consumption has grown from 5.4 billion tons of standard coal in 1965 to 18.7 billion tons of standard coal in 2015; nearly a 2.5-fold increase. As the world's second largest economy, China has witnessed more rapid energy consumption growth in recent decades. China's primary energy consumption has increased from 2.6 billion tons of standard coal in 2005 to 4.3 billion tons of standard coal in 2015: up 70% in the last 10 years. Figure 1.1 shows global energy consumption in the past 50 years. Figure 1.2 shows China's energy consumption in the past 35 years.

Global energy development has gone through a course of evolution from firewood to coal and further to oil and gas. At present, over 80% of the world's energy consumption relies on coal, oil, natural gas, and other fossil energy resources. Figure 1.3 shows the trend of energy consumption structure through history. Figure 1.4 compares the world's energy consumption structure in 1980 and 2015. Figure 1.5 shows the world's primary energy consumption structure by region in 2015.

The imminent exhaustion of fossil energy resources and the pollution and emission caused by the use of fossil resources places tremendous pressure on energy supply and ecological environment. The predicted shortage and exhaustion of fossil energy resources, current threats of global climate change and environmental pollution

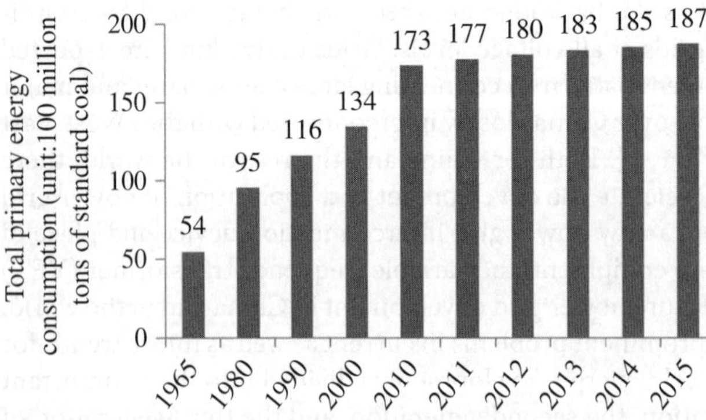

Figure 1.1 World primary energy consumption growth. Data source: BP, Statistical Review of World Energy 2016.

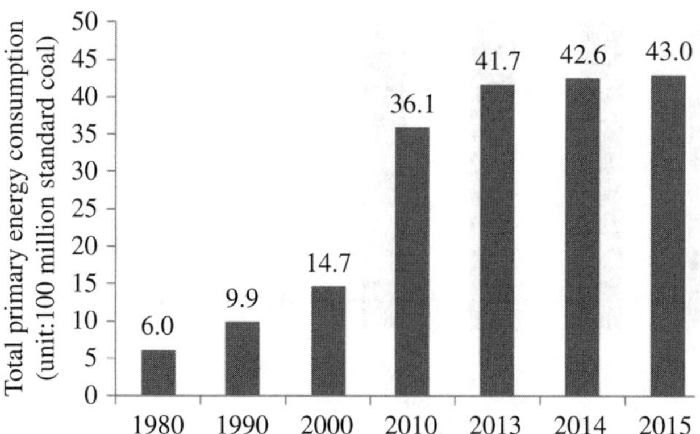

Figure 1.2 China's primary energy consumption growth. Data source: National Bureau of Statistics of China, *China Statistical Yearbook 2016*.

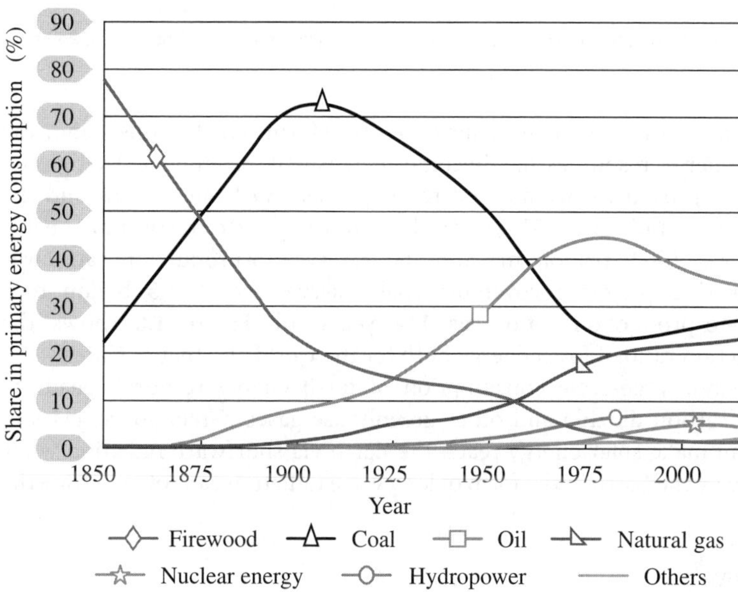

Figure 1.3 Changes in the composition of world energy consumption [36].

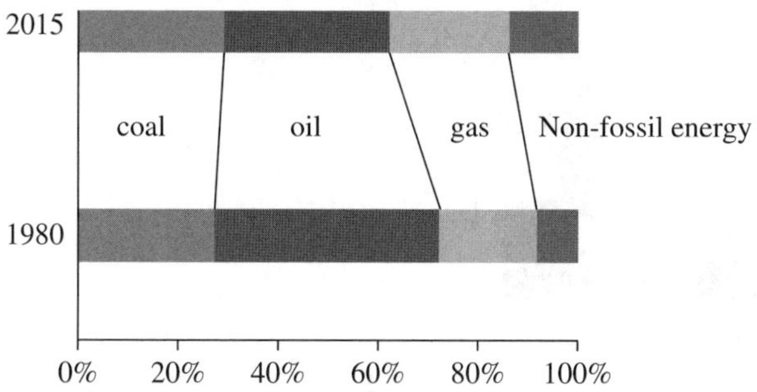

Figure 1.4 Structure of global primary energy consumption in 1980 and 2015. Data source: IEA energy balances of non-OECD countries 2016.

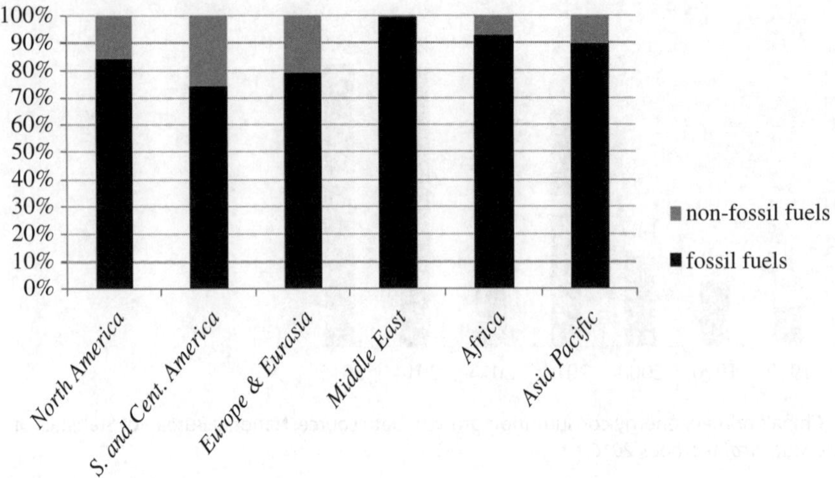

Figure 1.5 2015 Primary energy consumption structure by region. Data source: BP, Statistical Review of World Energy 2016.

have become the main driving force of new energy reform. On the one hand, the shortage of fossil energy resources has become increasingly prominent. By the end of 2015, the world's proved recoverable oil reserves was 239.4 billion tons and the reserves-to-production ratio was 51 years; the world's proved recoverable natural gas reserves was 186.9 trillion m^3 and the reserves-to-production ratio was 52.8 years; the world's proved recoverable coal reserves was 891.5 billion tons and the reserves-to-production ratio was 114 years [6]. Figure 1.6 shows the reserves-to-production ratio of fossil energy both for the world and that of China.

On the other hand, large-scale consumption of fossil energy resources leads to high emissions of carbon dioxide and other greenhouse gases. Greenhouse gases in the atmosphere can make solar energy reach the Earth via shortwave radiation while the energy radiated by the Earth in the form of long wave radiation cannot penetrate the

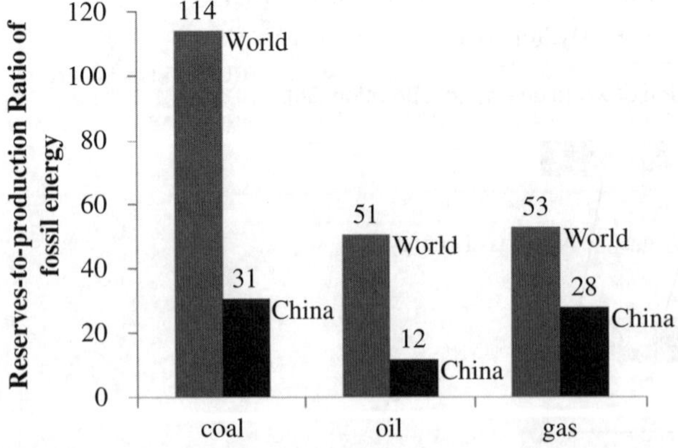

Figure 1.6 Reserves-to-production ratio of fossil energy. Data source: BP, Statistical Review of World Energy 2016.

greenhouse gas layer, resulting in the greenhouse effect that causes global warming and other abnormal changes in global climate. Research shows [7] that 60% of greenhouse gas emissions are from energy consumption. At the same time, if the average temperature rises over 2°C compared to that of pre-industrial times, this will have a disastrous impact on the Earth. To prevent the atmospheric temperature from rising by no more than 2°C, the corresponding atmospheric carbon dioxide concentration cannot exceed 450 ppm (1 ppm = 1×10^{-6}). To solve the problem of greenhouse gas emissions, the key is to optimize the energy structure and control the use of fossil energy resources. Figure 1.7 shows the amount of global carbon dioxide emissions from 1980 to 2014. Figure 1.8 shows the evolution of atmospheric carbon dioxide concentration.

Coordinated development of energy, ecology, environment, and economy is a major issue related to sustainable development of human society. Since the 1990s, the call for use of clean energy has become increasingly louder and its purpose is to gradually replace fossil energy with new energy sources (such as nuclear power) and renewable energy (including hydropower, biomass energy, solar energy, wind energy, geothermal

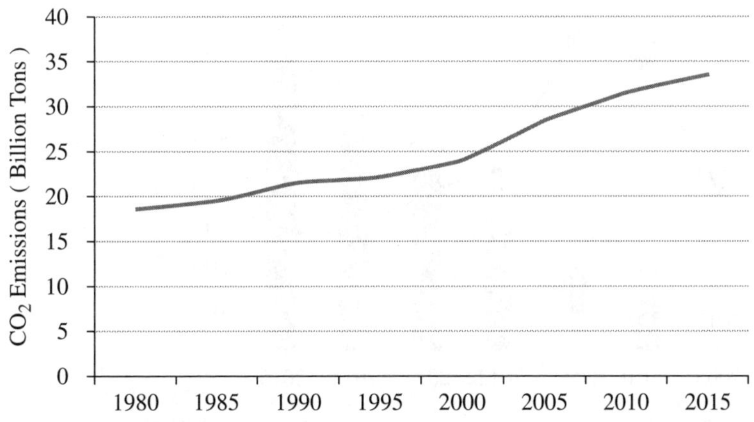

Figure 1.7 Total CO$_2$ emissions (1980–2014). Data source: BP, Statistical Review of World Energy 2016.

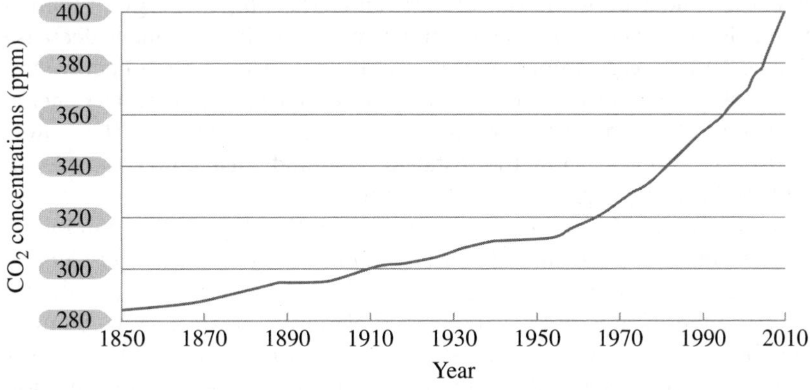

Figure 1.8 Atmospheric carbon dioxide concentrations [36]. Source: www.globalcarbonproject.org/activities/AcceleratingAtmosphericCO2.htm.

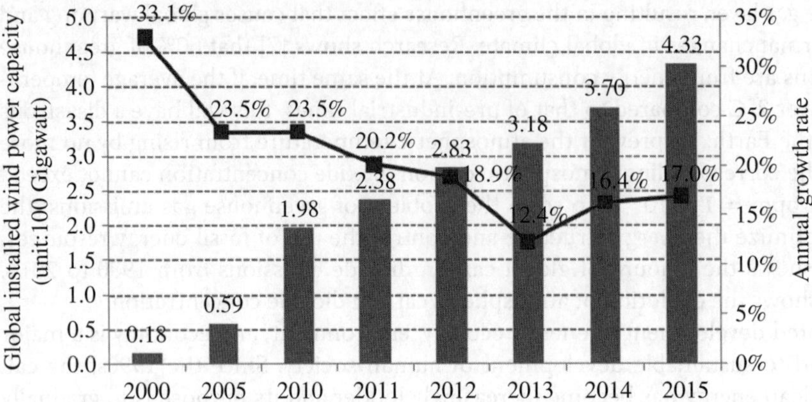

Figure 1.9 Global installed wind power capacity and growth rate. Data source: GWEC global wind report 2016.

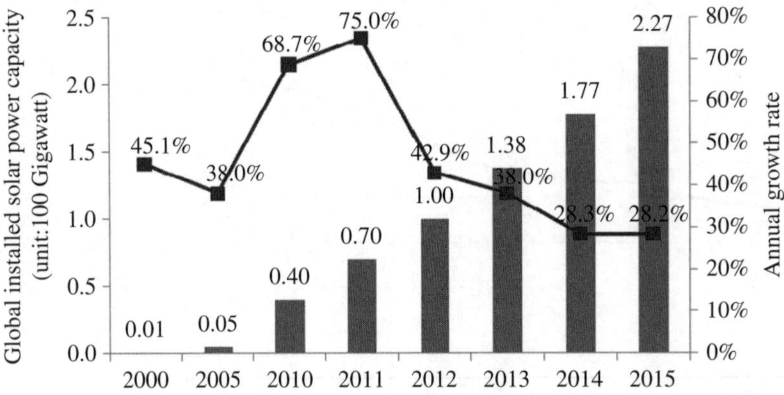

Figure 1.10 Global installed solar power capacity and growth rate. Data source: REN21 renewables 2016 global status report.

energy, ocean energy, and hydrogen) to ensure a sustainable supply of energy for human beings. This trend is called *new energy reform* and its main goal is to gradually decrease the proportion of fossil energy, improve clean and efficient use of fossil energy, and make renewable energy and nuclear energy account for a larger share in primary energy consumption. Figures 1.9 and 1.10 show the global installed wind and solar power capacity. Figures 1.11 and 1.12 show the installed wind and solar power capacity in China. Figure 1.13 shows per capita electricity consumption.

At the United Nations Conference on Environment and Development held in Rio de Janeiro, Brazil, in June 1992, the *United Nations Framework Convention on Climate Change* was formulated; in December 1997, the *Kyoto Protocol* on reducing greenhouse gas emissions was adopted at Kyoto. At the United Nations Climate Change Summit held in September 2009, the Chinese government promised that by 2020 China's non-fossil energy consumption would account for 15% in primary energy consumption and energy consumption per unit of GDP would drop by 40–45%. Vigorously developing clean energy is an effective way to solve the world's energy security and climate change

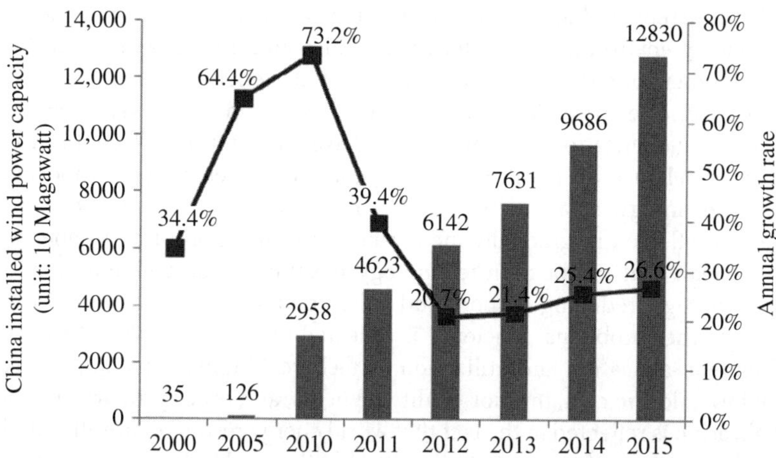

Figure 1.11 China's wind power capacity and growth rate. Data source: China Electricity Council.

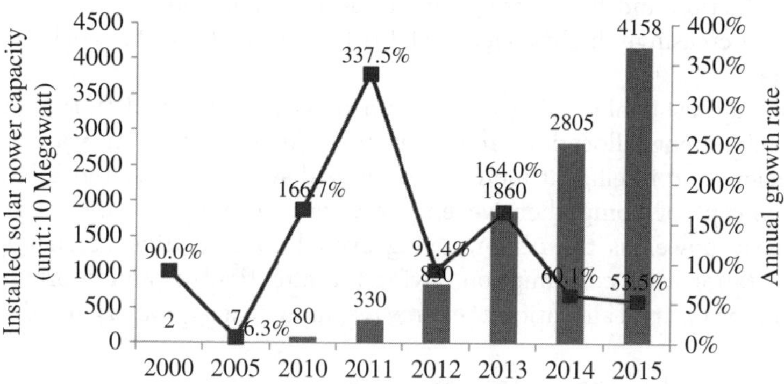

Figure 1.12 China's solar power capacity and growth rate. Data source: China Electricity Council.

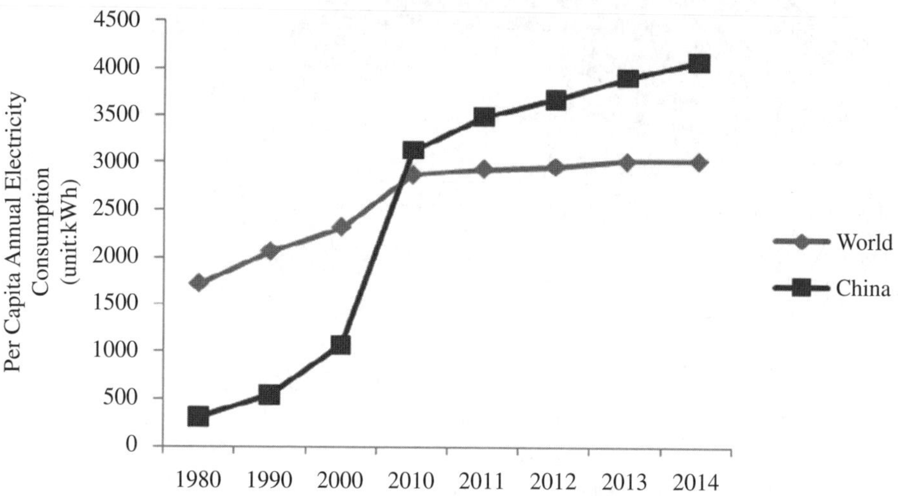

Figure 1.13 Comparison of per capita electricity consumption. Data source: China Electricity Council.

problems and play an irreplaceable role in meeting energy demand, improving energy structure, and reducing environmental pollution. It is also an important direction for world energy development and the main characteristic of new energy reform.

At present, the world energy development shows a trend for diversification, cleanliness, efficiency, globalization, and marketing. Diversification is reflected in the change from an over-reliance on fossil energy to vigorously developing hydropower, nuclear energy, wind energy, solar energy, and biomass energy in energy development. Cleanliness is reflected in gradually increasing the proportion of clean energy, actively deploying clean coal combustion technology, desulfurization technology, and denitrification technology, reducing carbon, and sulfur emissions to deal with haze, climate change, and other problems. Efficiency is reflected in improving the efficiency of energy development, allocation and utilization by technical innovations, improving the utilization of low calorific coal, inferior quality oil fields, and other primary energy resources. Globalization is reflected in the fact that world energy resources are allocated to an increasingly larger range of areas and transmitted farther and farther, representing an important part of economic globalization. Marketing is reflected in the regulation of global energy production and flow through price mechanisms, and world energy marketing levels are increasingly higher. Figures 1.14–1.16 show the global coal, oil, and natural gas trade.

The key to achieving sustainable energy development is to build a platform for efficient energy development, allocation, and utilization. An important measure is to construct an efficient and intelligent modern power grid system. In Reference [1], the development concept of comprehensive energy view is proposed, namely the view of taking electric power as the center, giving overall consideration to energy production, transportation, and consumption, giving full attention to the important role of power grids in the optimal allocation of energy resources in a large areas in order

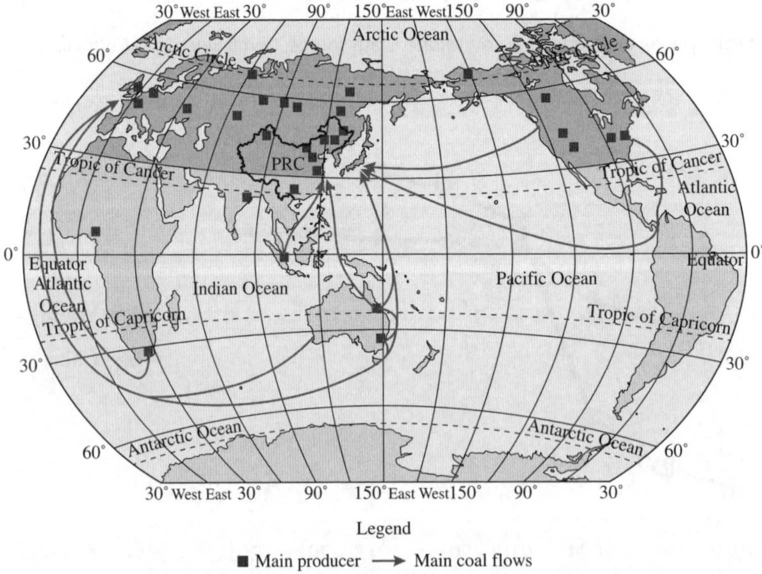

Figure 1.14 Global coal trade. Source: BP, Statistical Review of World Energy 2016.

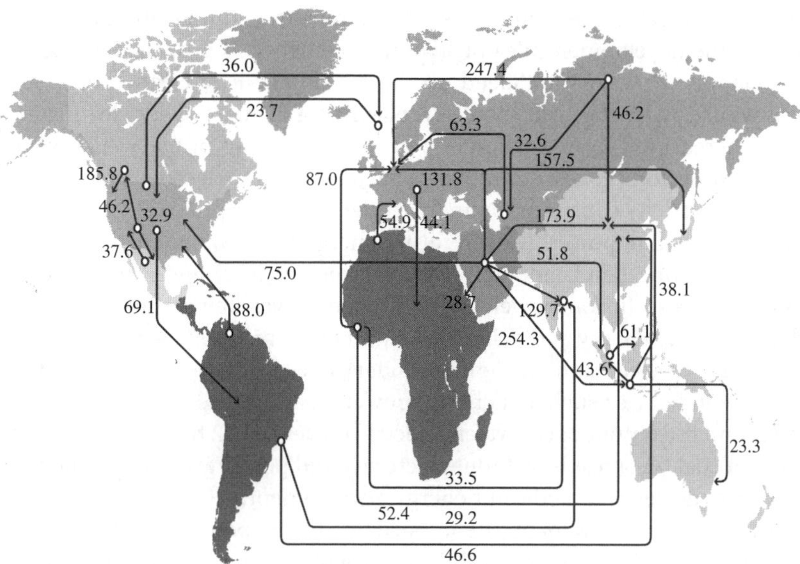

Figure 1.15 Global oil trade 2015 (million tonnes). Picture source: BP, Statistical Review of World Energy 2016.

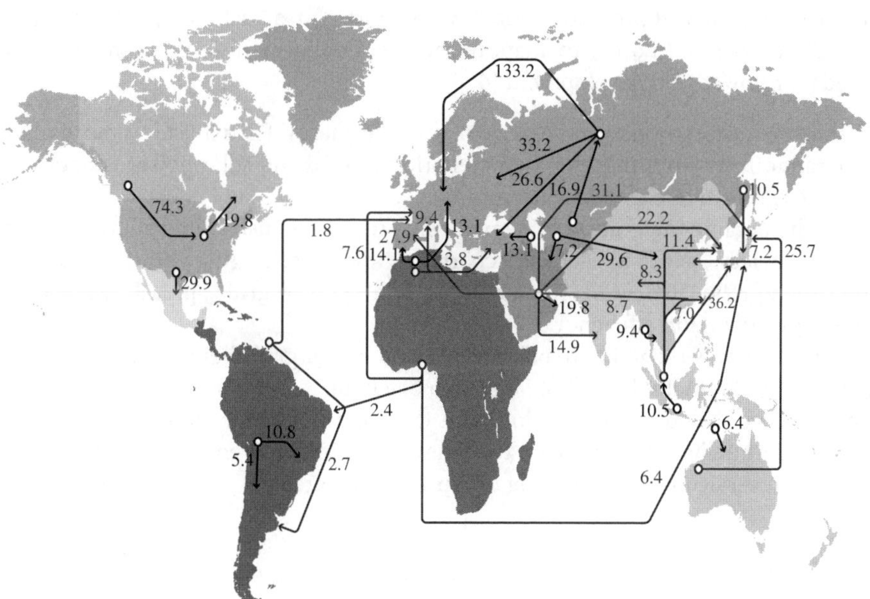

Figure 1.16 Global natural gas trade in 2015 (billion cubic meters). Picture source: BP, Statistical Review of World Energy 2016.

to ensure safe, economical, efficient and green development, and utilization of energy. In Reference [37], the development idea of "two replacements" is proposed, namely in the energy development process, to speed up the development of clean energy such as solar power, wind power, and hydropower, replacing fossil energy with clean energy alternatives and increasing the proportion of non-fossil energy in primary energy; and in the energy consumption process, to replace coal and oil with electricity to increase the proportion of electricity consumption in final energy consumption.

The role of electricity and power grids in future energy development is obvious and irreplaceable. On the one hand, not only hydropower, nuclear power, wind power, and solar power, but also fusion power and hydrogen power being studied must be transformed into electric energy and delivered to users to achieve efficient application. On the other hand, as a clean and efficient secondary energy, electricity can meet the demands for many industries such as lighting, power, heating, and transportation. In particular, the economic output of equivalent electric power is 3.22 times that of oil and 17.27 times that of coal. As a result, without electricity and power grids, it is difficult to achieve the diversified, clean, and efficient energy development.

In the new round of energy reform, the importance of the power grid has become increasingly prominent. China's successful development and application of UHV AC and DC transmission technology [2] as well as the maturation of large power grid security and stability control technology has laid a good foundation for large-capacity, long-distance, low-loss transmission, and large-scale optimal allocation of electric power. Meanwhile, it will provide an effective platform for distributed generation and micro-grid integration. In order to adapt to the energy reform, new requirements have been made for power grid development [4, 8].

1. The ability to accommodate large-scale renewable energy. Particularly, it must adapt to the random and intermittent characteristics of wind and solar power generation, and increase clean energy integration capacity.
2. Realize the organic integration of demand side response, distributed source, energy storage devices, electric vehicle charging and discharging power stations, energy comprehensive utilization systems with power grids, greatly improving end energy utilization efficiency.
3. High power supply reliability to effectively avoid power grid chain reaction accident and basically eliminate the risk of large area blackout.
4. Extensive and in-depth combination with the communication information system and modern intelligent technology to achieve an integrated energy, electricity, and information service system covering urban and rural areas.

1.2.2 Development and Upgrading of Power Grids

In 1831, the British physicist Michael Faraday discovered the law of electromagnetic induction. On this basis, the original AC generator, DC generator, and DC motor soon emerged. Early motor manufacturing and power transmission technologies were mainly for DC and power lines were 100–400 V low-voltage DC lines characterized by short transmission distances and small capacities. Then, with the application of alternating current and continuously improved grid voltage levels and through more than 100 years of development, power grid technology innovations, scale expansion, and function improvement have been constantly achieved. The world's power grid development can generally be divided into three stages; the first, second, and third generations of power grids [4].

The first generation of power grids are referred to as AC-dominated power grids that were developed from the end of the nineteenth century to the end of the Second World War. Transmission voltage reached 220 kV and stand-alone capacity was no more than 100–200 MW. Major milestone events: in 1882, Edison built the world's first commercial power plant (using the 1.6 km 660 kW, 110V DC cable for power transmission) in New York; between 1885 and 1886, Westinghouse built the first AC power transmission system, and built the 35 km AC transmission line from Niagara Falls (three 3675 kW hydropower units) to Buffalo in 1895. From then on, AC transmission has dominated the world's power grids. In 1916, the United States built the first 132 kV AC transmission line, built and began to use the 239 kV AC transmission line in 1923, and built the 287 kV AC transmission line in 1937. In 1918, they made the first 60 MW turbine generator and, in 1929, the first 200 MW turbo generator. In 1932, the Soviet Union built the Dnieper Hydropower Station with a unit capacity of 62 MW. In 1934, the United States built the Grand Coulee hydropower station with a unit capacity of 108 MW and in 1935, they built the Hoover Hydropower Station with a unit capacity of 82.5 MW.

Since the 1950s, with large hydropower development, developed countries in Europe and America have begun to develop large-capacity units, EHV, and large interconnected power grids that characterize the second generation. Major milestone events include: in 1952, Sweden built the first 380 kV EHV AC transmission line with a total length of 620 km and transmission power of 450 MW. In 1954, the United States built the 345 kV AC transmission line. In 1956, the Soviet Union put into operation the Kuibyshev-Moscow 400 kV AC transmission line with a total length of 1000 km. In 1959, the voltage was increased to 500 kV and it was the first time 500 kV AC transmission line was used. In 1965, Canada built the first 735 kV AC transmission line. In 1969, the United States achieved 765 kV EHV transmission. In 1985, the Soviet Union built the 1150 kV UHV transmission line that underwent industrial trial operation for about 5 years and later on adopted reduced-voltage operation due to dissolution of the Soviet Union. By the end of the twentieth century, all developed countries had built a second generation of power grids characterized by EHV AC and DC power transmission and large interconnected power grids. Major equipment and hardware technology in the second generation of power grids include: supercritical and ultra-supercritical coal-fired units (600 MW and 1000 MW), 1000 MW nuclear power generating units, 700–800 MW hydropower units; EHV transmission and transformation equipment as well as line technology; high-speed relay protection and automatic safety devices; and optical fiber communication. The main system technologies involved include: overvoltage, insulation coordination, reactive power balance, secondary arc current, electromagnetic interference, and other EHV transmission system technologies; low-frequency oscillation of interconnected power grids/dynamic stability, transient stability, voltage stability, power system stability control, SCADA/EMS dispatching automation and other large power grid technologies; power system analysis and simulation technology such as detailed dynamic modeling, electromechanical/electromagnetic transient analysis, and reliability analysis.

For the third generation of power grids, in the 1990s, developed countries began to study distributed power generation, renewable energy, microgrids, and the power system market. Technologies such as the application of power electronic devices in electric power systems, high-speed optical fiber communication, 1000 kV UHV AC transmission, ±800 kV (and above) UHV DC transmissions, and smart grid equipment were studied. The development and construction of the third generation power grids is characterized by higher voltage level, large-scale utilization of renewable energy, and high intelligence techniques. The third generation power grids involve innovative change in five major areas:

1. New materials and new elements and devices. These form the basis of the development of modern power grids providing efficient, energy-saving, and environment-friendly equipment.
2. New power transmission technology. Large capacity, low loss, environment-friendly transmission technology (UHV transmission, superconducting transmission, underground cable transmission, etc.) are the main goals of R&D.
3. Large-scale renewable energy (centralized and distributed) integration is the key technology of modern power grids.
4. Smart dispatching and operation control based on advanced sensing, communication, control, computation, and simulation technology ensures the construction of efficient and reliable power grids.
5. A smart power distribution and consumption system helps realize demand side response as well as the integration of distributed sources, electric vehicles, and energy storage devices, forming a comprehensive energy, electric power, and information service system covering urban and rural areas.

China began to construct first generation of power grids in 1882 (Shanghai Electric Power Company). In 1949, China's capacity of power generation was 1.85 GW with an annual electricity generation level of 4.31 billion kWh. In 1966, China's generation capacity reached 17.02 GW with 82.5 billion kWh of electricity generation and the maximum voltage level of the power grids was 220 kV. China began to build the second generation of power grids in the 1970s, 20 years later than developed countries. In 1971, the Liujiaxia Hydropower Station and Liujiaxia-Guanzhong 330 kV AC transmission line (535 km; 420 MW transmission capacity) were completed, forming China's first inter-provincial power grid (Gansu, Shaanxi, Qinghai). In 1981, China built the first 500 kV AC transmission line (Pingdingshan-Wuhan), heralding the start of construction of large regional power grids with 500 kV as the backbone. Between the end of the twentieth and beginning of the twenty-first century, the completion of the Three Gorges Power Transmission Project accelerated national power grid interconnection. In 2005, the Northwest Power Grid 750 kV AC transmission line was put into operation. In January 2009, China's Southeastern Shanxi-Nanyang-Jingmen 1000 kV UHV AC test and demonstration project was completed and put into operation within just 28 months. It was the first commercial UHV AC transmission project in the world (see Figure 1.17). In July 2010, China's Xiangjiaba-Shanghai ± 800kV UHV DC demonstration project was put into operation within just 30 months. It is 1891 km long and has a rated transmission power of 6.4 GW. In Dec. 2012, China's Jinping-Sunan ±800 kV 7.2 GW UHV DC project (see Figure 1.18) was put into operation.

Since the founding of new China, China's power grid has developed into one of the world's largest interconnected power grids and has become the most important energy transmission and distribution network in the country. Between the 1970s and

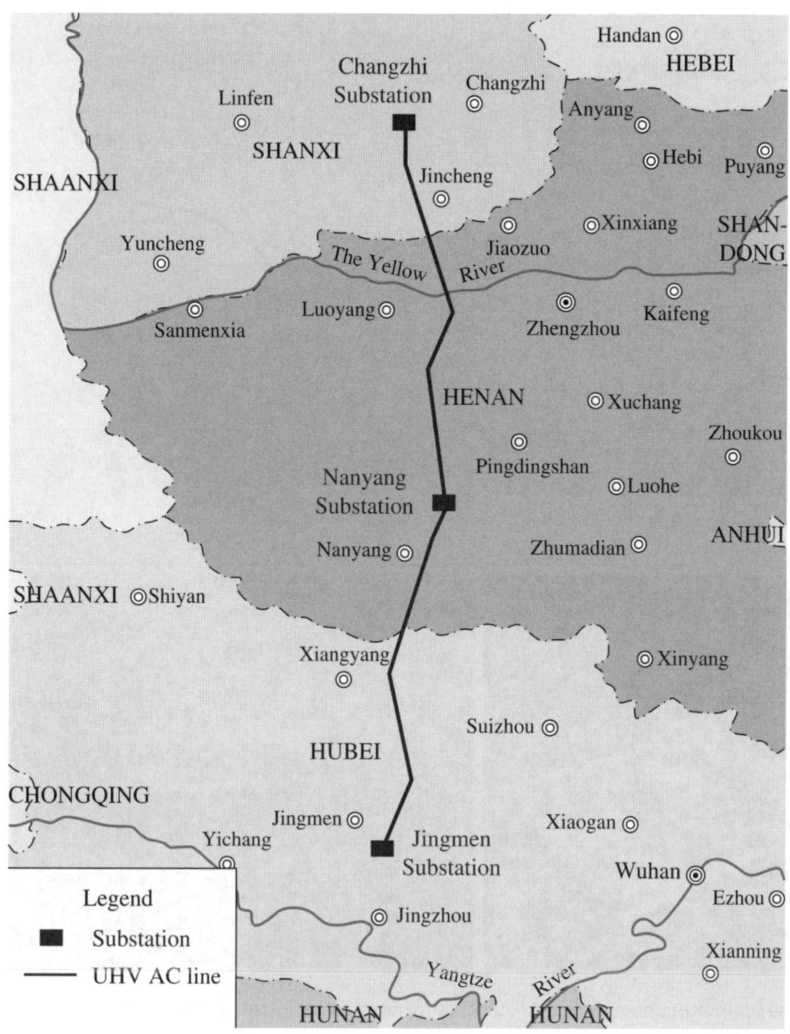

Figure 1.17 Shanxi-Nanyang-Jingmen 1000 kV UHV AC transmission line. Picture source: SGCC.

the beginning of the twenty-first century, based on the construction of the 330, 500, and 750 kV power transmission systems as well as the practice of large regional power grids and national grid interconnection, China has achieved rapid development and progress in technology (such as equipment, power system, control, protection, safe and stable operation, and simulation analysis) and comprehensively grasped the technology of second generation power grids, reaching the international advanced level. Since the beginning of the twenty-first century, with the development and construction of the UHV AC and DC transmission and smart grids, China has started the construction of the third generation power grids and gradually joined the top ranks of the world, catching up with developed countries. Table 1.1 compares the main characteristics of the first, second, and third generations of power grids.

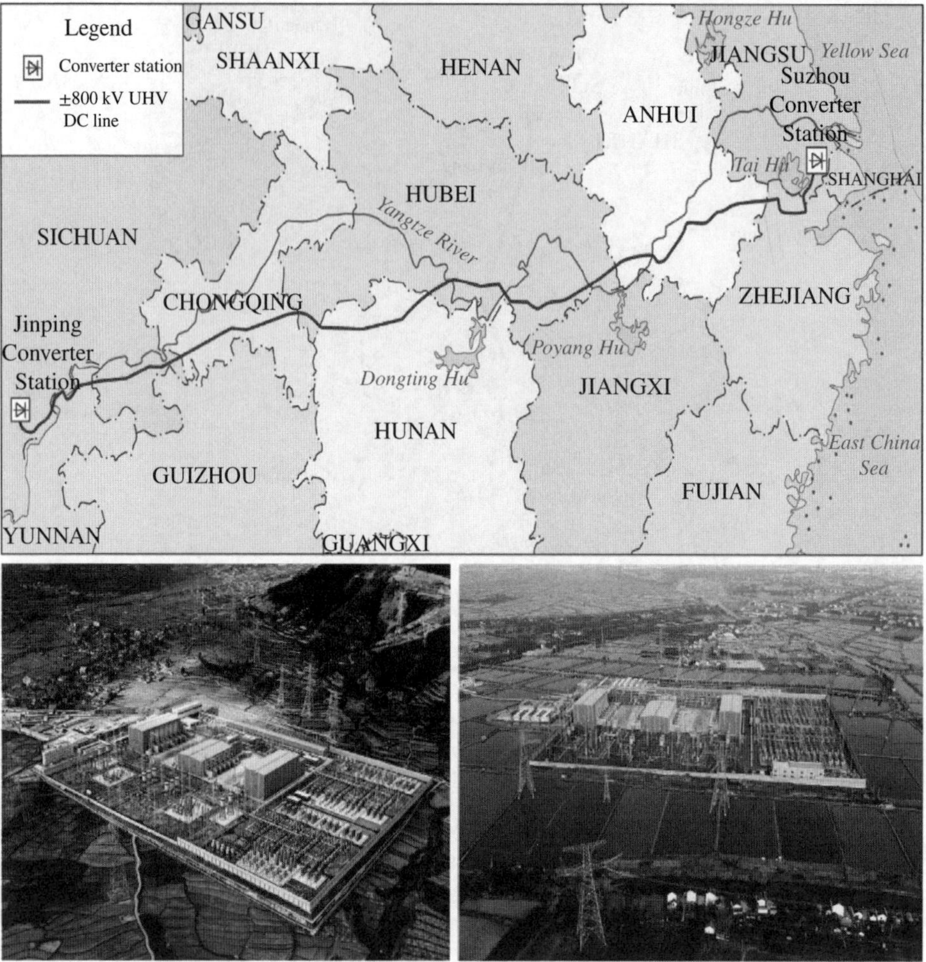

Figure 1.18 China's Jinping-Sunan ±800 kV UHV DC transmission line. Picture source: SGCC.

1.3 Large-Scale Power Allocation and Large Power Grid Interconnection

1.3.1 The Necessity and Importance of Large Power Grid Interconnection

With the increase in power grid voltage level and the expansion in scale, the installed capacity, load level, transmission distance, and allocation efficiency of power grids are greatly improved. Since the discovery of electricity more than 100 years ago, the world's power grids have undergone development from small power grids, isolated power grids, and urban power grids to large-scale interconnected power grids. Large-scale interconnection and formation of the layered and zoned power grids are the basic features of the second generation of power grids. On this basis, the third generation of power grids will form the power grid mode of organically combining backbone grids, local grids, and microgrids. This grid development mode is commensurate

Table 1.1 Comparison of characteristics between the first, second, and third generations of power grids.

	First Generation	Second Generation	Third Generation
Energy form	Dominated by low efficiency small coal-fired generating units and small hydropower generating units.	Efficient fossil energy, nuclear power and large hydropower (large hydropower development promotes the EHV transmission and grid connection) account for a large proportion.	Non-fossil energy generation accounts for a large proportion (e.g., 40~50% or above); combination of large centralized and distributed clean energy power.
Transmission mode	220 kV and below power transmission and distribution.	330 kV and above EHV AC and DC transmission, mainly in the overhead transmission mode.	Large capacity, low loss and environment-friendly transmission mode (UHV, superconducting power transmission, etc.).
Unit scale	Small units with the stand-alone capacity below 200 MW.	Large units with the stand-alone capacity between 300 MW and 1 GW.	Large units with unit capacity between 300 MW and 1 GW including large thermal power, hydropower, nuclear power, wind power generator clusters.
Grid model	Small power grids, urban power grids, isolated power grids, and small power grids.	Large-scale interconnected power grids, layered, and zoned power grids	Combination of backbone power grids, local power grids, and microgrids.
Control and protection	Simple protection and discrete relay protection.	Fast protection and optimal control; rapid removal of power transmission and transformation equipment fault; microcomputer-based protection.	Smart control and protection of grid; self-healing of power transmission, and transformation equipment and network; intelligent protection.
Dispatching mode	Empirical dispatching.	Analysis-based dispatching; power supply side energy management system that can adapt to the load changes.	Intelligent dispatching; integrated energy management system that can adapt to the changes of renewable energy power and load.
Power consumption mode	Passive power consumption.	Passive power consumption; single power service.	Active power consumption; widespread participation of users in power grid regulation; providing users with integrated energy and information services.

(Continued)

Table 1.1 (Continued)

	First Generation	Second Generation	Third Generation
Management model	Extensive management.	Vertical centralized management of power generation, transmission and distribution; introducing the power market mechanism.	Market oriented management model, fully mobilizes the initiative of power grids, users, and other participation parties.
Efficiency of electrical energy	Power plants' energy consumption rate and line loss rate are high.	Both power generating efficiency and power grid efficiency have been gradually improved.	Efficiency of power generating, grid, and consumption has been greatly improved.
Environmental impacts	High pollutant discharge of power plants.	Conventional pollutant emissions (SOx, NOx, etc.) are basically solved.	Pollutant and carbon emissions have been well controlled.
Reliability	Low power grid security and low power supply reliability.	Power grid security and power supply reliability have been significantly improved, but large power grid accident risks still exist.	Power supply reliability has been significantly improved; users' unexpected power outage risk has been basically ruled out.
Economic efficiency	Small units and small power grids have low economic efficiency and poor resource optimization allocation capacity.	Making full use of the scale economy of large units and large power grids; the ability to optimize the allocation large-scale resources.	Making full use of the scale economy of large units and large power grids; the ability to optimize the allocation large-scale resources; sustainability of high efficiency and adaptability of smart grids.
Sustainability	It is difficult for power supply to adapt to the needs of rapid economic development.	Highly dependent on large-scale use of fossil energy and nuclear energy resources; in the face of the two major challenges: exhaustion of fossil energy resources and global warming, this model of power grid is non-sustainable.	Fossil energy consumption is greatly reduced and carbon emissions are greatly reduced. It is a sustainable development model of power grids.

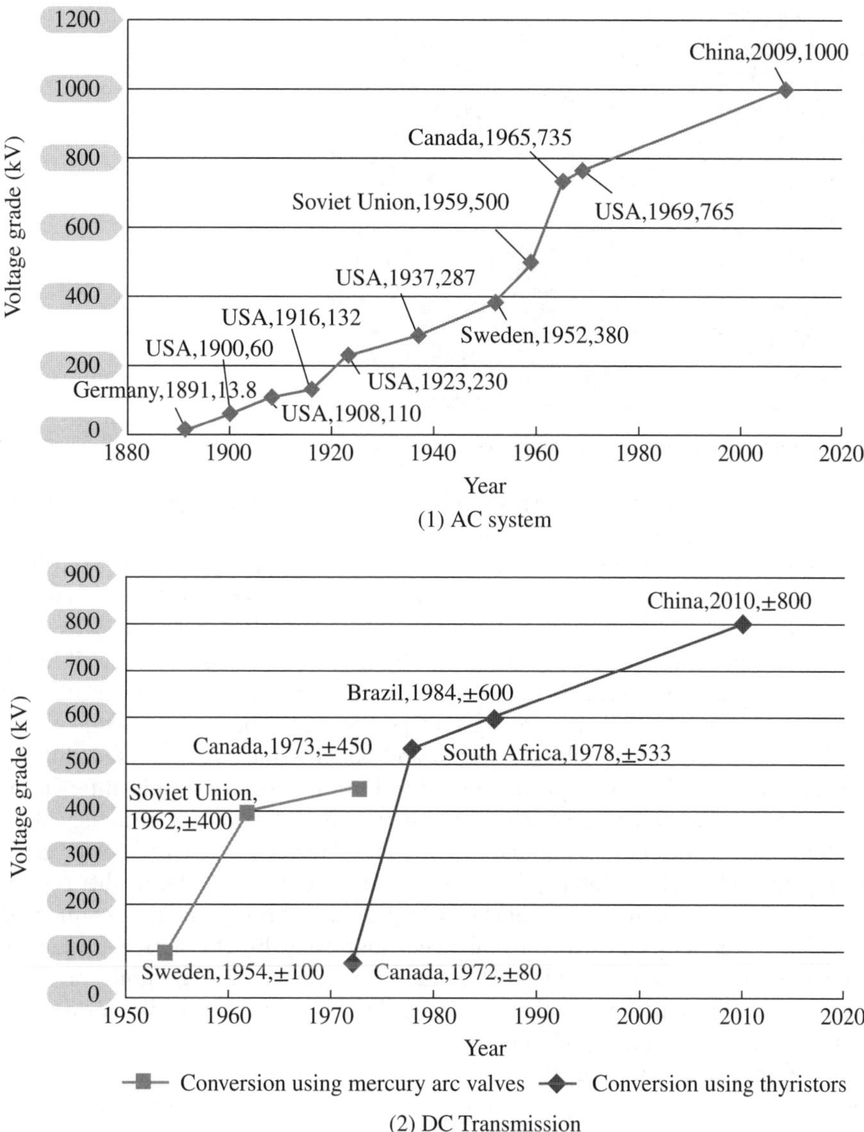

Figure 1.19 Voltage level of power system.

with the basic attributes of power grids and realistic demands of economic and social development. Figure 1.19 shows the development of AC/DC transmission voltage grades in the world. Figure 1.20 shows the interconnection of power grids in the world.

1.3.1.1 Power Grid Attributes

The power grid is a closely linked physical network that has a strong natural monopoly attribute. Strengthening the interconnection of large power grids can significantly improve the economic efficiency, security, and market competitiveness of power grids.

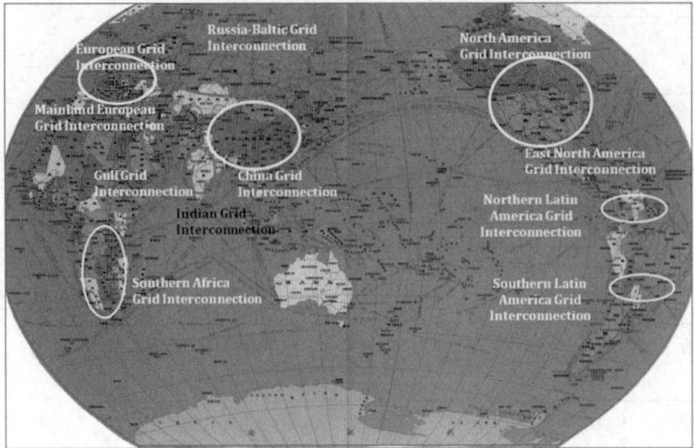

Figure 1.20 Interconnected power systems in the world.

Economically, larger-scale power grids can help better utilize power plants connected to power grids and make the operation mode of various power plants more efficient and flexible. For example, by giving the role of base load to nuclear power and coal-fired power and the role of load regulation to hydropower, the system can better adapt to the random characteristics of wind and solar generation. As larger power grids have more users, it is beneficial to make full use of the load characteristics from all kinds of users, forming complementary advantages and improving the efficiency of load management. Meanwhile, with the increase of power grid capacity, the average cost of per unit transmitted power will be reduced and benefits such as peak load shifting and averting, mutual aid of hydropower supply and thermal power supply, inter-basin complementation, and reduction of reserve capacity of power grids can be achieved.

From the prospect of security, compared with small power grids, large power grids have stronger anti-disturbance ability, higher reliability, and supply better quality power. When system faults occur, power support can be provided through the interconnected power grids. Power shortage can be balanced adaptively by the active frequency characteristics of large power grids that reduce the impact of large units or large capacity transmission channels' faults on system load. Judging from the world's and China's grid operation practices, the reliability of power supply is always improved with the expansion of scale of power grids.

At the market level, with the expansion of the grid coverage and the improvement of the large-scale resource allocation ability, more extension and reliable platforms will be provided for all kinds of power sources and loads to take part in competition, promoting fair trade in the electric power market. Therefore, large power grid interconnection has become the common trend in the world's power grid development and gradually the interconnection of local, regional, national, and even transnational and intercontinental power grids has been achieved.

1.3.1.2 Grid Interconnection

Adapting to the large-scale optimal allocation of electric power requires a lot of power grid interconnection. From a worldwide perspective, due to geological formation and for historical reasons, energy resources and energy consumption are unevenly distributed. Primary energy resources including hydropower, coal, and wind power are usually far away from the load center, which makes large power grid interconnection highly necessary. At present, more than 70% of the world's oil and natural gas resources are located in Middle East, Europe, and the former Soviet Union. To be specific, 57% of oil resources are located in the Middle East; 40 and 34% of the natural gas resources are located in Middle East, Europe, and the former Soviet Union, respectively; coal resources are distributed in North America, Asia Pacific, Europe, and the former Soviet Union [6]. At present, some developed countries, especially large cities, have a great demand for energy, but energy resources in the surrounding areas of these countries or cities have been exhausted. As a result, it is a matter of urgency to introduce energy and power from other areas. Figure 1.21 shows the fossil energy distribution around the world.

In China, 76% of the coal resources are distributed in North China and Northwest China; 80% of the hydropower resources are located in Southwest China; most land-based wind energy and solar energy resources are concentrated in Northwest China, Northeast China, and northern areas of North China. On the other hand, more than 70% of the energy demand is concentrated in East and Central China. Coal reserves and wind energy in Xinjiang account for 40 and 20.4% of China's total, respectively.

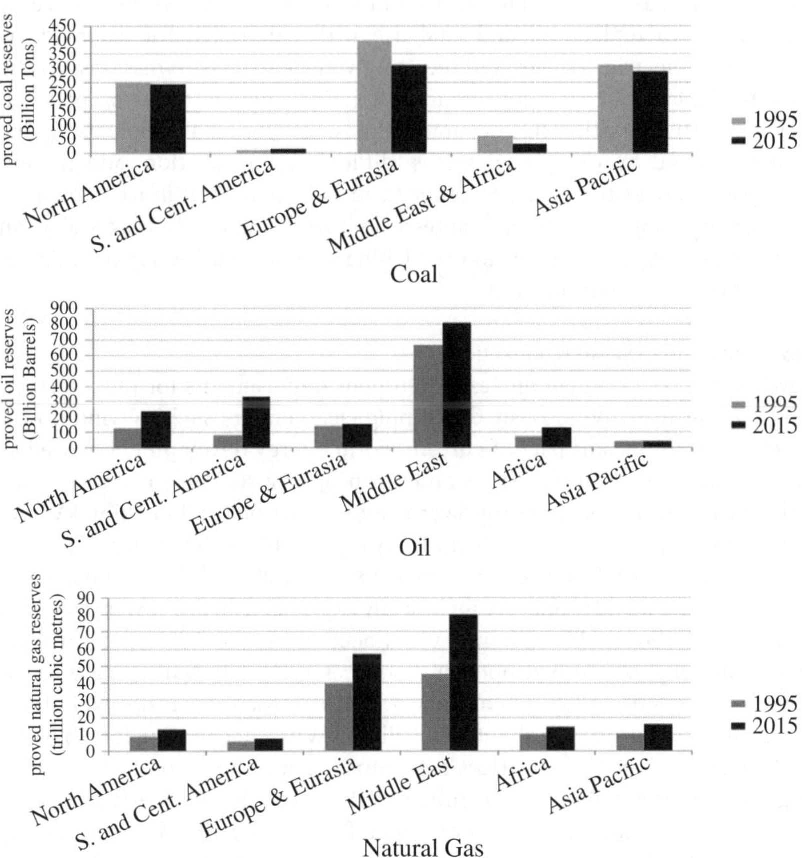

Figure 1.21 Distribution of proved fossil energy reserves in 1995 and 2015 (percentages). Data source: BP, Statistical Review of World Energy 2016.

Judging by China's future energy demand growth forecast, the load growth in East and Central China will still account for two-thirds of China's total energy demand growth. As a result, East and Central China will form China's energy consumption center. It can be seen from it that China's energy bases and load centers are 800 ~ 3000 km and even farther away from each other. With the shifting of China's future energy development focus westward, speeding up the large grid interconnection and long-distance power transmission will become more urgent [9–12].

From the perspective of solving the problem of environmental pollution, it is difficult to maintain China's long-term electric power development mode focusing on local balance. At present, the installed thermal power capacity in East and Central China has reached 630 million kW. The pollutant emissions are very high and the SO_2 emission per unit land area is 5.2 times that in West China. The SO_2 emission per unit land area in the Yangtze River Delta is even 20 times as much as China's national average. Therefore, due to the restrictions of land, environmental protection, transportation, and other factors, large-scale development of coal-fired power plants in East China is no longer suitable. Particularly, in order to solve the problem of PM2.5, we must optimize the layout of power sources, control the installed capacity in East China and reduce the emissions of pollutants. Meanwhile, we should vigorously develop the electricity replacement industry to replace oil and coal use with electricity: for example, by vigorously developing electric vehicles and reducing vehicle exhaust emissions. At the same time, due to the lack of coal resources in East China, the contradiction between coal and electricity transportation has become increasingly tense. The downtime due to coal shortages caused by rising coal prices, difficult transportation, and natural disasters has increased over the years. In order to meet China's continued growth in demand for electricity, long-distance and large-scale power transmission as well as an optimal allocation of energy resources all over China is required. See Figure 1.22 for the distribution of coal resources in China.

1.3.1.3 Clean Energy and Grid Interconnection

The rapid development of clean energy leads to urgent requirements for the interconnection of large power grids. Vigorously developing clean energy such as wind power and solar energy is an important part of current world energy development as well as a permanent solution to carbon emissions and ensuring a sustainable energy supply. The world's theoretical reserves of hydropower resources are about 39 trillion kWh/a, of which 16 trillion kWh/a, or 42%, were technically exploitable hydropower resources. The technically exploitable hydropower resources in Asia are about 7.20 trillion kWh/a, accounting for 46% of the world's total; the technically exploitable hydropower resources in South America are about 2.87 trillion kWh/a, accounting for 18% of the world's total; the technically exploitable hydropower resources in North America are about 2.42 trillion kWh/a, accounting for 16% of the world's total; the technically exploitable hydropower resources in Europe are about 1.04 trillion kWh/a, accounting for 7% of the world's total. Global wind energy resources are abundant and the theoretical reserves of wind energy resources are about 2000 trillion kWh/a. Affected by the atmospheric circulation, topography, land, and water and other factors, the world's wind energy resources are unevenly distributed. In terms of wind energy resource distribution in various continents, the theoretical reserves of wind energy resources in Africa, Asia, North America, South America, Europe, and Oceania account for 32, 25, 20, 10, 8,

Figure 1.22 Distribution of coal resources in China. Picture source: SGCC.

and 5% of the world's total, respectively. Solar energy is derived from solar radiation and is the most abundant and widest distributed resources in the world. The annual solar radiation in the world equates to about 116 trillion tons of standard coal, which is equivalent to 6500 times of the world's total primary energy consumption (18.19 billion tons of standard coal) in 2013; higher than the world's reserves of fossil energy resources. The distribution of world's hydropower resources are shown in Table 1.2. See Figures 1.23 and 1.24 for the distribution of world's wind and solar resources. Global distribution of hydro, wind, and solar resources is shown in Figure 1.25.

China has the world's richest hydropower resources. There are more than 3800 rivers with theoretical reserves of hydropower resources of 10,000 kW and above in China. China's theoretical annual power generation totals about 6.08 trillion kWh; the technically exploitable installed capacity reaches 570 million kW, equivalent to annual power generation of 2.47 trillion kWh, which is mainly attributed to the huge hydropower resources in the Yangtze River, Yarlung Zangbo River, and Yellow River. The technically exploitable hydropower resources in these three rivers account for 47, 13, and 7% of China's total, respectively. By the end of 2017, China's installed hydropower capacity was 340 million kW, accounting for about 60% of China's technically exploitable hydropower resources. As a result, there are great potentials for hydropower development in the future. China also has a land area of about 200,000 km^2 where the wind power density 10 m above ground is 150 W/m^2 and above, and theoretical reserves of wind power are above 4000 GW. According to the results of wind energy resource assessment [13] released by the China Meteorological Administration (CMA) at the beginning of 2010, China's potential exploitable wind energy resources above Grade 3 and 50 m above ground in onshore areas are about 2400 GW; China's potential exploitable offshore wind energy resources above Grade 3, 50 m above the ground

Table 1.2 Distribution of the world's hydropower resources.

Region	Nation	Theoretical Reserves	Technologically Developable Capacity (TWh/year)
Asia	China	6080	2470
	Russia	2300	1670
	India	2640	660
Europe	Norway	600	240
	Turkey	430	220
	Sweden	200	130
North America	USA	2040	1340
	Canada	2070	830
	Mexico	430	140
South America	Brazil	3040	1250
	Venezuela	730	260
	Columbia	1000	200
Africa	Democratic Republic of Congo	1450	780
	Ethiopia	65	260
	Cameroon	29	120
Oceania	Australia	270	100
	New Zealand	210	80
	Papua New Guinea	180	50

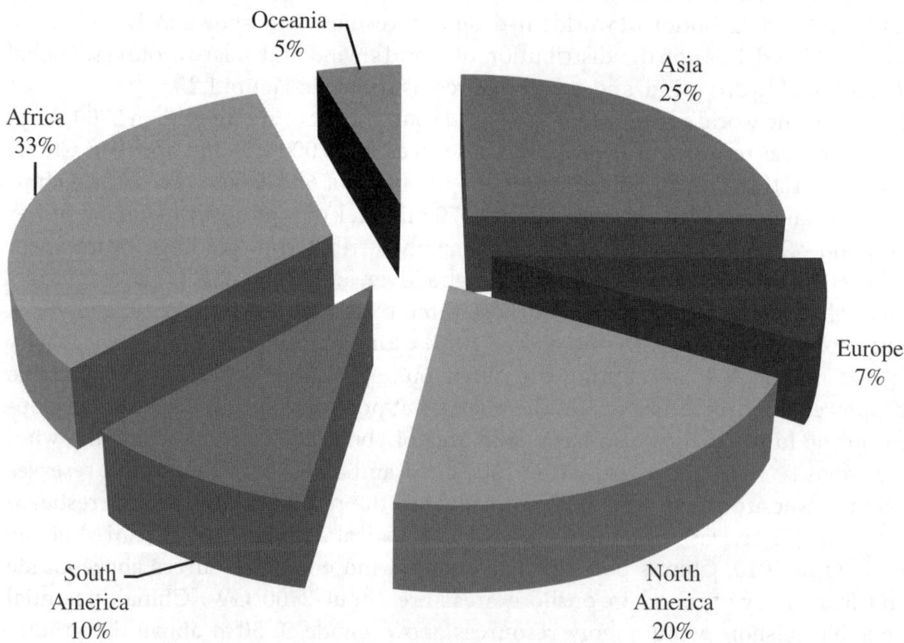

Figure 1.23 World wind power resource distributions [37].

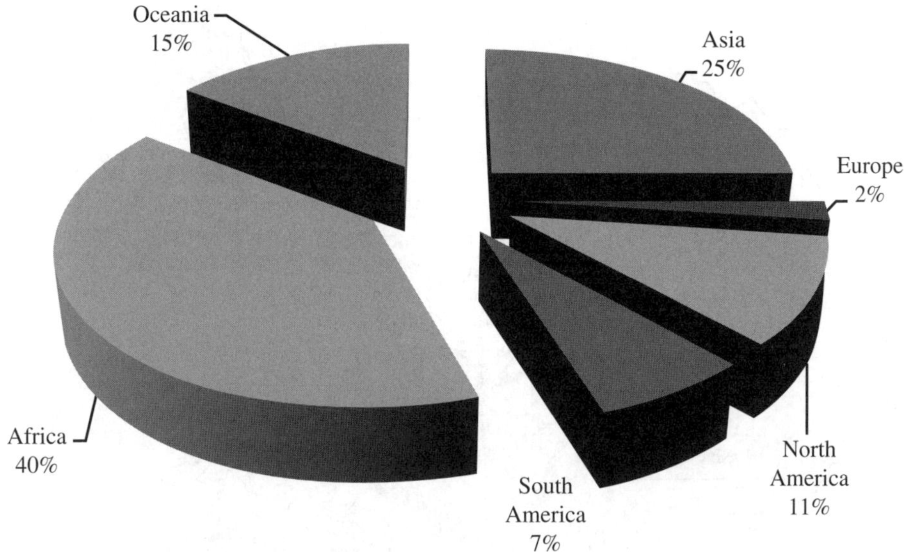

Figure 1.24 World solar power resource distributions [37].

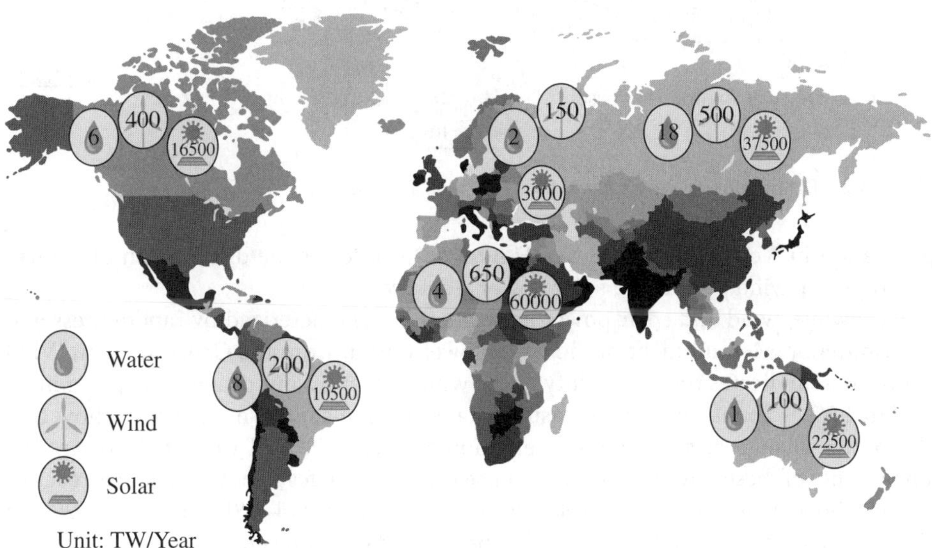

Figure 1.25 Global distributions of hydro, wind and solar resources [37].

and 5~25 m in depth are about 200 GW. In addition, China's solar energy resources are very rich and theoretical reserves reach 1700 billion tons of standard coal per year. According to planning by the National Energy Administration (NEA), China has plans to construct nine 10-GW wind power bases in Jiuquan, the Gansu province, coastal areas in the Jiangsu province, Inner Mongolia, Hebei, and other areas to create the "Wind Power Three Gorges Project" and develop large-scale photovoltaic power

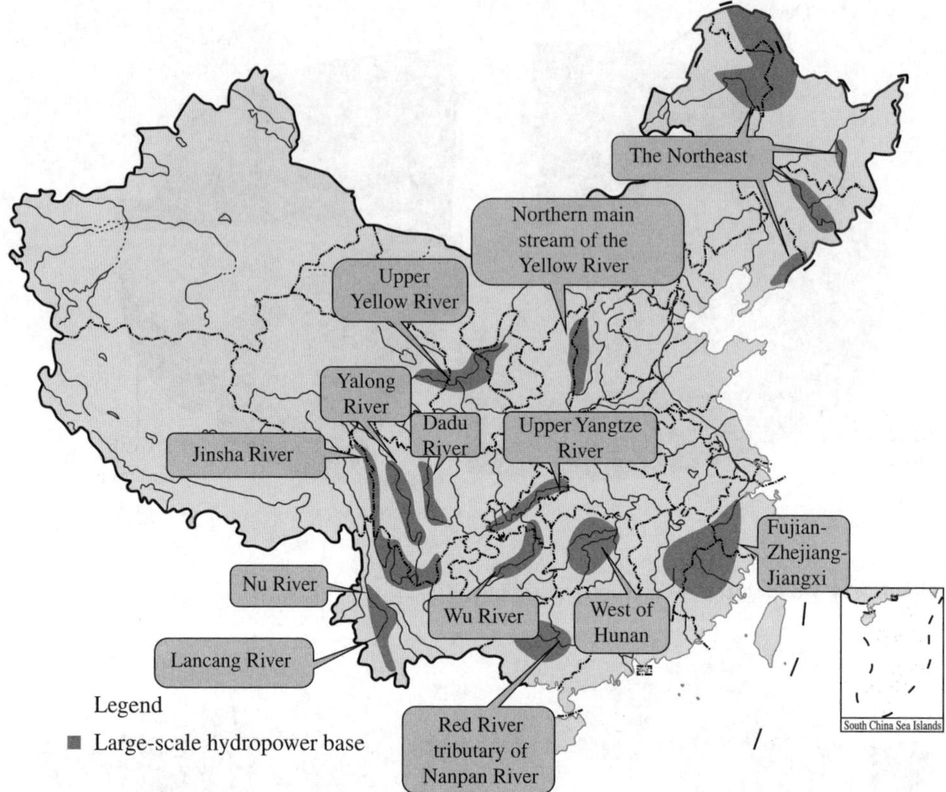

Figure 1.26 Distributions of hydropower bases in China. Picture source: SGCC.

generation in Northwest China. See Figures 1.26–1.28 for the distribution of China's Hydropower, wind power, and solar power resources.

Meanwhile, wind and solar power generation are characterized by randomness and intermittency. Measurement of the wind power output from the Gansu Jiuquan Wind Power Base indicates the probability of the wind power output change rate per minute within 1.5% is about 99%. The measured data in some southern cities also show that the power change of solar energy per minute reaches 70% of the rated power. The uncertainty of these power sources will pose great challenges to power balance, reactive power, and voltage control of the electrical power systems [14–18]. See Figure 1.29 and Figure 1.30 for the intermittency of wind power and solar power.

In the European Blackout on November 4, 2006, wind turbines were disconnected and reconnected in a disorderly manner during fault and recovery periods, respectively, which made fault control more complex [19–21]. In the first half of 2011, wind turbines in Gansu Jiuquan Wind Power Base experienced two large-scale disconnection accidents and security issues of wind power grid integration attracted attention from all parties. Judging from the analysis results, the problem was attributed to non-unified wind turbine standards and inadequate low-voltage ride through capability, but weak interconnection of the power grid will also affect the full utilization of wind power. Due to these characteristics of new clean energy, the interconnection of large power grids

Figure 1.27 Distributions of wind power bases in China. Picture source: SGCC.

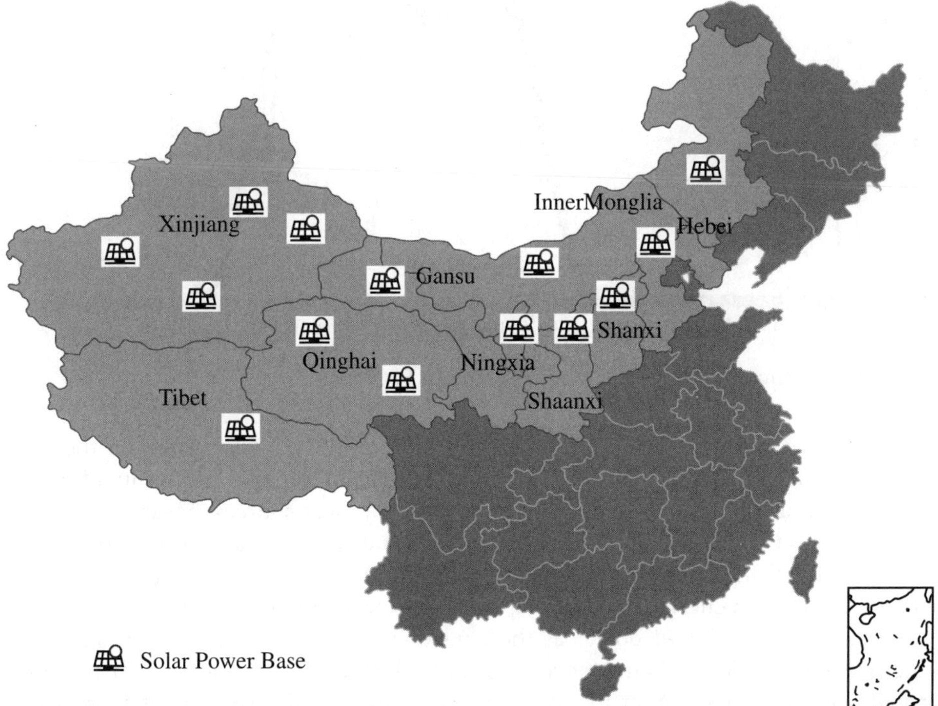

Figure 1.28 Distributions of solar power bases in China. Picture source: SGCC.

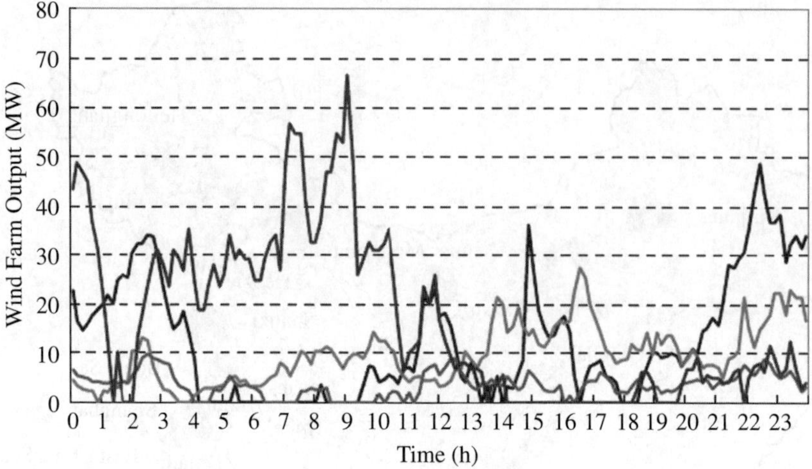

Figure 1.29 Intermittency of wind power. Picture source: SGCC.

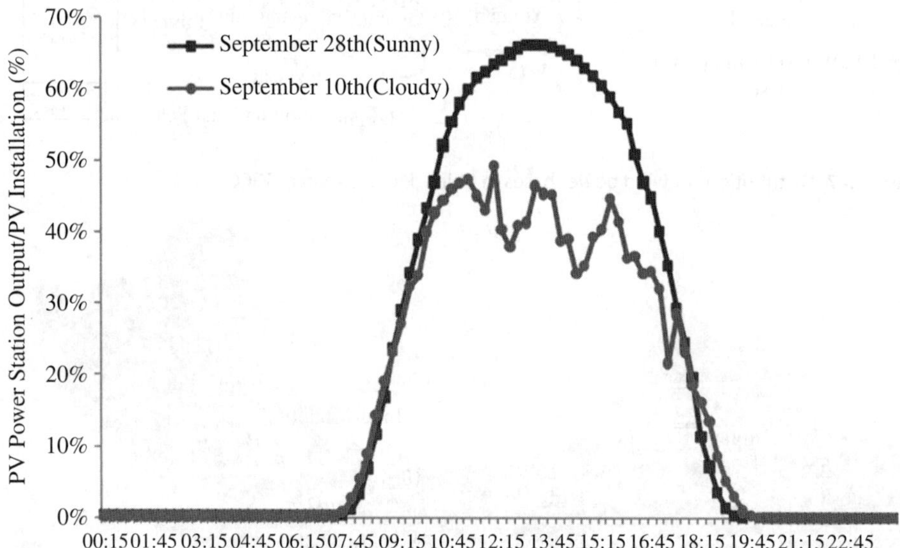

Figure 1.30 Intermittency of solar power. Picture source: SGCC.

is required as well as control of the proportion of wind and solar power in the total installed capacity. In addition, there is a requirement to improve the ability of power source integration through mutual coordination of power sources. Measures such as bundled outward transmission (consumption) of wind power, hydropower, photovoltaic power, and coal-fired power through large power grids can better facilitate the effective utilization of clean energy. In addition, the rapid development of electric vehicles, the connection of distributed power at the consumers' side, and the application of the micro-grid technology and the form of power grids (especially the form of distribution grids), will be quite different to that of existing power grids. The operation mode of power grids will also undergo significant changes. Therefore, this poses new challenges to the construction, operation, and management of power grids. What is more, the development of large power grids must speed up to satisfy consumers' demand and ensure grid security and the economic efficiency of grid operation.

1.3.1.4 Large Power Grid Interconnection is Required to Adapt to the Needs of Development of the Third Generation of Power Grids

To further strengthen the interconnection of power grids and improve their ability to optimize the allocation of energy resources is an important part in the construction of the third generation of power grids [22–27]. The national condition that China's energy resources and productivity are reversely distributed means that China must speed up the interconnection of power grids all over China and even on a larger scale. At present in China, with the expansion of the scope of the optimal allocation of energy resources, the scale of power grids has been expanded from small urban power grids, provincial power grids, and regional power grids to interconnected national power grids. In the 1950s, China's highest voltage level was 220 (110) kV and hundreds of small regional power grids gradually came into being. Between the 1970s and the 1980s, with continuous growth of power demand and installed capacity, almost 30 provincial power grids with 220 kV as the backbone network were gradually formed. Between 1970s and the 1990s, with the emergence of 330 and 500 kV voltage levels, inter-provincial power grids developed rapidly and finally formed six regional power grids; the Northeast China Grid, North China Grid, East China Grid, Northwest China Grid, and South China Grid. Between the end of the 1990s and the beginning of the twenty-first century, regional power grids such as the Central China Grid and East China Grid, Central China Grid and South China Grid, Central China Grid and Northwest China Grid achieved asynchronous interconnection through HVDC transmission projects (including back-to-back DC transmission system). Meanwhile, the Northeast China Grid and North China Grid, Central China Grid, and North China Grid achieved synchronous interconnection through AC transmission projects. In 2010, Xinjiang Grid and Northwest Power Grid were interconnected through the 750 kV AC power transmission project; in 2011, the Tibet Grid was interconnected with the Northwest Power Grid through the Qinghai-Tibet ±400 kV DC transmission project and the supporting 750 kV AC transmission project. Meanwhile, the Hainan Grid was inter-connected with the South China Grid through submarine cables. Figure 1.31 shows the synchronous grids in China by 2015. At present, except for the Taiwanese province, all China's power grids are basically interconnected with each other.

In January 2009 the world's first 1000 kV UHV AC demonstration project, the Southeastern Shanxi-Nanyang-Jingmen transmission line, was completed and put into operation; in July 2010, the ±800 kV Xiangjiaba-Shanghai UHVDC demonstration project was completed and put into operation; in December 2012, the ±800 kV Jinping-Southern Jiangsu UHV DC demonstration project was completed and put into operation. With the rapid development of the UHV transmission technology, by 2020 SGCC will have built a backbone network consisting of five vertical and five-horizontal UHV AC transmission lines and 27 inter-regional UHV DC power transmission projects connecting large energy bases and main load centers to form the energy distribution pattern of large-scale "West-East electricity transmission" and

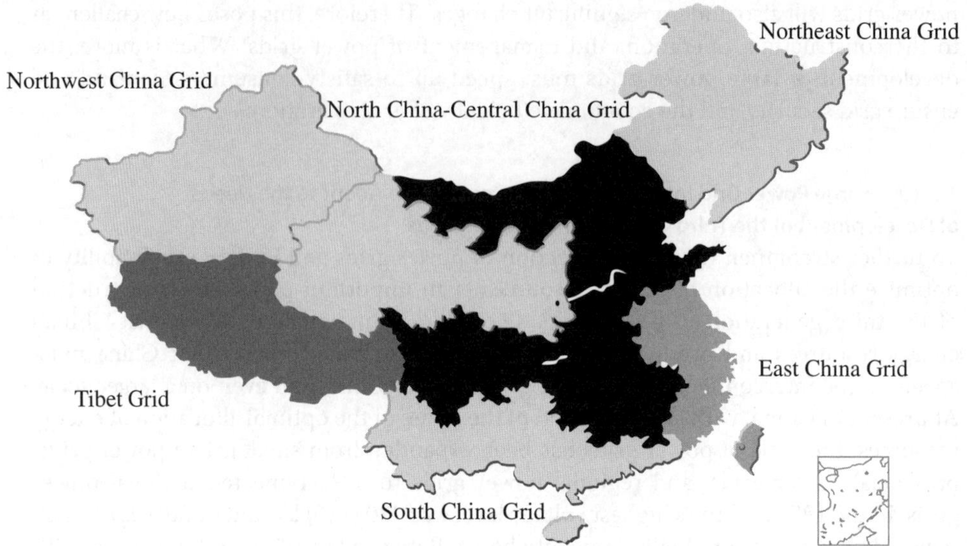

Figure 1.31 Schematic diagram of the interconnection of China's power grids in 2015. Picture source: SGCC.

"North-South electricity transmission." Figure 1.32 shows a schematic diagram of the interconnection of China's power grids in 2020 [1].

1.3.1.5 Large Power Grid Interconnection is an Important Trend in World Power Grid Development

The scale of interconnected synchronous power grids in most areas in the world is gradually expanding. At present the total installed capacity of the synchronous power grids in eastern North America exceeds 760 GW, covering an area of about 5.2 million km^2. With 500 kV as the backbone network and containing some 750 kV power grids, the synchronous power grids in eastern North America have the largest installed capacity in the world. Figure 1.33 shows a schematic diagram of interconnected power grids in North America. The installed capacity of synchronous power grids in Western Europe reaches 690 GW, covering an area of about 4.5 million km^2. The synchronous power grids are dominated by 400 kV power grids, but also have a few 330 kV power grids. After the American Blackout in 2003, the USA accelerated the development of grid technology and put forward the concept and idea of Grid2030, a super-power grid; namely, constructing the national backbone network [28], connecting the East and West Coast of the USA with Canada and Mexico. Figure 1.34 shows a schematic diagram of America's Grid2030 Plan. In order to meet the needs of clean energy development, European countries proposed to construct a "Europe-Mediterranean-Middle East" super-power grid by 2050 to connect renewable energy power generation bases in North Africa, the Middle East, and areas along the coast of the North Sea in Europe. Figure 1.35 shows a schematic diagram of the European SuperGrid Plan. Figure 1.36 shows a schematic diagram of the Desertec plan. Figure 1.37 shows a 2050 projection of the major powerhouses and super producers in Europe and North Africa. According to the plan, by 2020 at least 35% of the power in Europe will be supplied by distributed and

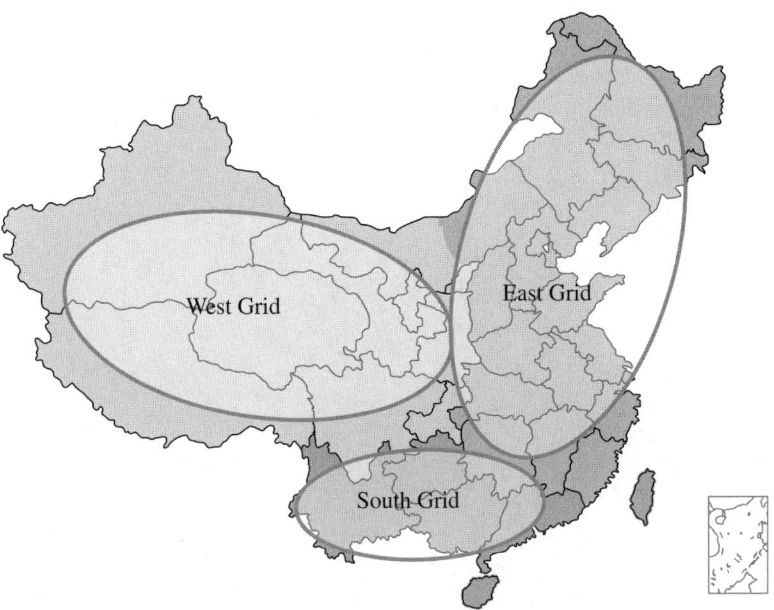

Figure 1.32 Schematic diagram of the interconnection of China's power grids in 2020. Picture source: SGCC.

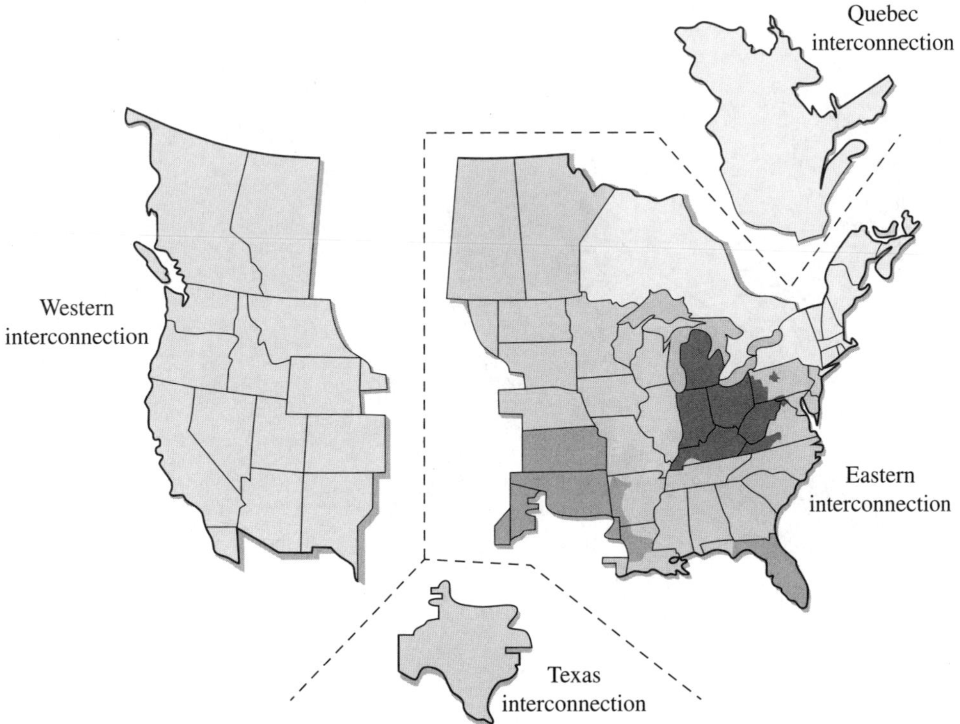

Figure 1.33 Schematic diagram of interconnected power grids in North America in 2012. Picture source: EIA.

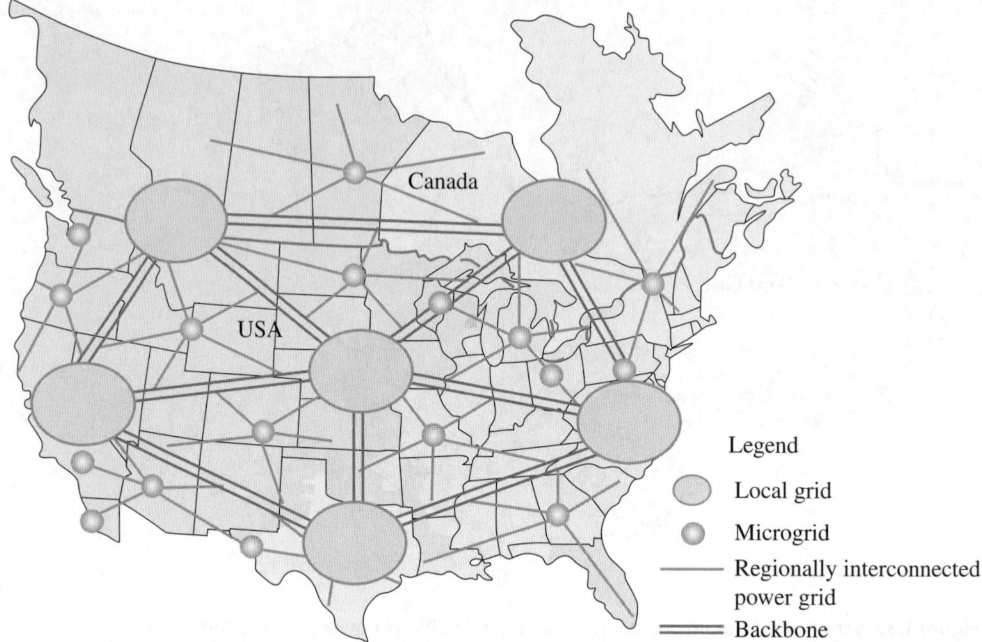

Figure 1.34 Schematic diagram of America's Grid2030 plan. Picture source: United States Department of Energy.

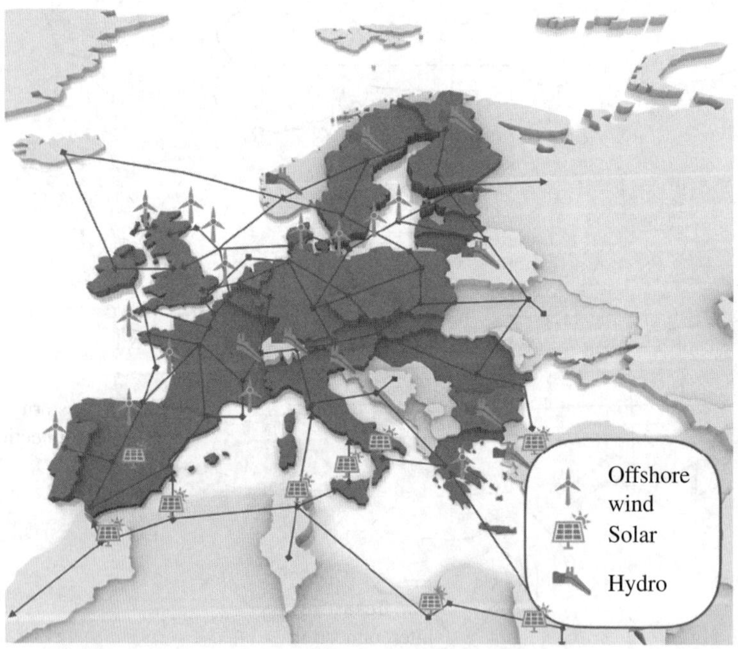

Figure 1.35 Schematic diagram of European SuperGrid plan. Picture source: Europagrid.

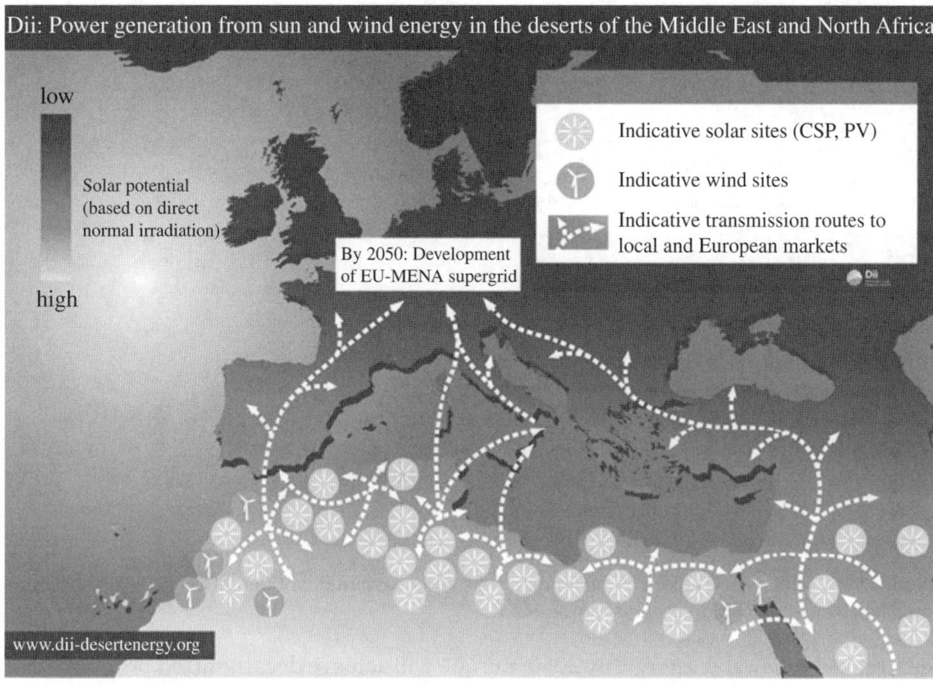

Figure 1.36 Schematic diagram of desert planning. Picture source: Dii-desertenergy.

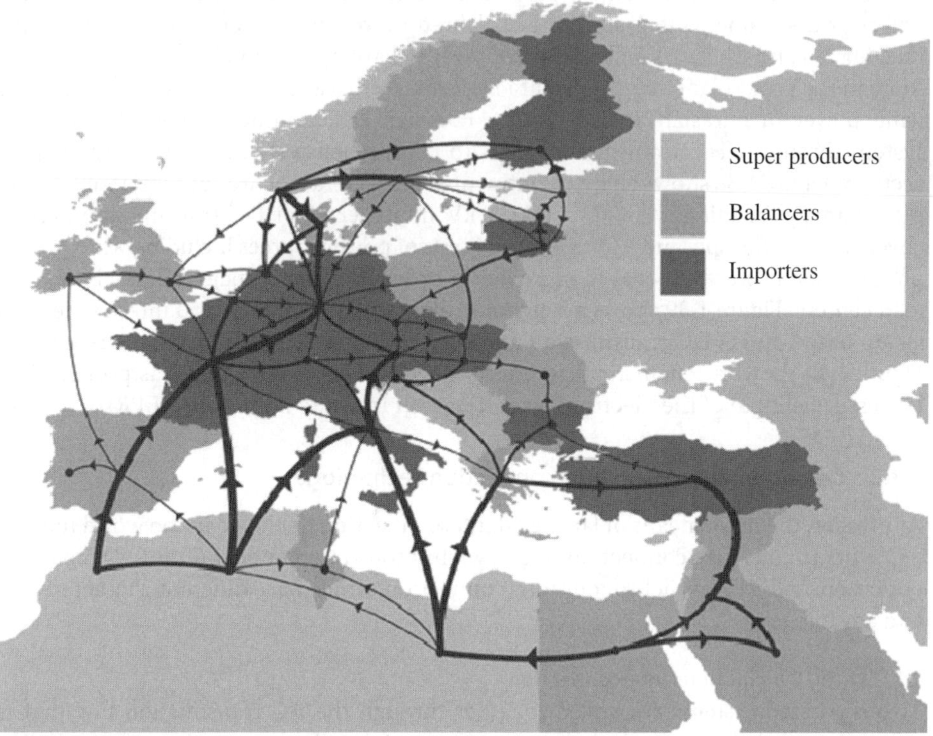

Figure 1.37 North and South – EUMENA's Powerhouses in 2050. Picture source: DESERTEC [38].

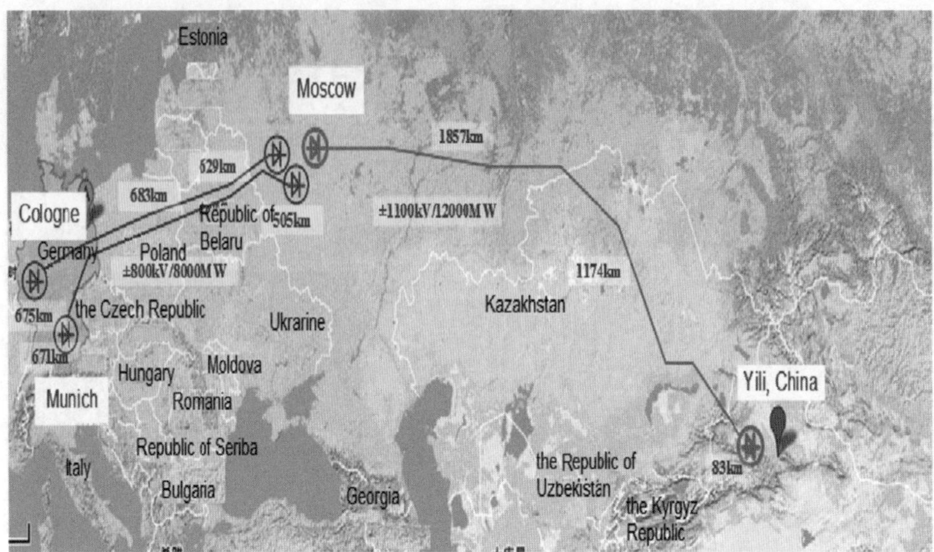

Figure 1.38 Schematic diagram of the Europe, Asia, and Africa intercontinental UHVDC Asynchronous grid interconnection vision. Picture source: SGCC.

centralized renewable energy. By 2050, Europe will achieve decarbonized production of all power. Driven by these new demands, the smart grid will come into being [29, 30].

Faced with the common worldwide problem of energy security, at the G-SEP Leaders' Summit held in 2012, the SGCC proposed the vision of building an intercontinental power grid – namely integrating the coal-fired power in Central Asian countries like Kazakhstan and China's Xinjiang, hydropower in Yarlung Zangbo River, hydropower in Russian Far East, solar energy in Sahara Desert in Africa and onshore as well as offshore wind power in Eurasian areas – into a Eurasian intercontinental energy allocation system through UHV transmission lines to comprehensively improve energy supply security. Figure 1.38 shows an intercontinental UHV DC asynchronous interconnection plan: namely, building a ±800 or ±1100 kV multi-terminal DC transmission system covering Asia, Europe, and Africa with all kinds of power sources including wind power, photovoltaic power, and coal-fired power in order to connect major energy bases and load centers. Figure 1.39 shows a schematic diagram of the global grid interconnection vision based on HVDC transmission and connection of all kinds of clean energy [31]. Figure 1.40 shows a schematic diagram of the global super-grid that was proposed by Professor Jielingth of the Electric Power Research Institute in the USA (EPRI).

1.3.2 Development of Grid Interconnection Technology

At present, the main means of large grid interconnection deployed in the world include AC synchronous interconnection, DC asynchronous interconnection, AC/DC parallel operation, and VFT asynchronous interconnection, which have different characteristics and advantages.

1.3.2.1 AC Synchronous Interconnection

Two separated systems are linked together through the AC transmission line that is generally called the network tie line (see Figure 1.41). The power flow of the tie line

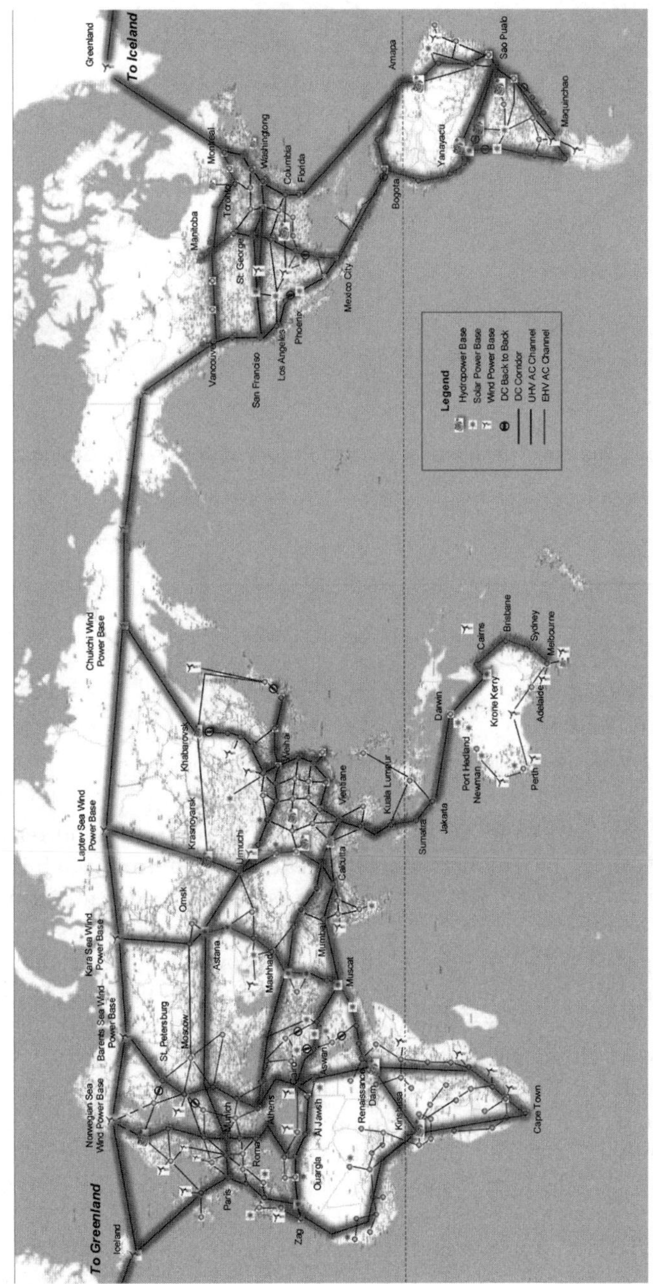

Figure 1.39 Schematic diagram of the global grid interconnection vision based on HVDC transmission and connection of all kinds of clean energy.

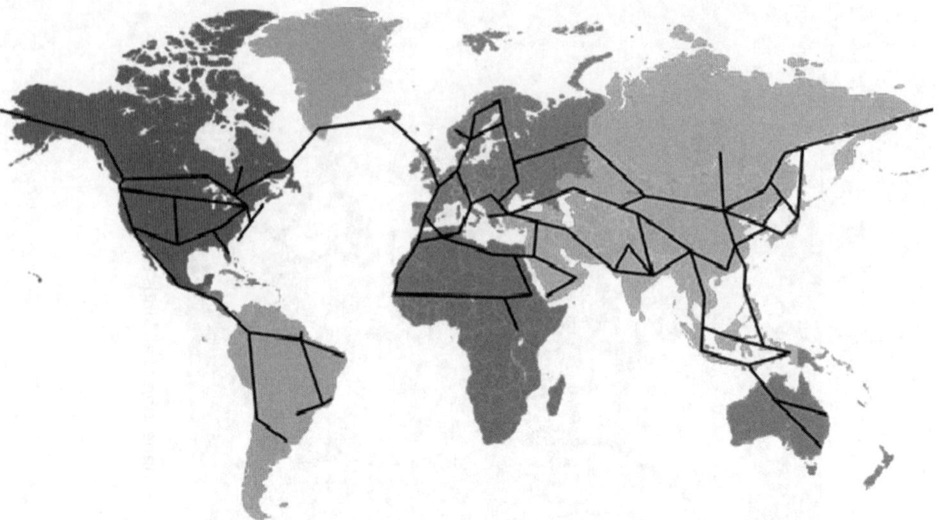

Figure 1.40 Schematic diagram of the global super-grid. Picture source: Clark W. Gellings, Global Supergrid.

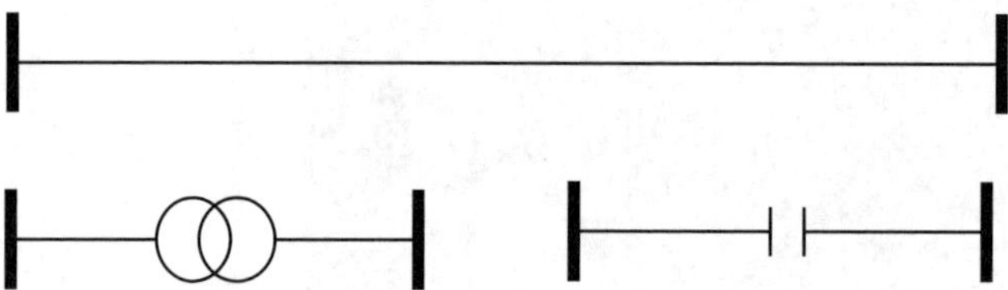

Figure 1.41 Typical AC synchronous interconnections in a power system.

is determined by the law of the AC circuit. In order to control the power of the tie line, the control strategy of regional automatic generation control (AGC) is generally adopted. Through AC interconnection, the systems on both sides in a synchronous power grid must control the power-angle of generator in the whole system to ensure safe and stable operation and to prevent the tie line power from exceeding the limit and becoming unstable; AC synchronous interconnection can not only achieve active power exchange, but also reactive power exchange. In addition, when the fault occurs in the system, a short-circuit current will be increased on the other side.

From the perspective of the world power grid development history, the scale and scope of the AC interconnection depends on the power grid voltage level and the load growth needs. The transmission capacity and distances of AC lines of different voltage levels are shown in Table 1.3. The higher the voltage level is, the longer the AC transmission line length and the larger the scale of the interconnected power grid. The success of China's 1000 kV UHV AC transmission demonstration project not only provides technical support for the construction of Central China-North China-Eastern China UHV synchronous power grid, but also provides a reference for other countries and regions in the world in power grid interconnection. A strong transmitting end grid or receiving end grid is built through the AC interconnection. The greater the number of generators

Table 1.3 Transmission capacity and distance of AC lines at different voltage levels.

Voltage Level (kV)	Natural Power (10 MW)	Transmission Capacity (10 MW)	Transmission Distance (km)
110	3	3–6	30–120
220	16	10–20	100–250
330	36	20–50	200–500
500	95	40–100	250–800
750	230	100–250	500–1200
1000	500	200–600	1000–2000

connected with the synchronous power grid, the larger the inertia of the synchronous power grid, which can significantly enhance the synchronous power grid's ability to deal with faults and accommodate external inflow.

At the same time, with the expansion of the scale of the synchronous power grid, the risk of low-frequency oscillation and accident diffusion has attracted much attention. For example, there is a low-frequency oscillation risk in China's "long chain" AC interconnected system of Central China-North China-Northeast China, which restricts the exchange power level of regional power grids and results in a small transmission scale of the Central China-North China, North China-Northeast China 500 kV AC tie line. In order to ensure the safe operation of large power grids, we need to construct AC synchronous power grids in a scientific way. On the one hand, we need to adopt safe and stable automatic control, PMU, other technologies, and measures to improve the large power grid state perception and controllability; on the other hand, in the power grid construction process, we should reasonably design the grid architecture to ensure a clear interface of the interconnected system, in order to effectively control key lines and important sections as well as actively prevent the risk of cascading failures caused by an electromagnetic loop network.

1.3.2.2 DC Asynchronous Interconnection

Two separate AC systems are linked through the HVDC transmission project (see Figure 1.42). This HVDC project can be a conventional HVDC project, a back-to-back DC project, or a flexible DC device. A DC project generally includes a DC line and converter stations on both sides of the line. The converter station mainly consists of a converter transformer, converter valve, AC filter, DC filter, smoothing reactor, overvoltage protection facilities, and other devices. The DC transmission power is decided by the DC control system and generally the constant power control mode or constant current control mode is adopted. Through DC interconnection, the systems on both sides can be asynchronous grids or even power grids of different frequencies, and there is no requirement for the power angle of generator in the interconnected systems. DC interconnection can only transmit active power, thus the systems on both sides need to provide a certain proportion of reactive power support. DC asynchronous interconnection can effectively isolate the interaction between the interconnected power grids and realize the effective and controllable transmission of active power. However, large-scale DC power exchange requires the support of

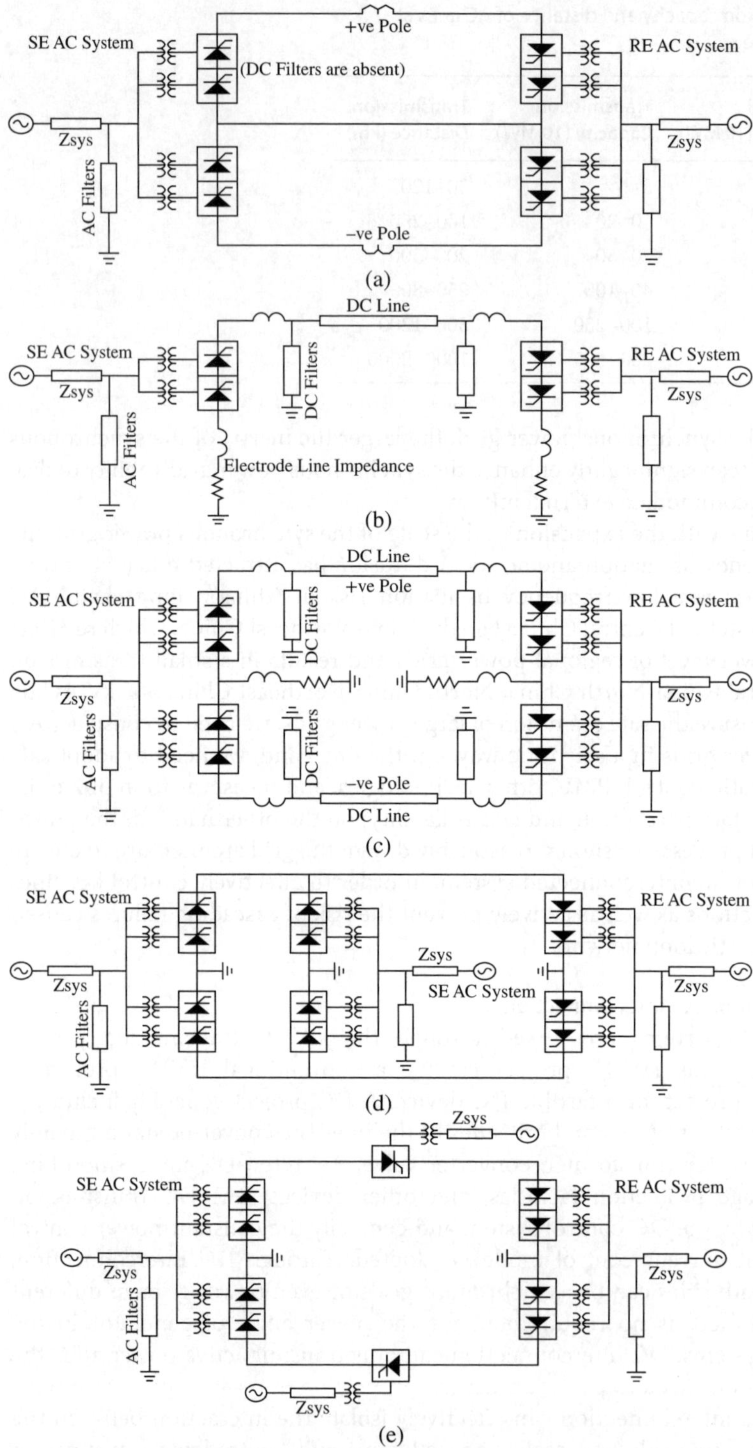

Figure 1.42 Typical DC asynchronous interconnections. (a) Back-to-back, (b) monopolar HVDC, (c) bipolar HVDC, (d) multi-terminal in series, and (e) multi-terminal in parallel.

strong AC grids; otherwise DC faults will lead to power supply accidents, and AC system faults may also cause DC commutation failures and even unipolar or bipolar blocking. Due to the limitation of the DC transmission characteristics, the normal operation of the conventional DC system (including back-to-back DC system) needs the support of the voltage of the systems on both sides to achieve normal commutation and rectification. Therefore, in supplying power to a weak power grid, the conventional DC connection has some limitations. In addition, as the DC transmission system will lead to certain harmonics, it should be equipped with a filter of a certain capacity.

1.3.2.3 AC/DC Parallel Operation

To be specific, there are both DC transmission channels and AC tie lines between two systems. This kind of interconnection, in essence, is an AC synchronous interconnection. The DC transmission power is determined by the DC control system; the power flow of the AC tie line is determined by the law of the AC power grid. The systems on both sides in a synchronous power grid must control the generator power angle of the whole system to ensure safe and stable operation and to prevent the tie line power from exceeding the limit and becoming unstable. The AC/DC parallel operation and control can significantly improve the transmission efficiency, safe, and stable performance of the interconnected systems and effectively suppress the low-frequency oscillation of interconnected systems. Figure 1.43 shows a AC/DC parallel operation power grid. In Reference [32] the operation and control of the AC/DC parallel system is studied and some control methods as well as ideas for flexible AC/DC parallel power transmission are proposed.

1.3.2.4 VFT Asynchronous Interconnection

Two separate AC systems are linked together through the VFT and step-up transformers on both sides (see Figure 1.44). Specific information about the VFT will be introduced in subsequent chapters. The power flow of the VFT is determined by its control system. Interconnection through the VFT has a similar effect as a back-to-back DC project. The

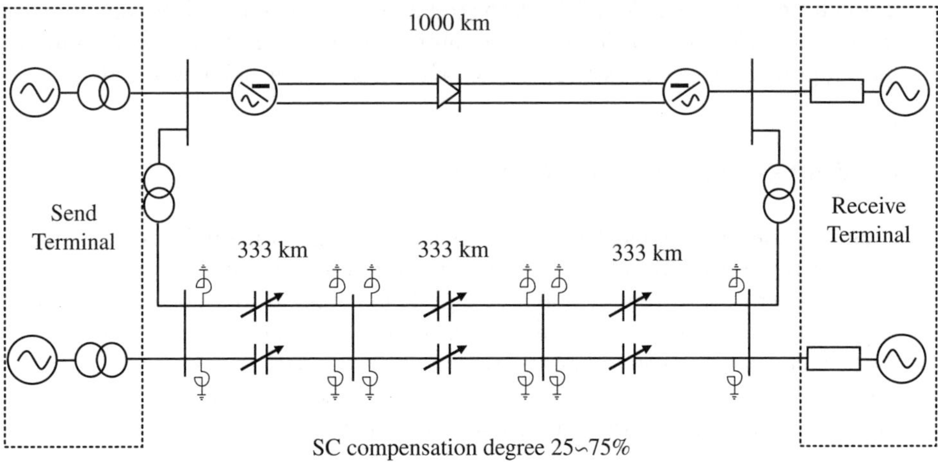

Figure 1.43 Schematic diagram of an AC/DC parallel operation power grid [32].

Figure 1.44 Typical VFT asynchronous interconnection.

systems on both sides can be asynchronous power grids and have no requirement for the power angle of generator. The VFT transmits active power as well as reactive power. In addition, in case of a system fault, this limits the increment of a short-circuit current. The VFT can also run in parallel with the AC channel. In this operation mode, the VFT is equivalent to a phase shifter capable of continuous and large-scale regulation.

These four kinds of interconnection methods have different characteristics and can be applied on different occasions. Generally speaking, the AC interconnection is applicable to countries and regions with the same management subject or unified management conditions and abilities. The interconnection scale depends on the voltage level and scale of energy exchange demand. The DC interconnection, to some extent, can isolate the systems on both sides and prevent their mutual influence on each other. So it is applicable to occasions when the systems on both sides belong to different management subjects or different countries, when the interconnected power grids have different voltage frequencies, and when power of an ultra-large capacity is transmitted over an ultra-long distance. The VFT is suitable for the marginal interconnection between large power grids and the interconnection of weak or small power grids to main grids. It can play an important role to achieve an organic combination of the backbone power grids and local power grids as well as microgrids.

At the same time, in order to ensure safe and stable operation of power grids, flexible power flow control, and development of the power market, there must be a clear interface between interconnected power grids, which is even more important to AC synchronous interconnection. Otherwise, once an interconnected line fails, this may lead to large-scale power transfer and affect the safe operation of the whole interconnected system. A similar situation occurred in the USA and Canada Blackout of 2003 and led to serious consequences. Therefore, the safety of large power grids has attracted attention from various parties [33–35]. In strengthening the interconnection of power grids, we must select suitable interconnection modes based on the actual situation of the power grids to ensure the safe operation of the interconnected power grids.

1.4 Main Content of this Book

Starting from the development features of modern power grids, in this book we carry out comprehensive research on VFT theory, control and simulation, and engineering applications. Problems are solved, such as VFT simulation analysis tools, operation control strategy, and large system low-frequency oscillation. A comparative study is made here on the technical economy of VFT and phase shifter as well as the back-to-back DC transmission system leading to systematic and innovative research results [5]. This book is divided into eight chapters. The main content of each chapter is as follows:

Chapter 1: Power Grid Development and Interconnection. On the basis of analyzing the development trend for world energy, electric power, and the power grid development trend, we summarize the basic characteristics of the first, second, and third generations of power grids. Starting from the needs of large-scale optimal allocation of electric power, we introduce the development of the power grid interconnection technology and compare the characteristics and performance of different types of interconnection.

Chapter 2: Proposal and Application of VFTs. In this chapter, we focus on the physical composition, technical characteristics, basic functions, foreign research, and engineering applications of VFTs, discuss the necessity of developing them and their application prospects in China, and elaborate on the problems related to research on VFTs that have to be solved urgently.

Chapter 3: Basic Equations and Simulation Models of VFTs. Based on the principle of general motors and circuits, we derive and establish the power flow steady-state equation, electromechanical transient equation, and electromagnetic transient equation that can effectively represent the static and dynamic characteristics of the VFT. These equations reflect the relationship between the main parameters of the VFT such as rotor speed, transmission power, driving torque, voltage and current of the main circuit, control driving circuit ignition angle, and the electromagnetic coupling and power exchange law of VFTs. We reveal the basic principle of asynchronous grid interconnection transmission and flexible operation control by using VFTs, and prove that the VFT can improve the stability level of an interconnected electric power system. On this basis, based on the derived basic equations of the VFT and the basic functions provided by programs such as the Power System Analysis Software Package (PSASP), EMTP, and PSCAD/EMTDC, we study and establish the digital simulation model of VFTs, which are applicable to power flow calculation, electromechanical transient analysis, and electromagnetic transient research of electric power systems. Simulation modules in relevant simulation programs of electric power systems are realized so as to create conditions for analysis of VFTs in a large electric power system.

Chapter 4: VFT Control System Research and Modeling. In this chapter, we study the overall control strategy of VFT, construct the DC driving circuit model, and put forward and establish block diagrams and parameters of various control systems that can realize functions such as frequency control, power angle control, synchronous grid connection control, power control, reactive power voltage control, system power flow optimization, and suppression of low-frequency oscillation. These control modules are mainly based on PID control, have strong robustness, and can adapt well to different system conditions and operation modes providing support for the implementation of various control targets of VFT.

Chapter 5: Analysis of Operation Characteristics and Application of VFTs in the Electric Power System. We use the simplified electric power system, a typical four-generator power system, and a complex practical system to build the all-digital simulation system including VFTs based on software such as PSASP, EMTP, and PSCAD/EMTDC and carry out an all-digital simulation including processes such as VFT startup, grid integration, power regulation, reactive voltage control, fault response, power supply to weak systems, system frequency regulation, and damping low-frequency oscillation. Simulation results show that the developed VFT simulation model and control system are effective; the VFT can be used to achieve power bidirectional transmission and smooth regulation between asynchronous power grids, effectively prevent the impact of the fault on the interconnected power grids, realize basic functions such as system frequency adjustment, power flow optimization, damped oscillation, power support, and power supply to islands. With the power step response time being about 400 ms, the VFT can meet the needs of power control and regulation and has sound system stability.

Chapter 6: Design of an Adaptive Low-Frequency Oscillation Damping Controller Based on a VFT. We study the low-frequency power oscillation problem of a large interconnected electric power system and propose the self-adaptive low-frequency power oscillation damping control method based on the VFT and Prony method. This method can actively identify the system oscillation mode and adjust the damping controller parameters. The simulation of the typical four-generator power system and complex practical system simulation proves this method can better adapt to the system changes and fault interference so as to effectively suppress low-frequency power oscillation of the synchronous interconnected systems.

Chapter 7: Technical and Economic Characteristics of VFTs. We compare the differences and similarities of VFT and different types of power phase shifter, DC power transmission devices, and other devices. Characteristics and advantages of VFT are analyzed. Compared with a phase shifter, a VFT can realize asynchronous interconnection as well as flexible and smooth power regulation of two power grids with different frequencies, it has better adaptability, controllability, and flexible regulation performance, and it does not produce harmonics. Compared with the back-to-back system, the VFT has advantages such as simple structure, fewer components, low loss, high reliability, small footprint, easy maintenance, lower reactive power configuration needs, strong system adaptability, convenient expansion, and no harmonics. It can be used as an alternative technique for asynchronous interconnection and has broad development and application prospects.

Chapter 8: Summary and Prospects. In this chapter, we summarize the main conclusions of VFT research and propose key issues to be studied next.

1.5 Summary

1. At present, human society is facing the threat of imminent fossil energy depletion and global climate change. In the new energy reform characterized by clean energy and smart grids, the energy structure will undergo dramatic changes in the future and electric power is expected to become the most important energy form. As a result, there are higher requirements for safety, adaptability, interaction, and ability to optimize the allocation of energy resources of the power grid.

2. Power grids and grid interconnection technology have entered the era of the third generation power grids characterized by the large-scale utilization of renewable energy and smart power grids. In the next 40 years, the world will complete the transition from the second to third generation power grids. Therefore, the promotion and application of new technologies and new devices will become more urgent.

3. Third generation power grids will gradually evolve to the network pattern of combining large power supplies with distributed generation and backbone power grids with microgrids. As a new smart grid device, the VFT can play a positive role in large power grid interconnection.

4. In the new round of world energy reform, particularly driven by the large-scale optimal allocation of electric power, we must accelerate the innovation and development of grid technology. The VFT is an innovative technology for future grid interconnection and has broad prospects in the future.

References

1 L. Zhenya. *Electric Power and Energy in China*. Beijing: China Electric Power Press, 2012.

2 L. Zhenya. *UHV Grid*. Beijing: Economic Press, 2006.

3 S. Yinbiao. Research and application of 1000 kV UHV AC transmission technology. *Power Grid Technology*, 2005, 29(19): 1–6.

4 Z. Xiaoxin, C. Shuyong, L. Zongxiang. Review and prospect of the development of power grid and power grid technology. *Proceedings of the CSEE*, 2013, 33(23): 1–11.

5 C. Gesong. *Modeling and Control of VFT for Power Systems*. D. Phil Thesis, Beijing: China Electric Power Research Institute, 2010.

6 BP Statistical Review of World Energy, 2012.

7 D. Archer. *The Global Carbon Cycle*. Princeton, NJ: Princeton University Press, 2010.

8 Z. Xiaoxin. The development prospects of power grid and its technology during the new energy revolution era. *Annual Meeting of the Chinese Society of Electrical Engineering*, 2010.

9 National Bureau of Statistics. *China Energy Statistics Yearbook 2008*. Beijing: China Statistics Bureau Press, 2009.

10 State Grid Energy Research Institute. *Handbook of International Energy and Power Statistics* (2009). Beijing: SGCC, 2009.

11 SGCC. *Research on the Energy Base Construction and Medium and Long Term Development Plans for Electric Power*. Beijing: SGCC, 2006.

12 Chinese Academy of Engineering. *Research on the Medium and Long Term Development Strategy for Energy*. Beijing: Chinese Academy of Engineering, 2010.

13 Wind Energy Resource Assessment Center of China, Meteorological Administration. *Assessment of Wind Energy Resources in China* (2009). Beijing: Meteorological Press, 2010.

14 L. Feng, L. Yichuan. Influences of large-scale wind energy converters on transmission systems. *China Electric Power*, 2006, 39(11): 80–84.

15 H. Dongsheng, L. Yongqiang, W. Ya. Study of the shunt-connected Wind Power Generation System. *High Voltage Engineering*, 2008, 34(1): 142–147.

16 Z. He, X. Jianyuan, Z. Mingli, et al. Status and key issues of wind power technology development. *East China Electric Power*, 2009, 37(2): 314–316.

17 A. Vladislav. *Analysis of dynamic behavior of electric power systems with a large amount of wind power*. Dissertation. Denmark: Technical University of Denmark, 2003.

18 T. Geetha, V. Jayashankar. Stability assessment of power system models for higher wind penetration. *Power System Technology and IEEE Power India Conference*, 2008.

19 G. Rui, D. Yu, L. Yaochun. Analysis of large-scale blackout in UCTE power grid and lessons to be drawn to power grid operation in China. *Power System Technology*, 2007, 31(3): 1–6.

20 L. Chunyan, C. Zhou, X. Mengjin. Analysis of large-scale blackout in Western Europe Power Grid on November 4 and relevant suggestion to Central China Power Grid. *High Voltage Engineering*, 2008, 34(1): 163–167.

21 S.K. Salman, A.L.J. Teo. Improvement of fault cleaning time of wind power using reactive power compensation. *IEEE Power Tech Conference*, Porto, Portugal, 2001.

22 Z. Xiaoxin, G. Jianbo, S. Yuanzhang. *Basic Research on the Operational Reliability of Large-Scale Interconnected Power Grids*. Beijing: Tsinghua University Press, 2008.

23 Z. Baosen, G. Ricai. On development of interconnection of power networks in China. *Power System Technology*, 2003, 27(2): 1–7 [in Chinese].

24 H. Zhenxiang, X. Yusheng, Q. Jiaju. A review of CIGRE-2000 on power system interconnection. *Automation of Electric Power Systems*, 2000, 23(24): 1–4 [in Chinese].

25 Y. Yongkang. Research on inter-provincial power system interconnection andthe electric power exchange project. *Power System Technology*, 1996, 20(7): 14–18 [in Chinese].

26 P.A.S. Pegado. Large international interconnections in South America and Brazil. *2000 CIGRE 37*, Paris, France, 2000.

27 J.P. Dresbrosses. International interconnections of power systems: trends for the next decade. *2000 CIGRE 37*, Paris, France, 2000.

28 L. Zongxiang, J. Jinfeng. Interpretation of the vision of the United States "Grid2030." *China Electric Power Enterprise Management*, 2004.

29 A.S. Massoud, B.F. Wollenberg. Towards smart grid. *IEEE Power & Energy Magazine*, 2005,3(5):34–41.

30 M. Vadari. Demystifying intelligent networks. *Public Utilities Fortnightly*, 2006, 145(11):61–64.

31 GEIDCO. Global Energy Interconnection Backbone Grid Research. *International Conference on Global Energy Interconnection*, 2018.

32 C. Gesong. Discussion on the flexible AC/DC power transmission mode. *National Conference on Power System Technology*, 2001.

33 Z. Xiaoxin. Power system burst disaster and risk of blackout. *Power Industry Summit*, 2007.

34 Z. Xiaoxin, Z.J. Chao, S. Guorong, X. Yusheng. Draw lessons from the Northeast America-Canada power grid blackouts accident. *Power System Technology*, 2003, 27(9): 1.

35 B.G. Gher, P. Dusan, R. Dietmar, T. Erwin. Lessons learned from global blackouts lessons from the Northeast, 2007, 40(10): 75–81.

36 R.A. Hefner III, *The Grand Energy Transition* [Yuanchun M, Boshu L, Trans.]. Beijing: China Citic Press, Citic Publishing House; 2013.

37 L. Zhenya. *Global Energy Interconnection*. Beijing: China Electric Power Press, 2015.

38 S. Mludi, I.E. Davidson, Dynamic analysis of the southern Africa Power Pool (SAPP) Network, presented at *2017 IEEE PES-IAS Power, 2007*.

2

Proposal and Application of VFTs

2.1 Overview

In order to meet the demand for the large-scale optimal allocation of power and to adapt to the development trend of large-scale power grid interconnection, it is urgent to achieve a breakthrough in the power grid interconnection technology and equipment. Under this circumstance a new type of interconnection equipment, the VFT, has been developed [1–4]. It is also the second asynchronous interconnection technology after the DC power transmission system. A lot of the literature [1–34] shows that this device has unique advantages in solving problems related to power grid interconnection such as asynchronous interconnection, fault isolation, power supply to weak power systems, and low-frequency power oscillation. On the one hand, the VFT can realize interconnection of two asynchronous power grids as well as flexible and smooth control of transmission power. Meanwhile, it can supply power to weak power systems and even passive systems, providing active and reactive power support. To some extent it has more reliable interconnection power supply performance than the back-to-back system. On the other hand, the loss of the VFT is far lower than the VSC-HVDC device or the conventional back-to-back DC transmission system. As a result, it can achieve higher interconnection efficiency.

In this chapter, we mainly introduce the basic concept, physical structure, workflow, and basic functions of the VFT as well as its application in the world's engineering practice, discuss the necessity of developing VFT use and its application prospects in China, and summarize the major problems that have to be solved urgently in order to apply this technology. Meanwhile, we analyze the relation equation between VFT control torque and transmission power. In addition, we take the single-machine infinite system as an example to conduct quantitative analysis of the basic principle of VFTs to improve system stability performance [31].

2.2 VFT System Constitution

Integrating mechanical motion, electromagnetic theory, and electric energy conversion technologies, the VFT is a creatively conceived and ingeniously designed energy conversion device used in electric power systems. Specifically, this device integrates the technologies of transformer, phase-angle regulator, hydro-generator, doubly fed induction motor, DC drive control, and so on. It is a new device for interconnections between

Variable Frequency Transformers for Large Scale Power Systems Interconnection: Theory and Applications,
First Edition. Gesong Chen, Xiaoxin Zhou, and Rui Chen.

asynchronous grids and some experts have incorporated it into smart grid and flexible AC transmission system equipment [8, 28]. Table 2.1 compares the characteristics of the main flexible AC transmission system equipment including the VFT.

The core technology of VFTs lies in controlling the capacity and direction of the active power transmitted via the VFT by setting a three-phase winding rotary transformer on both the stator and rotor side, ensuring the synchronization of the rotor and the stator

Table 2.1 Flexible AC transmission system equipment and characteristics [35].

Category	Device	Functional Characteristics	Power Electronic Device Characteristics
Series type	TCSC, Thyristor Controlled Series Compensator	Power control; series impedance compensation; improving transient stability; suppression of subsynchronous resonance and low-frequency power oscillation; regulating line power flow; increasing synchronous torque	Based on semi-controlled devices
	IPC, Interphase Power Controller	Power control; limiting short-circuit current	Based on fully controlled devices
	GCSC, GTO Controlled Series Capacitor	line impedance control; regulating line power flow; increasing synchronous torque; damped oscillation	Based on semi-controlled devices
	TCSR, Thyristor Controlled Series Reactor	Regulating line reactance	Based on semi-controlled devices
	FCL, Fault Current Limiter	series reactor control; limiting short-circuit current	Based on semi-controlled devices
	SSSC, Static Synchronous Series Controller	Series injection current; regulating line power flow; increasing synchronous damping; damped oscillation	Based on fully controlled devices
	VFT, Variable Frequency Transformer	Asynchronous networking; power control; low-frequency oscillation damping; improving transient stability	
Parallel type	SVC, Static Var Compensator	Bus voltage control; reactive power compensation; oscillation damping	Based on semi-controlled devices
	STATCOM, STATic synchronous COMpensator	Regulating voltage control; Dynamic voltage support; reactive power compensation; oscillation damping; improving transient stability	Based on fully controlled devices
	SMES, Superconducting Magnetic Energy Storage	Generator rapid braking; suppression of low-frequency oscillation; active and reactive power compensation; improving power quality	

Table 2.1 (Continued)

Category	Device	Functional Characteristics	Power Electronic Device Characteristics
series parallel type	TCDB, Thyristor Controlled Dynamic Brake	oscillation damping; improving transient stability;	Based on semi-controlled devices
	CSR, Controlled Shunt Reactor	Limiting the power frequency and switching overvoltage; reactive power and voltage regulation; suppression of secondary arc current; oscillation damping	Based on semi-controlled devices
	DVR, Dynamic Voltage Restorer	Parallel injection current; regulating bus voltage; dynamic voltage support	Based on fully controlled devices
	UPFC, Unified Power Flow Controller	Power control; voltage control; reactive power compensation; phase-angle control; series impedance compensation, oscillation damping; improving transient stability	Based on fully controlled devices
	PAR, Phase-Angle Regulator	Regulating the voltage phase angle; the line power flow control	Based on semi-controlled devices
	MSSC, Multimode Static Series Compensator	Regulating the voltage phase angle, line power flow, and oscillation damping	Based on semi-controlled devices
	IPFC, Interline Power Flow Controller	Series injection voltage; parallel injection current; regulating the voltage phase angle, line power flow, and oscillation damping	Based on semi-controlled devices
	CSC, Convertible Static Compensator	Series injection voltage; parallel injection current; realizing the power flow regulation between multiple lines; oscillation damping; dynamic voltage support	Based on fully controlled devices

magnetic field in the rotating space through the DC motor drive system and adjusting the phase shift of the rotor magnetic field corresponding to the stator magnetic field. The system constitution of VFT is shown in Figure 2.1. In an actual power system, VFT system includes devices such as a VFT, DC rectification and driving module, DC motor, step-down transformer module, reactive power compensation capacitor module, and corresponding circuit breaker [1–5, 31]. The main functions of all modules and devices are as follows:

2.2.1 VFT Device

The cutaway view of the VFT is shown in Figure 2.2. It is mainly composed of rotary transformer, DC driving motor and collector ring [1]. Main parts of a VFT are shown in Figure 2.3.

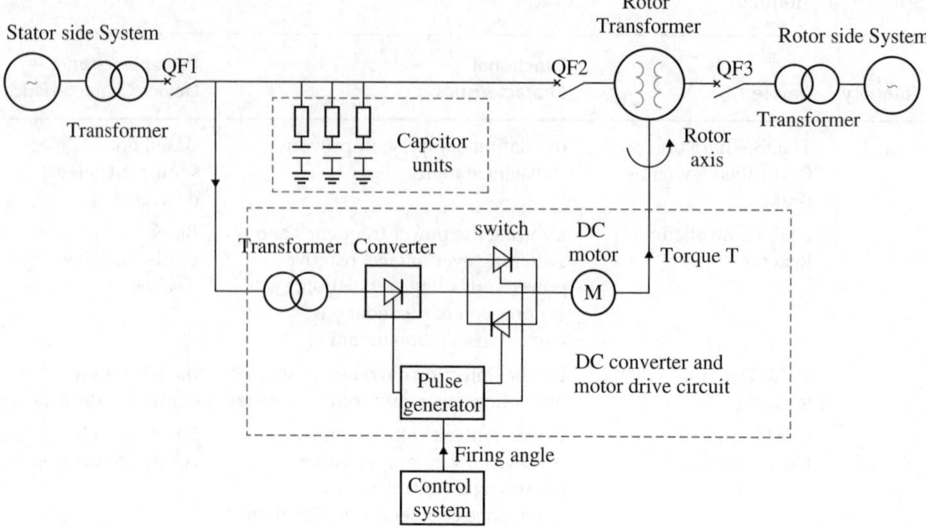

Figure 2.1 VFT system constituents [1].

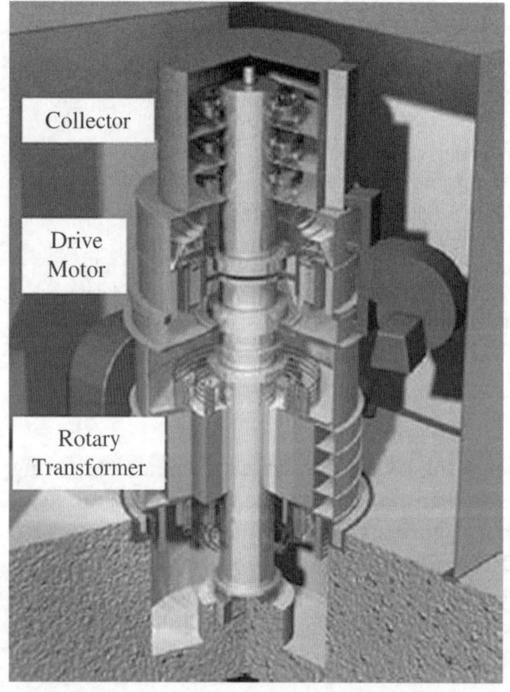

Figure 2.2 Cutaway view of a VFT [1].

1. The collector ring is located in the upper part of the VFT. In general, it is of a carbon brush structure in order to conduct current between rotor three-phase winding and the external system (see Figure 2.4). The electrical connection of the rotor winding and the external AC system is achieved by this method. The design, service life and

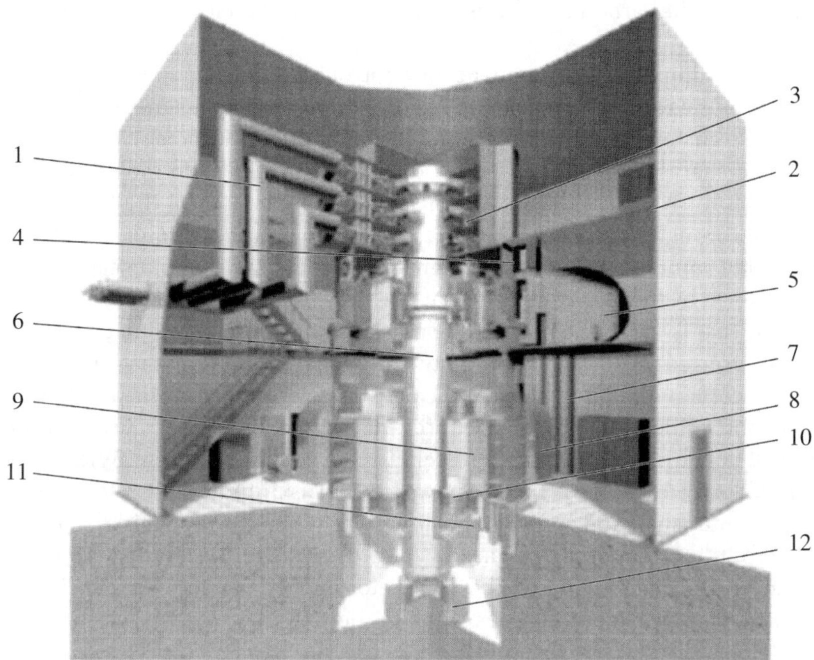

Figure 2.3 Main parts of a VFT [2]. Where: 1. Rotor bus duct; 2. Air hood; 3. Three-phase collector; 4. DC drive motor; 5. DC motor fan; 6. Upper support; 7. Stator bus duct; 8. Rotary transformer fan; 9. Stator core; 10. Rotor core; 11. Winding connection; and 12. Lower thrust support base.

Figure 2.4 VFT collector brusher [20].

current capacity of the carbon brush are key issues in the capacity and maintenance cost of the VFT.

2. A DC motor is an important component for controlling the rotor speed, adjusting the phase shift of the rotor winding, as well as the stator winding magnetic field, and controlling the transmission power of the VFT. Generally, it is located in the middle of the VFT and shares the shaft with the rotor (see Figure 2.5). On the rotor shaft a DC driving motor rotor winding is also installed that can control the DC motor driving torque through a DC rectifying circuit so as to adjust the rotor speed and phase shift of the rotor as well as stator magnetic field. The transmission capacity of VFT is determined by the torque on the rotor of rotary transformer that is exerted by the DC motor. That is to say, the power transmitted by the VFT is the function of the driving torque exerted by the DC motor on the rotor shaft. Assuming that the torque is exerted in a certain direction to make the active power flow from the stator side system to the rotor side system, when the torque is exerted in the opposite direction, the power will flow from the rotor side system to the stator side system. Within the design capacity of VFT, the transmission power of VFT is proportional to the torque exerted on the rotor shaft and so also to its direction. When the transmission power of VFT is zero, the rotor torque required is also close to zero. Different from the conventional high-speed DC motor, the DC motor applied in the VFT usually has a low speed and can provide sufficient torque in the stationary state or close to the stationary state.

3. The rotary transformer is located in the lower part of the VFT. As the most important part of the VFT, it is composed of three-phase winding of the stator, three-phase

Figure 2.5 Magnet frame of VFT DC motor [20].

winding of the rotor, stator core, and rotor core. The rotor winding is connected with the system on one side through the collector ring while the stator winding is connected with the AC system on the other side. The phase shift of the rotor and stator winding current magnetic field is dynamically adjusted through the rotation of the rotor and the control of the speed to achieve the controlled power exchange between the systems on both sides based on the change in the frequency of the systems on both sides and the given power control targets. In the steady-state, that is when the sum of rotor rotation (electrical) frequency and rotor current frequency is equal to the stator current frequency, the stable power transmission between the two systems can be achieved. In power control, the VFT generally controls the power transmission capacity through a closed-loop power control system. The control system compares measured power value with expected set value, and then the deviation value of power will be used to adjust the output torque of DC motor. The step response rate of the power conditioning is generally about hundreds of milliseconds, which can satisfy the fast response to the system disturbance and maintain the stable power transfer. In absorbing the reactive power, the VFT follows the law of AC circuit, which is determined by leakage reactance of rotary transformer and the square of the current of VFT. At the same time, the VFT can also absorb reactive power from the system on one side and transmit some of the reactive power to the system on the other side.

In short, in a VFT (see Figure 2.3), the rotary transformer is the main body that achieves power exchange between systems on both sides of the rotor and stator through the magnetic coupling principle; the DC motor is the key device for controlling rotor speed and transmission power of the VFT; and the collector ring is the device that forms a direct electrical connection between the rotor winding and external system. The rotor inertia of the VFT is usually quite large, which can enhance the ability of VFT to cope with external shock or disturbances, and provide quite good damping characteristics, but meanwhile increases the response time of the VFT in the power step regulation process.

2.2.2 DC Rectification and Motor Drive

This module consists of power transformer, six-pulse bridge rectifying circuit, thyristor ignition control circuit, thyristor bridge commutation circuit, and DC driving motor. Its main function is to change the driving torque of DC motor through regulation of input current and direction of DC motor winding so as to achieve the goal of transmission power control. It can be said that there are a lot of methods to realize a DC drive. For a converter valve, in addition to the semi-controlled thyristor, an insulated gate bipolar transistor (IGBT) and other fully controlled power electronic devices can also be used. Figure 2.6 shows the DC motor drive system model, which is an example based on the thyristor valve rectification and IGBT commutation [31].

2.2.3 Step-Down Transformer

The voltage of interconnected systems is generally 110 kV (or above), limited by mechanical design, and due to the insulation design of rotating components and other factors, the VFT cannot be directly connected. The step-down transformer is needed to reduce

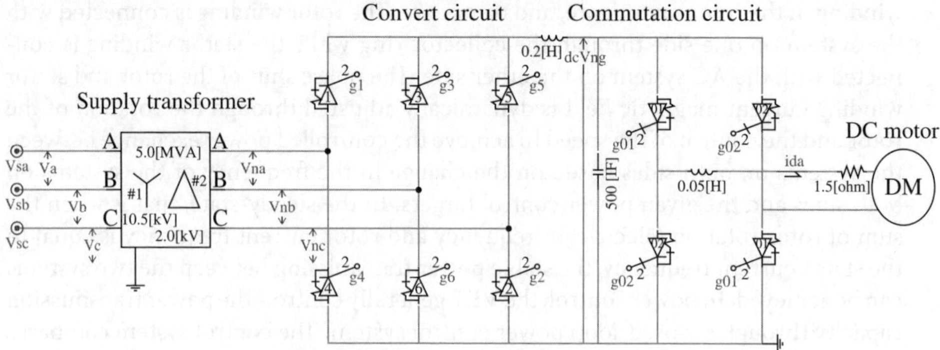

Figure 2.6 Simplified model of a DC motor driver.

the voltage of the systems on both sides to 15–25 kV, which is equivalent to the voltage of hydraulic turbine generator. The conventional generator step-up transformer is usually used as the step-down transformer. In selecting capacity of the transformer, the active power the VFT has to transmit as well as the power factor has to be taken into consideration.

2.2.4 Reactive Power Compensation Capacitor Bank

Based on the change of VFT current and bus voltage, corresponding capacity of capacitor banks are switched automatically to compensate for the reactive power absorbed by the leakage reactance of the VFT and the step-down transformer to optimize the system reactive power balance and improve the system voltage performance.

2.2.5 Circuit Breaker

Based on the system instructions, the VFT is switched on and off by the circuit breaker to achieve synchronous connection and islanding of systems on both sides. Meanwhile, it can quickly clear faults based on system measurement and protection requirements to ensure the safety of systems and equipments as well as create the conditions for system operation and maintenance.

2.3 Basic Functions of VFTs

The VFT has functions similar to those of the back-to-back DC transmission system. The basic means for the realization of various functions of the VFT are speed regulation and power regulation. Speed regulation is the basis for all of the control, namely, the control of the rotor speed to ensure that the magnetic field formed by the rotor winding current is synchronous with the magnetic field formed by the stator winding current, which is the basic premise of the stable operation of the VFT. Power regulation is the basic operation mode. From the perspective of mechanical motion, in essence it adjusts the rotor winding equivalent magnetic field and the phase shift of the stator winding magnetic field. Generally, the closed-loop negative feedback method is used to control the power of the VFT at a given level. In case of the disturbance or the abnormal operation state of the power grid, the command power value is changed to achieve different control targets such as suppression of power oscillation so as to improve the operation state and stability of the system. Generally speaking, the VFT mainly has the following functions [1–3, 6, 31].

2.3.1 Asynchronous Interconnection Function

To be specific, this achieves the synchronization of the rotor winding current magnetic field and the stator winding current magnetic field by controlling the motion of the rotor to form stable interconnection between the asynchronous power grids.

2.3.2 Transmission Power Control

This controls the exchanged power of the interconnected power system based on the exchanged power requirements in the dispatching plan to achieve the controlled transmission of the electric power between interconnected power grids. It is the basic function of the VFT as well as the basic function of the VFT as the series flexible AC transmission system device.

2.3.3 Frequency Regulation Function

When the frequency change of the system on either side exceeds the set value, the transmission power and direction of the VFT can be changed within the design capacity of the VFT and the scope the system permits to increase the frequency of the system with a lower frequency and reduce the frequency of the system with a higher frequency, which is equivalent to using the system on one side as the frequency modulation power supply of the system on the other side so as to maximally maintain the frequency of the systems on both sides at the permissible scope.

2.3.4 Power Supply to Weak Systems

Different from conventional DC based on the current commutation principle, the VFT does not need the support of grid voltage in operation. As a result, it can be used to supply power to weak systems and even passive systems. Take the Langlois Substation equipped with the world's first VFT as an example [1–5]. The local power grid on one side is islanded from other parts of the network, so whether this isolated system has local power supply or not, the VFT will continue to keep running. If there is no local power supply, the VFT will automatically supply all the power within its operating limit range to the isolated system. If there is a local power supply, the VFT will compensate the difference between local power supply and local load and assume the task of frequency regulation together with the governor of the local generator.

2.3.5 Black-Start Power

Black start is crucial to restore the safe operation of the power grid. When the system on one side stops operation due to failure and the power supply is required to be restored, the system on the other side can restore the power supply to the outage system through the VFT and provide black-start power. Meanwhile it can effectively isolate the shock that might be caused to the normal system in the black-start process.

2.3.6 Suppression of Low-Frequency Power Oscillation

When there is a weakly damped low-frequency oscillation mode between the interconnected systems, if the VFT is installed in the section, the power exchanged between the interconnected systems can be regulated dynamically through the power regulation function and the system low-frequency damping can be increased by selecting appropriate parameters to suppress low-frequency power oscillation [34].

2.3.7 Power Emergency Regulation

Generally, in case of large system disturbances (e.g., the system on one side loses the key line or generator), the output power of the VFT will be adjusted to the preset value in a very short period of time according to the previously set power control curve to provide emergency power support.

2.4 Startup and Control of VFTs

The startup, control, and exit are routine operations of the VFT. When two AC systems need to be interconnected through a VFT, the main working process is as follows:[6]

2.4.1 Switching No-Load VFTs

In Figure 2.1, first close QF1, and then close the circuit breaker on the stator side (e.g., QF2), which is equivalent to switching no-load VFT. As the core of the VFT is a saturated element, certain inrush current might be produced in switching on the no-load VFT. At this point, the phase angle can be used to control the switching on of the VFT and suppress the amplitude of the inrush current of VFT. Generally when the voltage reaches the peak value, the switching on of the VFT will produce the minimum inrush current. Therefore, the switching on command should be issued based on the circuit breaker operation delay and the output voltage waveform of the phase-locked loop [36].

2.4.2 Adjusting Rotor Speed

Detect the frequency of the systems on both sides, calculate the frequency difference, and then adjust the output torque of the DC motor and control the rotor speed to make the frequency of the rotor side winding induction voltage consistent with the frequency of the system on the other side in order to make the voltage frequency on both sides of the circuit breaker QF3 consistent with each other.

2.4.3 Synchronizing Close

Adjust the angle difference of the rotor magnetic field to make the phase difference in voltage between the systems on both sides of the circuit breaker QF3 within the given range. When necessary, adjust the system voltage to make the amplitude difference in voltage between the systems on both sides of QF3 within the given range. Then close the circuit breaker on this side to make the two power grids achieve the asynchronous interconnection at a small exchange power. Figure 2.1 shows the circuit breaker in the synchronizing close state.

2.4.4 Power Regulation

On the basis of ensuring the stable frequency of the VFT, adjust the angle difference between the rotor winding current magnetic field and the stator winding current magnetic field to adjust the transmission power of the VFT from close to zero to the preset power and achieve the stepless regulation and step regulation of the transmission power within the scope of design.

2.4.5 Capacitor Bank Switching

When the VFT transmits power, the total leakage reactance including the step-down transformers on both sides will absorb a certain amount of reactive power, resulting in voltage drop and increasing the system loss. At this point, measure the current of VFT and the system bus voltage and switch on the capacitor bank of a certain capacity in sets to optimize the system reactive power balance. In terms of selection of the input capacity of the capacitor, there are two methods. One is based on the principle of minimum reactive power exchange. In other words, the reactive power supplied by the capacitor bank should be equivalent to the reactive power absorbed by the leakage reactance of the VFT and step-down transformer. The other is to change the input capacity of the capacitor bank based on the bus voltage level to control the bus voltage within a reasonable range.

2.4.6 System Application Control

According to the actual needs of the electric power system operation control, use the power regulation as the basic means to achieve a series of functions such as system frequency regulation, suppression of low-frequency oscillation, optimization of the system power flow, and supplying power to passive systems and weak power systems.

2.4.7 Failure Cleaning

For the short-term external fault, generally there is no need to take any measures because after the fault is removed, the VFT system will resume normal operation. For the long-term or permanent external fault as well as the fault of the module of the VFT, the VFT should be removed through the circuit breaker according to the protection setting principle.

2.5 VFT Mechanism for Improving System Stability

In this section, we will use the simplified electric power system and its equivalent circuit diagram shown in Figure 2.7 as the research subject to discuss the impact of the VFT on the stability of the power angle of the electric power system.

In Figure 2.3, E'_1, E'_2 and G1, G2 are the transient electromotive force of the generator G1 and G2, respectively, and are approximately assumed to remain unchanged; δ_1 and

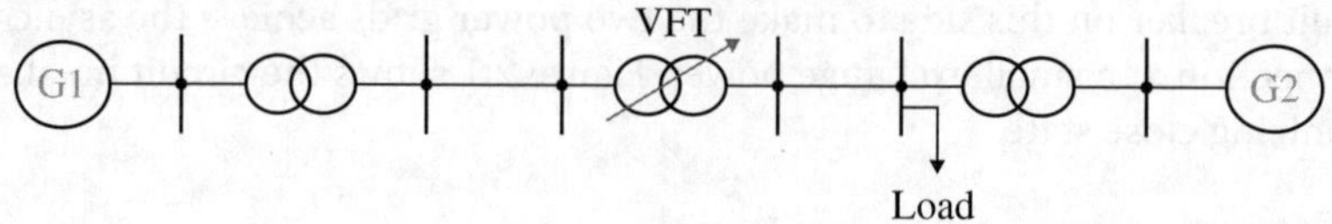

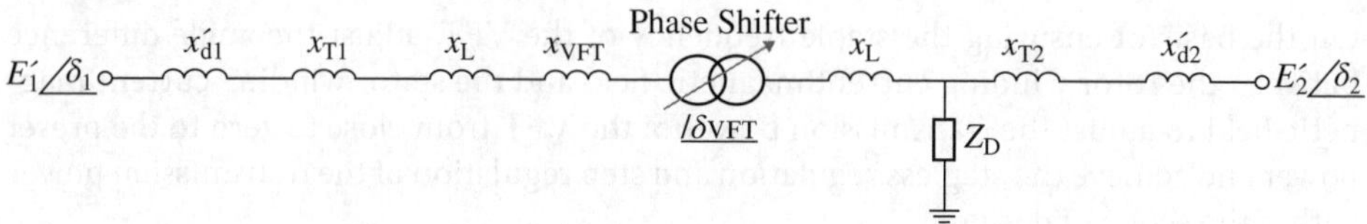

Figure 2.7 Simplified electric power system and its equivalent circuit diagram.

δ_2 are the transient electromotive force angle of the generator G1 and G2, respectively; constant impedance load is expressed as Z_D. The VFT is treated as the impedance X_{VFT} and is connected with the ideal phase shifter in series. The angle of the phase shifter is the stator and rotor phase shift of the VFT and expressed as δ_{VFT}. Conduct the Y-delta equivalent transformation on the equivalent circuit to eliminate the load nodes, the network will only contain generator nodes and the electromagnetic power of the generator can be calculated by Equation (2.1).

$$P_{Ei} = \mathrm{Re}(\dot{E}_i \hat{I}_i) = \mathrm{Re}\left(\dot{E}_i \sum_{j=1}^{G} \hat{E}_j \hat{Y}_{ij} \right)$$

$$= E_i \sum_{j=1}^{G} E_j(G_{ij} \cos \delta_{ij} + B_{ij} \sin \delta_{ij})$$

$$= E_i^2 G_{ii} + E_i \sum_{\substack{j=1 \\ j \neq i}}^{G} E_j(G_{ij} \cos \delta_{ij} + B_{ij} \sin \delta_{ij}) \tag{2.1}$$

where,

$Y_{ij} = (G_{ij} + jB_{ij})$ is the mutual admittance between the generator Node i and j.

$\delta_{ij} =$ angle difference between $\dot{E}_i$ and $\dot{E}_j$.

The electromagnetic power of the generator can be expressed as Equation (2.2) and Equation (2.3).

$$P_{E1} = E_1'^2 G_{11} + E_1' E_2'[B_{12} \sin(\delta_1 - \delta_2 - \delta_{VFT}) + G_{12} \cos(\delta_1 - \delta_2 - \delta_{VFT})] \tag{2.2}$$

$$P_{E2} = E_2'^2 G_{22} + E_1' E_2'[-B_{12} \sin(\delta_1 - \delta_2 - \delta_{VFT}) + G_{12} \cos(\delta_1 - \delta_2 - \delta_{VFT})] \tag{2.3}$$

The change of power relative to the generator power angle deviation is:

$$\Delta P_{E1} = \left(\frac{\partial P_{E1}}{\partial \delta_1} \right) \Delta \delta_1 + \left(\frac{\partial P_{E1}}{\partial \delta_2} \right) \Delta \delta_2 = K_{11} \Delta \delta_1 + K_{12} \Delta \delta_2 \tag{2.4}$$

$$\Delta P_{E2} = \left(\frac{\partial P_{E2}}{\partial \delta_1} \right) \Delta \delta_1 + \left(\frac{\partial P_{E2}}{\partial \delta_2} \right) \Delta \delta_2 = K_{21} \Delta \delta_1 + K_{22} \Delta \delta_2 \tag{2.5}$$

In which,

$$K_{11} = \left(\frac{\partial P_{E1}}{\partial \delta_1}\right) = -K_{12}$$

$$= E_1' E_2'[B_{12}\cos(\delta_1 - \delta_2 - \delta_{VFT}) - G_{12}\sin(\delta_1 - \delta_2 - \delta_{VFT})] \tag{2.6}$$

$$K_{21} = \left(\frac{\partial P_{E2}}{\partial \delta_1}\right) = -K_{22}$$

$$= E_1' E_2'[-B_{12}\cos(\delta_1 - \delta_2 - \delta_{VFT}) - G_{12}\sin(\delta_1 - \delta_2 - \delta_{VFT})] \tag{2.7}$$

Regardless of the damping of the system, Substitute ΔP_{E1} and ΔP_{E2} into the rotor motion equation, that is

$$\begin{cases} \dfrac{d\theta_{rm}}{dt} = \omega_{rm} = 2\pi f_{rm} \\[4mm] \dfrac{d\omega_{rm}}{dt} = \dfrac{1}{T_{Jr}}(T_d + T_r - T_s) \end{cases}$$

Where
θ_{rm} = rotor phase shift;
ω_{rm} = rotor angular velocity;
f_{rm} = rotor mechanical frequency;
T_{Jr} = moment of inertia of rotor;
T_d = mechanical torque exerted by the DC motor on the rotor;
T_r = torque exerted by the rotor winding current on the rotor;
T_s = torque exerted by the stator winding current on the rotor.

The obtained generator rotor operation equation is shown in Equation (2.8)

$$\frac{d\Delta\delta_1}{dt} = \Delta\omega_1\omega_0$$

$$\frac{d\Delta\delta_2}{dt} = \Delta\omega_2\omega_0 \tag{2.8}$$

$$\frac{d\Delta\omega_1}{dt} = \frac{-1}{T_{J1}}(K_{11}\Delta\delta_1 + K_{12}\Delta\delta_2)$$

$$\frac{d\Delta\omega_2}{dt} = \frac{-1}{T_{J2}}(K_{21}\Delta\delta_1 + K_{22}\Delta\delta_2)$$

Convert Equation (2.8) into matrix form to conduct eigenvalue analysis

$$\begin{bmatrix} \dot{\Delta\delta_1} \\ \dot{\Delta\delta_2} \\ \dot{\Delta\omega_1} \\ \dot{\Delta\omega_2} \end{bmatrix} = \begin{bmatrix} 0 & 0 & \omega_0 & 0 \\ 0 & 0 & 0 & \omega_0 \\ -\dfrac{K_{11}}{T_{J1}} & -\dfrac{K_{12}}{T_{J1}} & 0 & 0 \\ -\dfrac{K_{21}}{T_{J2}} & -\dfrac{K_{22}}{T_{J2}} & 0 & 0 \end{bmatrix} \begin{bmatrix} \Delta\delta_1 \\ \Delta\delta_2 \\ \Delta\omega_1 \\ \Delta\omega_2 \end{bmatrix} \tag{2.9}$$

Calculate the eigenvalue and the eigenvalue equation is shown in Equation (2.10)

$$\lambda^4 + \lambda^2 \omega_0 \left(\frac{K_{11}}{T_{J1}} + \frac{K_{22}}{T_{J2}} \right) + \omega_0{}^2 \left(\frac{K_{11}}{T_{J1}} \cdot \frac{K_{22}}{T_{J2}} - \frac{K_{12}}{T_{J1}} \cdot \frac{K_{21}}{T_{J2}} \right) = 0 \qquad (2.10)$$

As K has relations shown in Equation (2.6) and Equation (2.7), the eigenvalue equation can be changed into

$$\lambda^4 + \lambda^2 \omega_0 \left(\frac{K_{11}}{T_{J1}} + \frac{K_{22}}{T_{J2}} \right) = 0$$

$$\lambda^2 + \omega_0 \left(\frac{K_{11}}{T_{J1}} + \frac{K_{22}}{T_{J2}} \right) = 0 \qquad (2.11)$$

The eigenvalue can be obtained from Equation (2.11)

$$\lambda_{1,2} = \pm \sqrt{-\omega_0 \left(\frac{K_{11}}{T_{J1}} + \frac{K_{22}}{T_{J2}} \right)} \qquad (2.12)$$

In fact, the system has damping, so its static stability criterion is

$$\frac{K_{11}}{T_{J1}} + \frac{K_{22}}{T_{J2}} > 0 \qquad (2.13)$$

Substitute the coefficients K_{11} and K_{22} into Equation (2.13) to simplify

$$\frac{\cos(\delta_1 - \delta_2 - \delta_{VFT})}{\sin(\delta_1 - \delta_2 - \delta_{VFT})} > \frac{G_{12}}{B_{12}} * \frac{T_{J1} - T_{J2}}{T_{J1} + T_{J2}}$$

namely,

$$\tan(\delta_1 - \delta_2 - \delta_{VFT}) < \frac{B_{12}}{G_{12}} * \frac{T_{J1} + T_{J2}}{T_{J1} - T_{J2}} \qquad (2.14)$$

Discuss in $\left[-\frac{\pi}{2}, \frac{\pi}{2} \right]$, Equation (2.14) can be rewritten as

$$\delta_1 - \delta_2 < arctan \left(\frac{B_{12}}{G_{12}} * \frac{T_{J2} + T_{J1}}{T_{J2} - T_{J1}} \right) + \delta_{VFT} \qquad (2.15)$$

It can be seen from Equation (2.15) that $arctan \left(\frac{B_{12}}{G_{12}} * \frac{T_{J1}+T_{J2}}{T_{J1}-T_{J2}} \right)$ is a constant; changing the δ_{VFT} will change the stable operation range of the power system. If the disturbance leads to $\delta_1 - \delta_2$ exceeds the maximum value of the stability limit, increase δ_{VFT} to enable the system to restore stability. It can be seen from it that the VFT is equivalent to a movable balancer that can control and improve the stability of the system very well.

2.6 Existing VFT Applications in Power Systems

The variable frequency transformer was first developed by the GE Company in the US in the 1990s and succeeded in applying them to the power grid at the beginning of the twenty-first century. See the Appendix for the application of VFTs in North America.

Table 2.2 Summary table of VFTs put into operation in the world [1–5, 23].

No.	Name of Substation	Rated Parameters	Function	Operation Time	Notes
1	Langlois Substation in Canada	Phase I: one 105 MVA and 17 kV rotary transformer; 3000 horsepower (1 horsepower = 7355 W, the same below) DC motor and variable-speed drive system; 3 sets of switched capacitors with a capacity of 25 Mvar; 2 120/17 kV conventional step-up transformer.	Realize the interconnection of the power grids in Québec, Canada, and power grids in New York, America.	2003	Phase II: construction of VFTs of the same capacity.
2	Laredo Substation in America	One 100 MVA and 17 kV rotary transformer; 3750 horsepower DC motor and variable-speed drive system; 4 sets of switched capacitors with a capacity of 25 Mvar; 2 units of 142/17.5 kV conventional step-up transformer.	Used for the interconnection of the power grids in Texas, America, and power grids in Mexico.	2007	For this project, using the VFT is better than the back-to-back flexible DC transmission.
3	Linden Substation in America	3 units; each unit has one 100 MVA and 17 kV rotary transformer, one 230/17 kV step-up transformer and one 345/17 kV step-up transformer; one 17 kV circuit breaker and DC motor and variable-speed drive system, capacitor banks.	Strengthen the interconnection of PJM power grid and NYISO power grid.	2010	The compact design greatly reduces the used land area.

Listed in Table 2.2 are the parameters of VFT that have been put into operation in the world, their operation time, and other data. The world's first VFT was put into operation at Langlois Substation, Québec, Canada in October 2003, and used for the interconnection with the power grid in New York. During the project operation period, a large number of experiments were made to verify the adaptability of VFTs. For example, the power step response time of the VFT is about 0.4 s, mainly because the core equipment of the VFT is an electrical rotating machine with great inertia and the step response time is longer than the high-voltage DC response time. Meanwhile, the VFT can realize the stepless regulation of transmission power between asynchronous power grids and there is no power flow dead zone. In addition, VFTs can meet the needs of the island grid power supply.

The safe operation of the first VFT gradually attracted the attention of users and the electric power system researchers. Some electric power enterprises and researchers began to study VFTs comprehensively. With the deepening of the research over the time, the advantages of the VFT became increasingly prominent and its application scope was constantly expanded. The world's second VFT was funded by AEP and manufactured by GE. Installed at Laredo Substation in southwestern Texas, it was put into operation in 2007 and used for the interconnection of the Texas Power Grid with Mexico Power Grid. The world's third VFT project was installed between PJM power grid and NYISO power grid. A total of three 100 MW VFTS were installed and the project was put into use in early 2010. Thanks to its compact design and less land occupation, this project solved the problem of land resource shortage for the old power grid reconstruction.

2.7 VFT Applications in Global Energy Interconnection

2.7.1 Introduction of Global Energy Interconnection (GEI)

GEI is the advanced form of large power grid interconnection and the development idea for the interconnection of the global energy proposed by Mr. Liu Zhenya is based on China's successful practice of UHV power grids and the world's smart grid and clean energy development trends. The GEI is a globally interconnected strong and smart power grid with the UHV power grids as the backbone network, focusing on clean energy transmission. The GEI is composed of the backbone network of transnational, intercontinental UHV power grids, and smart grids at all voltage levels in all countries, connecting "wind power in the Arctic region and solar power around the equatorial region" and large energy bases in all continents. Adapting to the large-scale development of centralized and distributed power supply and supplying wind energy, solar energy, hydro-energy, ocean energy, and other forms of renewable energy to various users, it is a basic platform for the large-scale development, distribution, and utilization of clean energy in the world. In short, the essence of GEI is "Smart grid + UHV + Clean energy."

2.7.1.1 Smart Grid

The smart grid is an important foundation for building the Global Energy Interconnection. The smart grid integrates advanced transmission, intelligent control, access to new energy, new energy storage, information network, and other modern intelligent technologies. With strong flexibility and adaptability, it can meet the demands for integrating clean energy and distributed power, plug, and play of intelligent devices and intelligent interactive services. It is the result of innovative development of the power grid technology in the new century. The smart grid originated in Europe and the United States, but it is experiencing the fastest development and widest application in China. Figure 2.8 shows the conception of the smart grid.

Promotion of the smart grid: Since the 1990s, with the rapid development of information, communication, sensing, and electric power, smart grid technology has been gradually popularized. Faced with worldwide security pressure from power grids, energy, and economy, this new century is nurturing and developing the third Industrial Revolution featured by IT technology and renewable energies. Accelerated development for the smart grid is being ushered in.

Power system blackouts: 140 widespread blackouts have occurred since 1965, causing huge economic and social losses. The large-scale August 14 Blackout in 2003 in North America affected 50 million people and lost 61.8 GW of electricity, causing direct

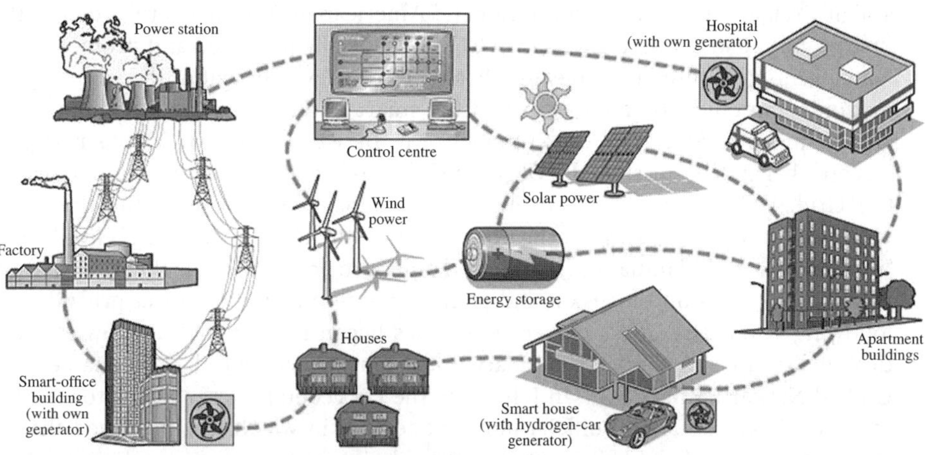

Figure 2.8 Concept of the smart grid. Picture source: CEPRI.

economic loss of 1 billion US dollars and indirect economic loss of 30 billion US dollars. The Indian blackout on 30 and 31 July 2012 took 48 GW of power offline, affecting 670 million people. Building a smart grid has become a major solution for utilities and experts to address grid security and prevent widespread blackouts.

Energy and environmental conflicts: World energy shortage and environmental pollution have become ever more prominent. New energies have to be developed and utilized to address the energy shortage. However, the intermittent and random features of wind and solar power pose a higher requirement on the security and adaptability of the power grid. The smart grid thus becomes the key to promote new energy and realize a coordinated development of energy and environment.

International financial crisis: Since 2007, the global financial crisis plunged the world economy into recession. The economic revitalization has become the core for current development, which requires new economic growth point. As a technology-intensive and capital-intensive industry, smart grid provides the solution for cracking down practical grid difficulties and energy problems. This strategic emerging industry can also boost the economic development and drive investment and employment. Therefore, the development of the smart grid has become the common choice among various countries, energy industries, and power companies. As countries may differ in their economic development stages, energy resource endowments, and electric technology levels, their philosophies and focuses on smart grid development are different. For example, the initial drivers for the US to develop a smart grid is to improve grid structure and enhance power grid security and stability, while Europe wants to develop new energies. The drivers for China are to upgrade grid technology, build robust grids, and improve the grids' capabilities to allocate resources optimally, as well as to improve their adaptability, and meet the demands for large-scale integration and accommodation of various kinds of energies. Relatively speaking, China has more to ask from smart grid developments, due to more pressure, and challenges.

The smart grid in the United States: The US has played a leading role in promoting smart grid. In 1998, Electric Power Research Institute (EPRI) carried out the research on the "complex interactive networks/system initiative" with the purpose to create a highly reliable and completely automatic American grid, which is regarded as the first prototype of a smart grid. In 2002, EPRI officially proposed and promoted the "Intelligent grid" research project dedicated to the development of the whole information communication architecture of the smart grid and the business innovation and technical research and development of it. Between 2003 and 2005, smart grid research began to flourish in the USA. The US Department of Energy (DOE) released the "Grid2030" and "national transmission technology roadmap," depicting the future vision and technical strategy of the power grids in the USA; it launched "Grid Wise," "modern power grid initiative (MGI)," and other projects, and made the whole nation reach a consensus on the vision and plan for modernization of power grids. In the following years, The USA's power enterprises began to carry out a series of practical applications in the field of smart grids. This smart grid plan was named the Unified National Smart Grid. In July 2009, the US Federal Energy Regulatory Commission (FERC) issued an official policy statement in which it is proposed that in the construction of the US smart grid should give priority to the making of standards including system interface, network security, wide-area measurement, demand response, energy storage technology, and electric vehicles in the hope of laying foundations for the formulation of other standards and promoting the construction of smart grid and driving the development of smart grid-related industries. In August 2009, President Obama delivered a speech in Indiana, in which he put forward new viewpoints on the development of smart grid and held that the United States should build "stronger and more intelligent power grids that can realize the transmission of power between the East and the West." Figure 2.9 shows the concept of the US

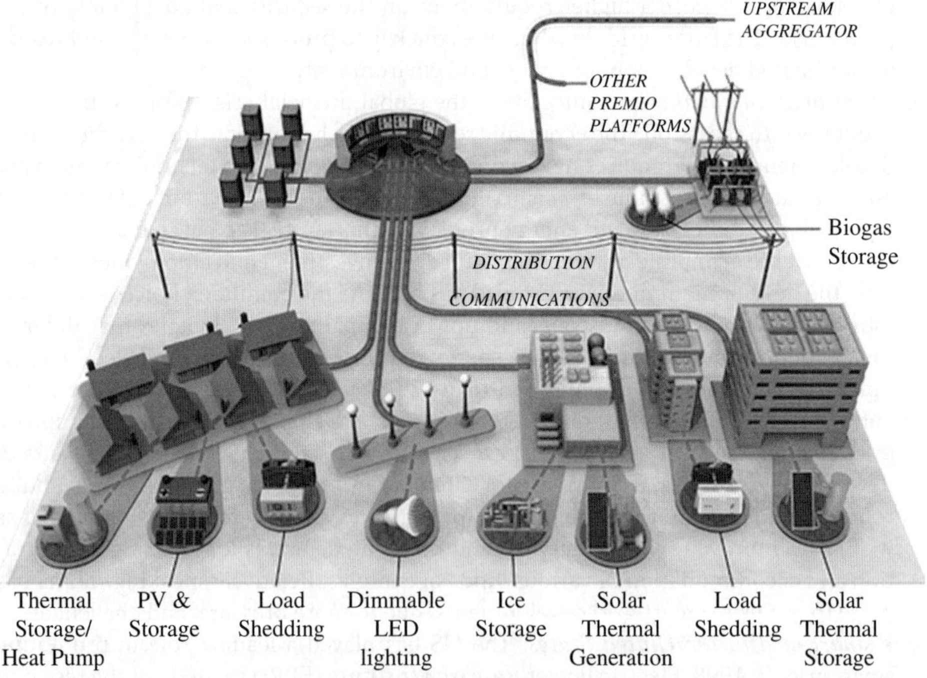

Figure 2.9 Concept of the smart grid in US Grid2030. Picture source: DOE, United States, Grid2030.

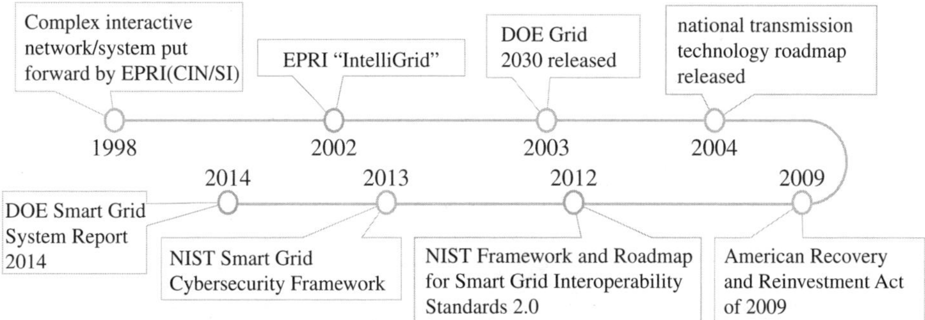

Figure 2.10 General roadmap for the development of smart grid technology in USA. Picture source: DOE, United States, Grid2030.

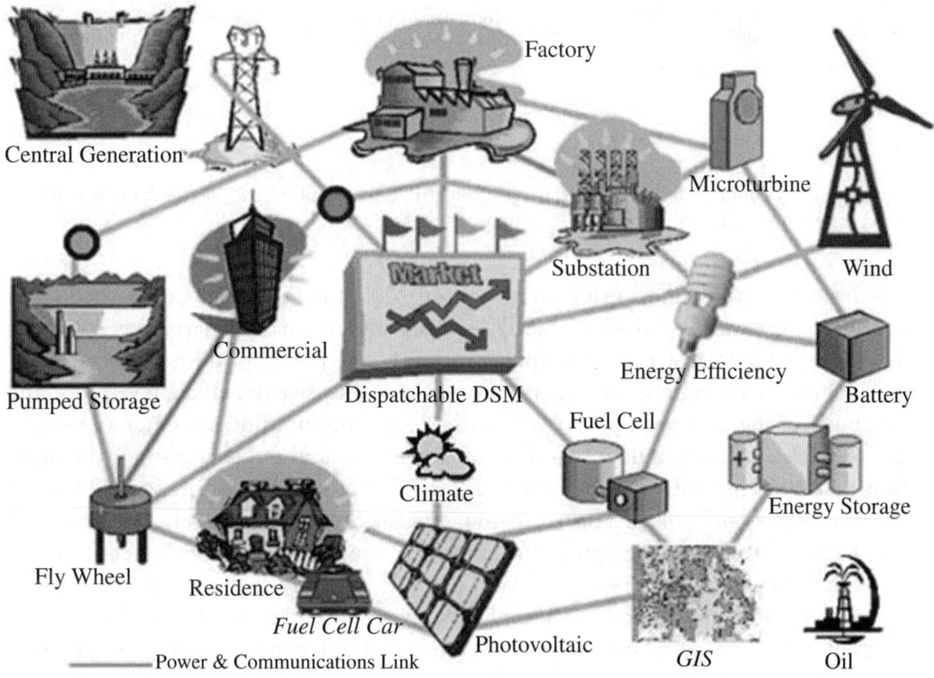

Figure 2.11 Schematic diagram of Pacific Northwest smart grid demonstration project. Picture source: Pacific Northwest National Laboratory, USA.

Smart Grid in Grid2030. Figure 2.10 shows the general roadmap for the development of smart grid technology in the USA. Figure 2.11 shows a schematic diagram of the Pacific Northwest Smart Grid Demonstration Project.

Europe's smart grid: EU attaches great importance to smart grids. Figure 2.12 shows the concept of the Smart Grid by Eurelectric. European countries have different power grid operation modes and their power demand tends to be saturated. The energy policies of these countries give priority to environmental protection and the development of renewable energy power generation. A low carbon economy is the main driver of

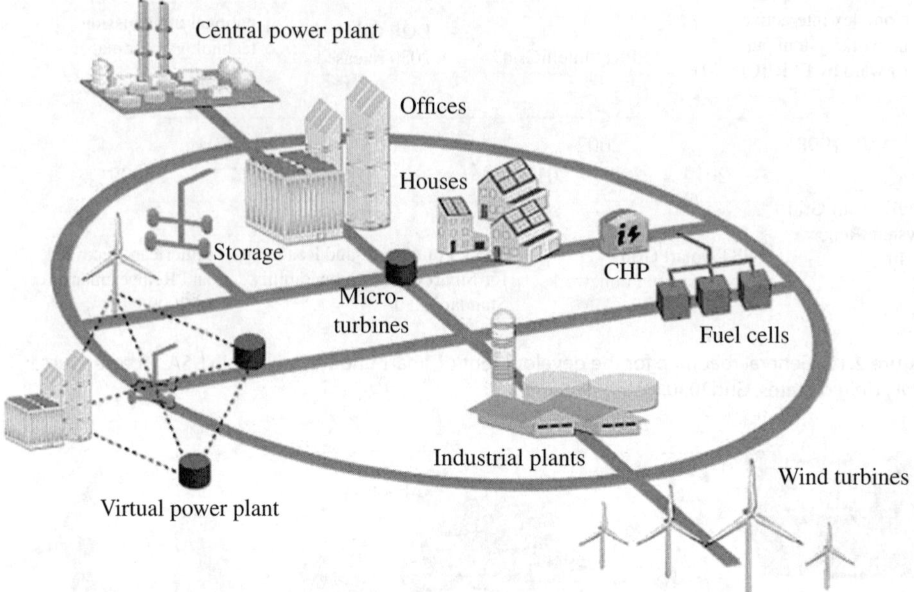

Figure 2.12 Conception of the smart grid from Eurelectric. *Source*: Eurelectric. Picture source: Federal ministry for economic affairs and energy, Germany.

the development of smart grids in Europe. In 2005, the European Commission first proposed the concept of the smart grid in Europe and set up the SmartGrids European Technology Platform with the purpose to change the power grid into a service network of interactions between users and operators, improve the efficiency, safety, and reliability of the European power transmission and distribution system, and remove the obstacles to large-scale integration of distributed and renewable energy generation. In 2006, the Platform released a report titled Vision and Strategy for Europe's Electricity Networks of the Future, formulated the Strategic Research Agenda (SRA) and Future Development Strategy for European Power Grids, and instructed the EU and other countries to carry out relevant projects to promote the implementation of smart grids. In September 2009, the EU proposed in the Energy Technology Development Strategy that it would select 30 cities to implement trial projects of smart grid to become the leader in the global green science and technology competition. EU described the initiative of using high technology to cope with climate changes and enabling European enterprises to seize the opportunity for a global low carbon economic transformation. Billions of Euros would be invested in the R&D and energy technologies to assist the EU in competing with the USA who invested $777 million. The energy in the smart grid pilot cities comes from the recycling of garbage, solar energy, and wind energy. This self-produced energy will be used to drive electric vehicles, trams, and other vehicles. These smart grid cities will become the core of the smart grid. The new generation of energy-efficient buildings and the transportation modes using renewable energy will become a reality in Europe.

China's smart grid: China started late in the development of smart grids, but they have developed very rapidly in China. China has surpassed European and American countries in smart grid plan formulation, technical research, equipment development, and engineering applications to take the leading position in the world. Since the beginning of the twenty-first century, the SGCC has carried out theoretical research and technological breakthroughs on smart grids, developed a complete smart grid development planning and technical standard system, completed and put into operation a large batch of smart grid pilot and demonstration projects covering new energy access, distributed power, intelligent transmission lines, intelligent substations, intelligent distribution networks, intelligent scheduling, electric vehicle charging and battery swapping, advanced measurement systems (AMI), PFTTH, and an intelligent community.

China's smart grid concept: At the International Conference on Ultra High-Voltage Transmission Technology held in Beijing in 2009, the SGCC proposed the construction of information-, automation-, and interaction-based strong and smart power grids with UHV as the backbone network and characterized by coordinated development of power grids at all levels [7]. The strong and smart grid will realize the full *integration* of "strong grid" and "smart technology," diverse comprehensive intelligence of power generation, transmission, transformation, distribution, consumption, and scheduling and overall integration of the grid information flow, power flow, and business flow. Information, automation, and *interaction* are important features of smart grids, including highly integrated grid topology information, power consumption information, control information, business information and management information, high automatic control of whole network collaboration, integrated regulation and control and fault self-removal, and a high interaction of power generation, power grid, consumers, resources, and the environment. Figure 2.13 shows the concept of the smart grid in China.

China's smart grid planning: The construction of a smart grid is a systematic project. Scientific planning is the prerequisite for the realization of the mutual connection and coordinated development of the power grid. Based on this consideration and in accordance with the principle of constructing strong and reliable, economic and efficient, clean and green, transparent and open, friendly, and interactive smart grids, SGCC has drawn up a development plan covering power generation, transmission, transformation, distribution, consumption, and scheduling and the construction of a communication and information platform to set up smart grid development goal and focus [8]. It is the first smart grid development plan covering all fields of the power grid in the world. According to the plan (Figure 2.14), China's smart grid development will go through three stages: the planning and pilot stage (kick-off 2009–2010), the comprehensive construction stage (roll-out 2011–2015), and the leading and promotion stage (scale-up 2016–2020). At present, China is in the comprehensive construction stage.

China's smart grid standards: There are a lot of institutions focusing on the research into smart grids and many smart grid equipment makers in China. As a great variety of smart grid technologies and devices are being developed and deployed, it is urgent to solve the problems of inconsistent standards and different and *incompatible* interfaces. SGCC has established a smart grid technology research system dominated by grid enterprises, participated in by scientific research institutions, equipment manufacturers, and electricity customers, while covering technological breakthrough, equipment development, test and inspection, engineering application,

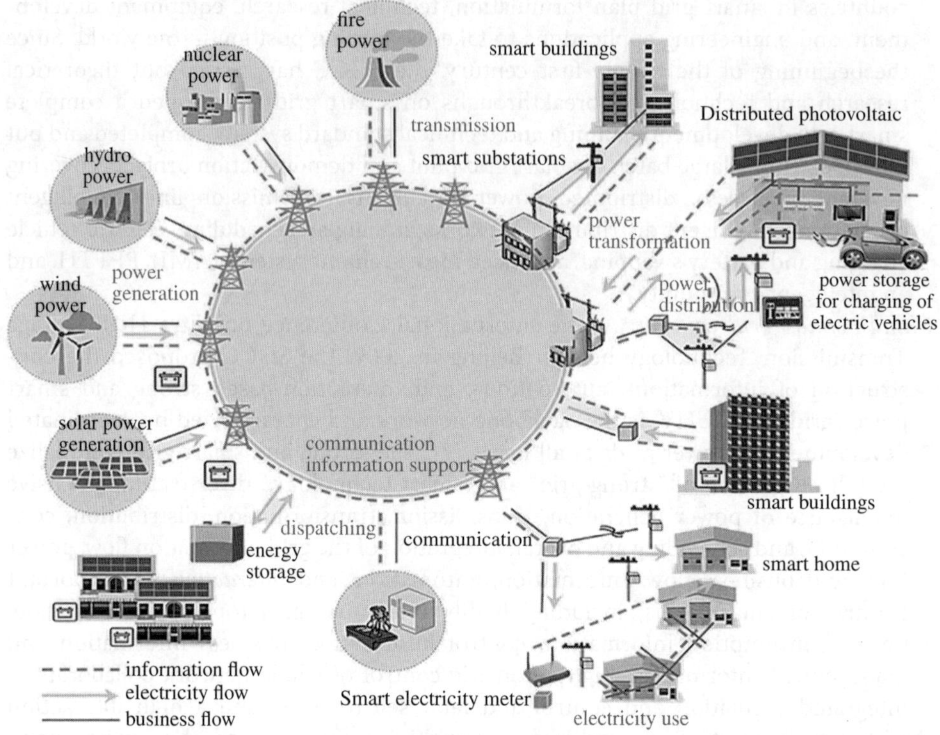

Figure 2.13 Concept of the smart grid in China. Picture source: SGCC.

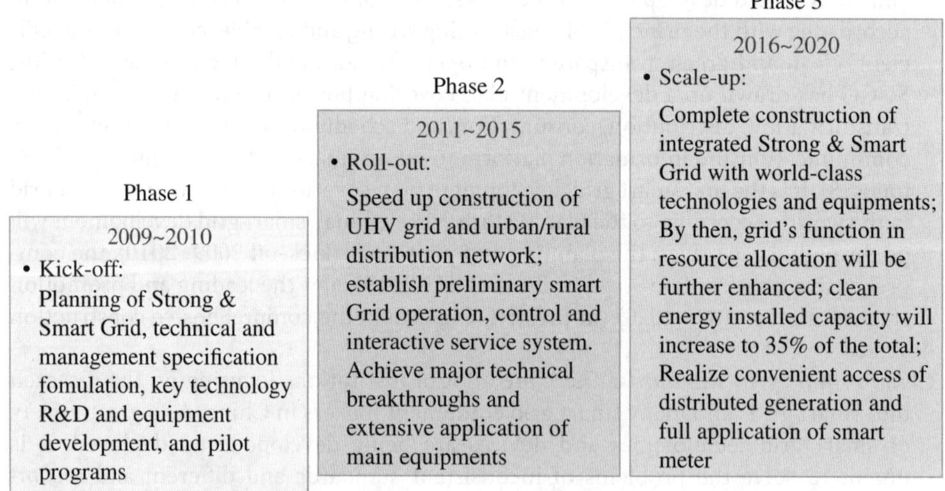

Figure 2.14 SGCC's smart grid development planning and construction goal. Picture source: SGCC.

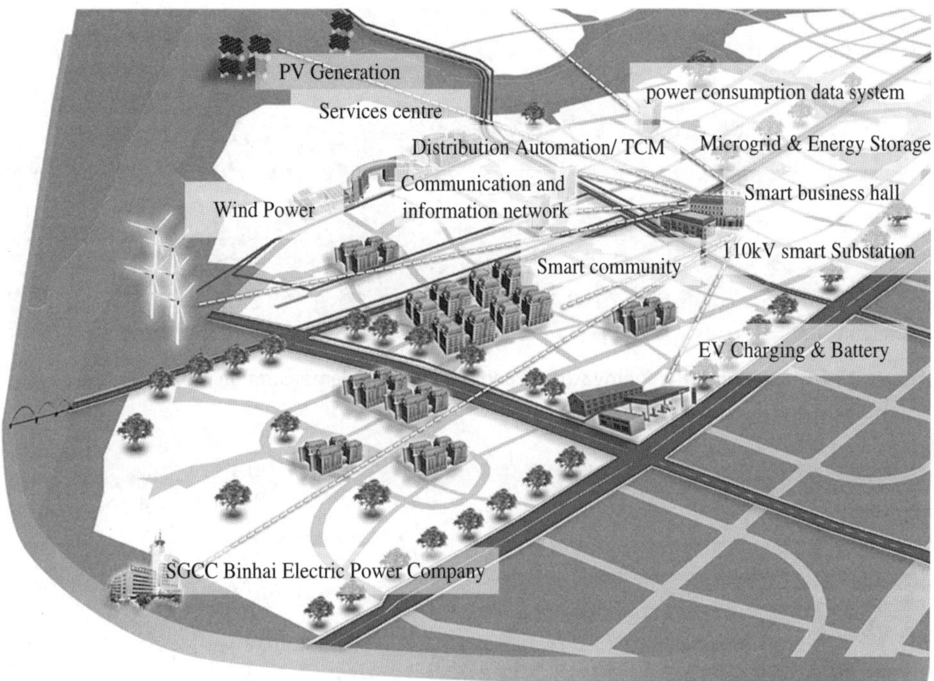

Figure 2.15 Sino–Singapore Tianjin Eco-City Smart Grid Comprehensive Demonstration Project. Picture source: SGCC.

and standard formulation: (1) It established the State Energy Smart Grid R&D Center; focusing on the research into customized power, intelligent power transmission and transformation, flexible transmission, microgrids, smart electricity consumption, energy storage, energy efficiency evaluation, information security technology, and so on. (2) It built the NESC and National Energy Large-Scale Wind Power Integration System R&D Center, possessing the perfect wind turbine and PV power plant testing abilities. (3) It promoted the IEC to establish three new technical committees including TC115 high-voltage DC, the PC118 smart grid user interface, and integration of high-capacity renewable energy into the power grid. Based on in-depth studies, the SGCC has released 363 UHV and smart grid enterprise standards, developed 145 industry standards, 66 national standards, and 19 international standards, forming a more perfect smart grid technology standard system.

Application of smart grids in China: Since 2009, the SGCC has established a batch of smart grid pilot and demonstration projects covering new energy integration, distributed power, intelligent transmission lines, intelligent substations, intelligent distribution networks, intelligent scheduling, electric vehicle charging and battery swapping, advanced measurement systems (AMI), PFTTH, and intelligent community, achieving remarkable results [10]. Figure 2.15 shows the Sino–Singapore Tianjin Eco-City Smart Grid Comprehensive Demonstration Project.

Power generation field: Serving new energy development is an important goal in the construction of smart grid and the key is to solve problems of random and intermittent integration and accommodation of energy forms such as wind and PV

Figure 2.16 China's wind/PV/energy storage and transmission demonstration project. Picture source: SGCC.

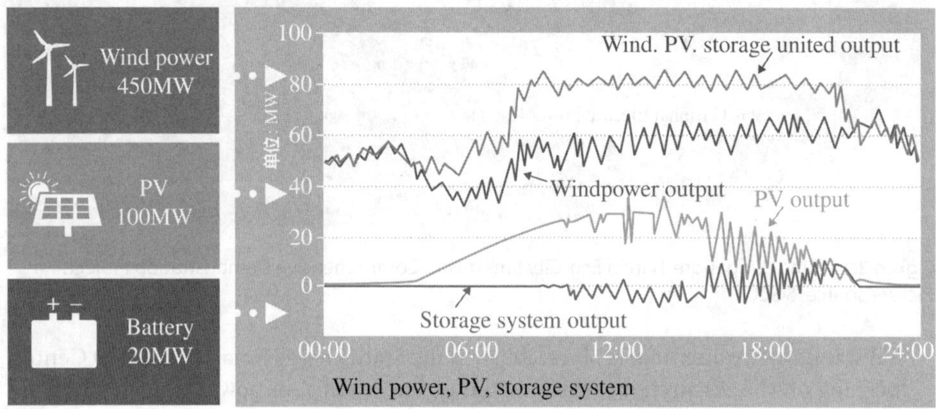

Figure 2.17 Performance of power output for China's wind/PV/energy storage/transmission demonstration project. Picture source: SGCC.

power. For this purpose, SGCC has carried out research on large-scale wind power forecasting and operation control, large-scale grid-connected PV power operation, conventional power plant coordination, and large-scale new energy power generation cluster control, and operational control of power generation from a great variety of clean energy sources including wind energy, PV energy and hydro-energy, distributed power, microgrid integration control, and applications of energy storage systems.

The SGCC has built the national wind/PV/energy storage/transmission demonstration project. After completion of construction, this project will have a total capacity of 670 MW. Phase I of the project involves 100 MW wind power, 40 MW PV power, and 20 MW energy storage (Figure 2.16). This project has come out in front in the world in terms of the wind turbine type, installed capacity of power-regulated PV power, scale of diverse chemical energy storage, and joint operational technology of new energy sources. It realizes the joint operation of wind power, PV power, energy storage, and transmission and significantly improves the power grid's new energy integration and accommodation capacities. Performance of power output for this project is shown in Figure 2.17. The application of these technologies has promoted China's new energy development. By the end of 2017 China's installed capacity of grid-connected wind power had reached 160 GW, ranking first in the world.

Breakdown: wind power generation: 24 sets of 2 MW doubly fed variable-speed wind turbines, 15 sets of 2.5 MW, two sets of 3 MW, and one set of 5 MW permanent magnet direct drive wind turbines, two sets of 1 MW vertical axis wind turbine; *PV power generation*: polycrystalline silicon PV module 37 MW, single crystalline silicon PV module 1 MW, amorphous thin film PV module 1 MW, back-contact PV module 0.5 MW, and HCPV cell module 0.05 MW; *energy storage device*: 14 MW ferric phosphate lithium cell, 2 MW vanadium redox flow battery, 1 MW sodium flow battery, 1 MW lithium titanate battery, and 2 MW gel battery.

Transmission field: Improving the power grid's transmission capacity and security is a key point in the construction of a smart grid. For China, the key is to solve the problems of long distance, large capacity, and high efficiency power transmission, the construction of EHV grids and improve the online monitoring and charged overhaul capacities of the transmission lines. For this purpose, SGCC has carried out research on UHV AC, UHV DC, flexible DC, flexible AC transmission, large section conductors, steel towers, disaster warning systems, line state detection systems in helicopters and UAV intelligent inspection, and high altitude transmission. At the same time, the SGCC has developed transmission technologies including environment-friendly transmission line construction, online monitoring, helicopter inspection, site maintenance, and ice and wind protection (Figure 2.18).

Power transformation field. The substation, especially the load-center substation, is an important part of the power system. Construction of a smart substation is the key to improving the intelligence level of the whole power system and the focus is to realize the digitization, equipment integration, business integration, and compact design of the whole substation. (1) *Digitalization*. To digitalize all kinds of signals, equipment, and control devices to form a digital substation model to set up a platform

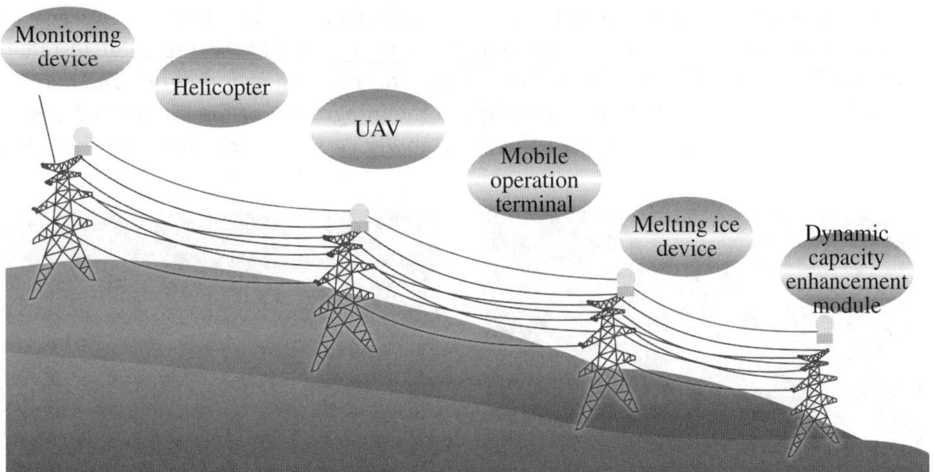

Figure 2.18 Main equipment in the intelligent transmission line. Picture source: SGCC.

for the comprehensive intelligent control and efficient management. (2) *Equipment integration*. To use new technologies, new materials, and new processes to optimize the design of key devices such as transformers and circuit breakers, and integrate relevant sensors and intelligent components to enhance equipment functions, control the size, and improve reliability. (3) *Business integration*. To integrate protection control, automation, and communication systems and integrate businesses such as online monitoring, site inspection, and operation maintenance to build an integrated business system, reduce crossing and repetition, realize the coordinated control, and improve the overall efficiency. (4) *Compact design*. To promote owner-dominated integrated design based on characteristics of different types of substations of different voltage levels to optimize the main wiring and station layout, and reduce area occupation and investment.

In order to achieve these goals, SGCC has carried out research on intelligent substation digital acquisition and control, equipment integration and system integration, equipment fault condition monitoring and self-diagnosis system, integrated business platform, and panoramic visual information system. By the end of 2013, SGCC had built 843 intelligent substations ranging from 110 (66) kV to 750 kV. At present, the SGCC is carrying out the development and construction of a new generation of intelligent substations in accordance with the principle of highly integrated system, reasonable structural layout, applicable advanced equipment, energy conservation and environmental protection, and the integration of support and regulation to realize the goals of smaller areas, lower cost, and high reliability. Let's take the 220 kV open-type substation as an example, the floor area is reduced by 40%; equipment investment is reduced by 20%; and the failure rate is reduced by 32% (see Figure 2.19 and Figure 2.20).

Distribution field. In recent years, with the popularization and application of distributed power, microgrids, electric vehicle charging, and discharging devices, the structure and functions of the distribution network has undergone important changes, changing from the original unidirectional radial power supply to multi-directional interactive power supply. According to China's planning, by 2020 the capacity of distributed power will reach 180 GW, accounting for 11% of the total installed capacity and electric vehicles in China will total more than 5 million. The core of intelligent power distribution is to adapt to the integration of the distributed power and microgrids, and to improve the adaptability of the distribution network. For

Figure 2.19 Interval optimization effect of a new generation of 220 kV open-type smart substations. Picture source: SGCC.

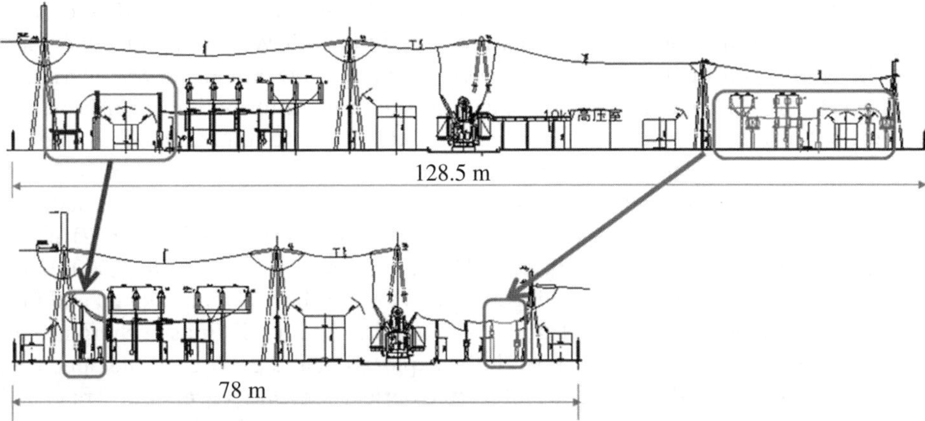

Figure 2.20 Optimization effect of the layout of new generation of 220 kV open-type smart substations. Picture source: SGCC.

this purpose, SGCC has carried out the research on and pilot construction of power distribution automation, smart parks, urban energy storage devices pilot projects, self-healing control systems, and emergency repair management systems. Compared with the traditional distribution network, the intelligent distribution network has significant advantages such as comprehensive information monitoring, automatic fault location, rapid emergency repair, line self-healing, rapid restoration of power supply, and friendly accommodation of clean energy (Figure 2.21).

Electricity consumption field. The key to intelligent electricity consumption is to meet the diverse needs of consumers and enhance interactivity. As a result, SGCC has carried out the research on and application of smart meters, electricity consumption

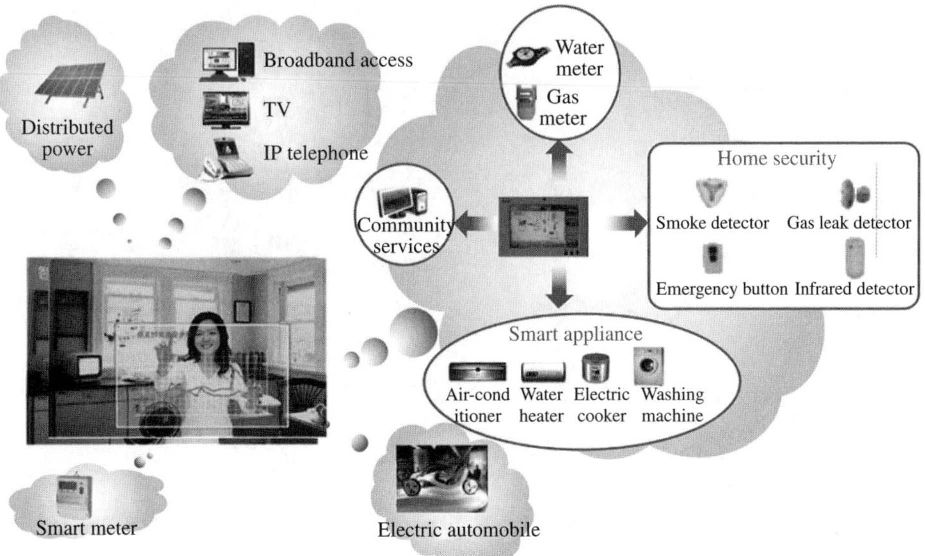

Figure 2.21 Comparison of intelligent and traditional distribution networks. Picture source: SGCC.

information acquisition system, interactive marketing services, demand side management, consumer side distributed power supply, electric vehicle charging facilities, bidirectional automatic settlement of electric charges, power quality monitoring, and PFTTH. (1) SGCC has built the electricity consumption information acquisition system for 202 million smart meters, which realizes the remote automatic meter reading, auto top-up, intelligent control, real-time power consumption control, line loss monitoring, and orderly electricity consumption management, and can meet the management of electric charges and price after the integration of distributed power. (2) SGCC has built 400 electric vehicle charging and battery swapping stations and 19,000 charging piles, forming the quick charging networks for four horizontally aligned and four vertically aligned expressways (Figure 2.22), to fully support all kinds of charging and battery swapping modes and carry out the exploration and practice of the operation mode. (3) SGCC has developed the integrated optical fiber and power optical fiber and power line, and completed the PFTTH of 270,000 households. In addition to transmitting power, it has spread Internet, telecommunications, radio and television signals for final consumers, provided various value-added services, innovated the grid operation mode, and provided society and the public with more convenient, abundant, and efficient services (Figure 2.23).

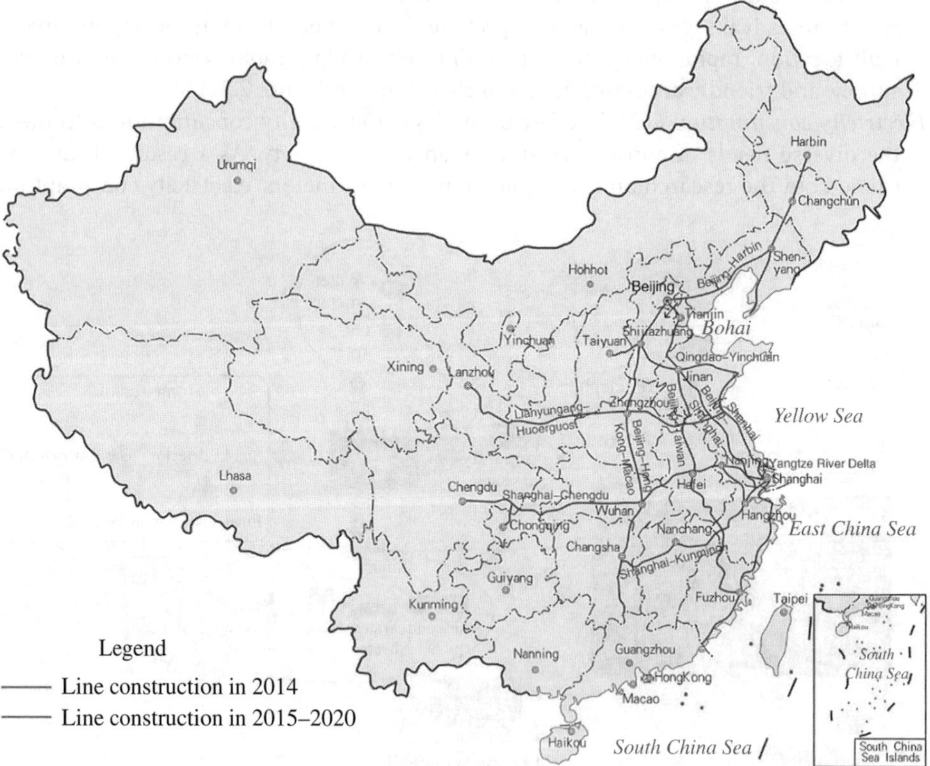

Figure 2.22 Quick charging networks for four horizontally aligned and four vertically aligned expressways in 2020. Picture source: SGCC.

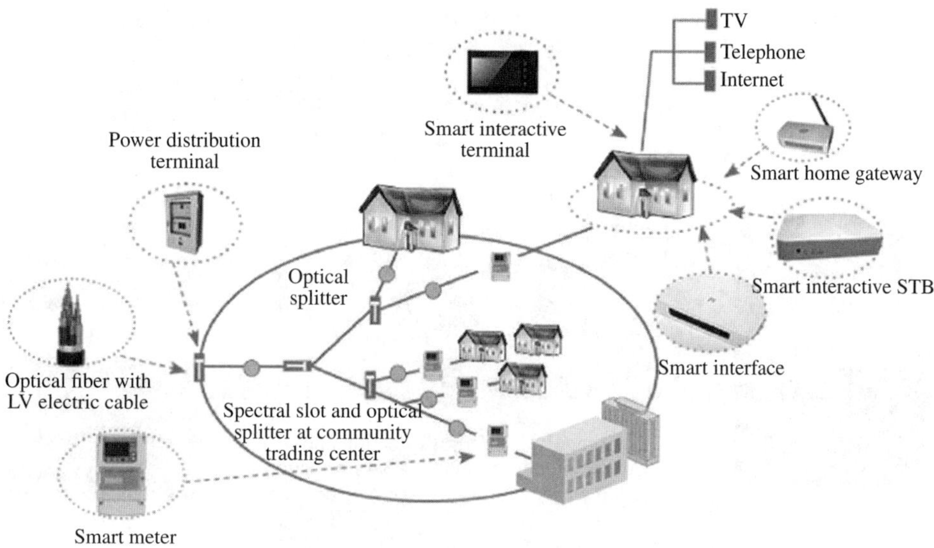

Figure 2.23 Schematic diagram of PFTTH. Picture source: SGCC.

Dispatching field. Dispatching is the control center of power grid operation. The dispatching of modern large power grids not only has to meet the requirements for the safety of the power grid, but also ensure the efficient utilization of clean energy and economic operation of the power system. In recent years, based on the unified dispatching of the whole network, SGCC has constructed the new generation of intelligent grid dispatching technology support systems covering the nation, provinces, and cities (counties) and gradually realizing the panorama of operation information, data transmission network, online security evaluation, refinement of dispatching decisions, operation control automation, optimization of coordination between power grids and power plants to ensure safe and reliable operation, flexible coordination, high quality and efficiency, economy and environmental protection of power grids, and maintaining a good record of safe operation of large power grids.

Up to now, SGCC has built the SCADA/EMS system covering the dispatching of all power grids at or above the municipal level, the PMU phasor measurement system covering the vast majority of power plants and substations of 500 kV and above, the large-scale wind power prediction system covering major wind farms, thunder, lightning, and other adverse weather monitoring systems covering major transmission channels and a large-scale grid simulation system with real-time simulation capacity. Through these systems, we can fully grasp the operation state, environment, and development of power grids, judge the emergencies in advance, and take corresponding measures in a timely manner. Through the joint operation of wind power, PV power, hydropower, thermal power, and other different types of power supply, we can realize the effective utilization of wind power, PV power, and hydropower. In 2013 China's clean energy power generation accounted for 29.3% of the total power generation. A panorama of power grid scheduling and operation information is shown in Figure 2.24.

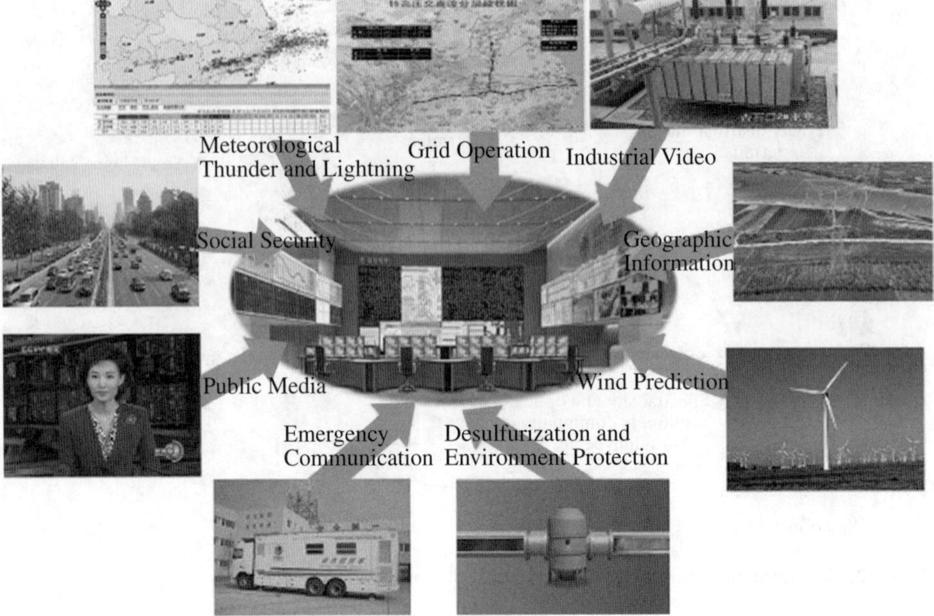

Figure 2.24 Panorama of power grid scheduling and operation information. Picture source: SGCC.

2.7.1.2 UHV Grid

The UHV power grid is the *key* to building the Global Energy Interconnection. UHV power grids are composed of AC systems of 1000 kV and above, and DC systems of ±800 kV and above. Characterized by long transmission distance, large capacity, high efficiency, low loss, land conservation, and sound safety, UHV power grids can transmit 10 GW power over thousands of kilometers and achieve the interconnection of trans-regional, transnational, and transcontinental power grids. The distances between continents and between continental energy bases and load centers are within the transmission range of UHV AC and DC power grids. UHV AC is mainly deployed for building strong national, continental, intercontinental synchronous power grids, and long-distance, and large capacity power transmission; UHV DC is mainly deployed for the super-long-distance and super-large capacity power transmission from large energy cases and the construction of transnational and transcontinental backbone transmission channels. Figure 2.25 shows the advantages of UHV transmission technology.

UHV transmission technology is efficient and highly cost effective. The transmission power, distance, and corridor efficiency of the 1000 kV UHV AC transmission line are 4-5, 2-3, and 2-3 times the transmission power, distance, and corridor efficiency of 500 kV AC transmission lines, respectively; the transmission loss and the unit cost of the 1000 kV UHV AC transmission line are 33 and 70% of the transmission loss and the unit cost of the 500 kV AC transmission line, respectively. The transmission power, distance and corridor efficiency of the ±800 kV UHV DC transmission line are 3.3, 3.0, and 1.5 times the transmission power, distance, and corridor efficiency of ±500kV DC transmission line, respectively; the transmission loss and the unit cost of the ±800 kV UHV DC transmission line are 40 and 70% of that of the ±500 kV DC transmission line, respectively. The transmission power, distance, and corridor efficiency of the

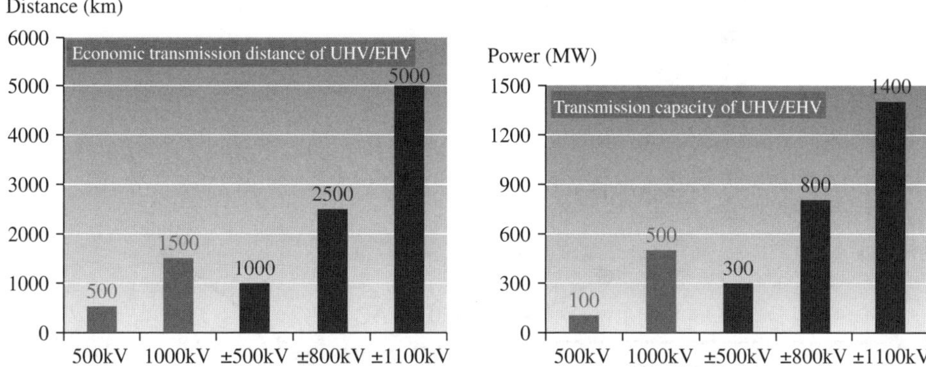

Figure 2.25 Advantages of UHV transmission technology. Picture source: SGCC.

±1100 kV UHV DC transmission line are 4-5, 5-6, and 1.9 times the transmission power, distance, and corridor efficiency of ±500 kV DC transmission line, respectively; and the transmission loss and the unit cost of the ±800 kV UHV DC transmission line are 25 and 40% of the transmission loss and the unit cost of the ±500 kV DC transmission line, respectively.

Since the 1960s, the former Soviet Union, Japan, the United States, Italy, and other countries have developed UHV transmission plans. Both the former Soviet Union and Japan have built 1000 kV-level transmission projects and constructed UHV transmission labs and test fields. In addition, they carried out a lot of research on many issues such as overvoltage, audible noise, radio interference, and ecological effect and made important progress.

The Soviet Union's research into UHV power grids. In 1980, the former Soviet Union began the construction of the 1150 kV AC transmission project connecting the power grids of Siberia, Kazakhstan, and Ural. It built 2362 km 1150 kV AC transmission lines and three UHV substations. A 900-km-long UHV transmission line within the territory of Kazakhstan started full voltage operation in 1985. After the collapse of the Soviet Union, for many reasons this transmission line began step-down operation in 1992. Figure 2.26 shows a line diagram of the 1150 kV UHV AC and ±750 kV UHV DC transmission line built by the former Soviet Union. In terms of UHV DC transmission technology, in 1990 the former Soviet Union began to construct a ±750 kV DC transmission project that had a design transmission distance of 2414 km and a design transmission capacity of 6 GW. All the equipment passed type test. A total of over 1000 km UHV DC transmission lines were constructed, but the construction stopped after the collapse of the Soviet Union.

Japan's research into UHV power grids. Since 1992 Japan has built 427 km 1000 kV parallel transmission lines on the same pole. In 1996 it put into operation the New Haruna UHV Equipment Verification Station and performed a number of tests on it. The equipment passed a total of 5 years of 1000 kV charged operation tests, through which the feasibility of the UHV transmission technology was preliminarily verified. After the completion of the UHV transmission lines, as Japan's electricity demand growth slowed down, it put off the nuclear power construction plan and the UHV transmission lines have ever since been in step-down operation at a voltage of 500 kV.

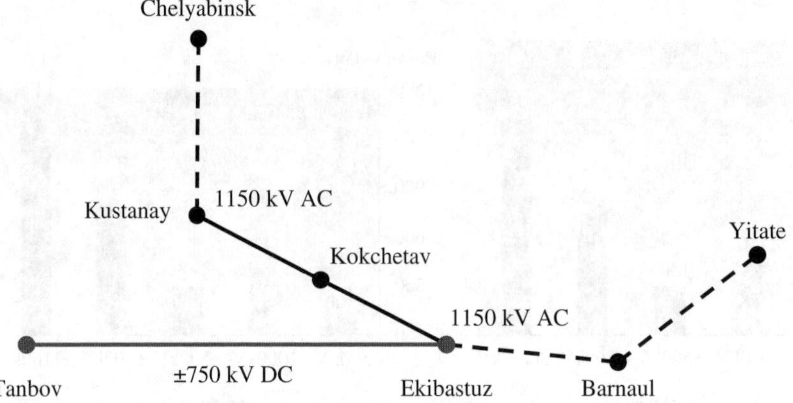

Figure 2.26 Line diagram of the 1150 kV UHV AC and ±750 kV UHV DC transmission line built by the former Soviet Union.

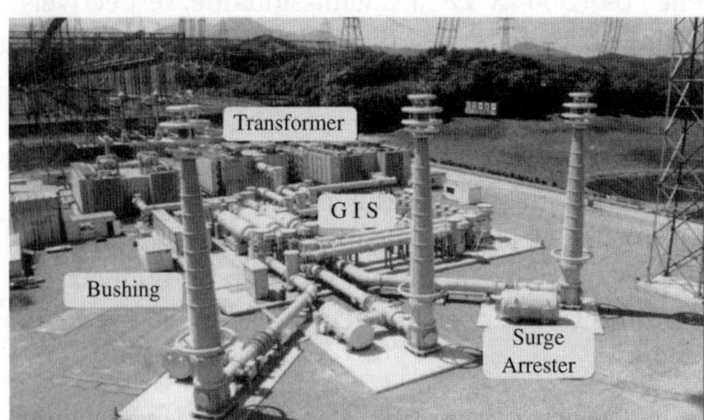

Figure 2.27 Japan's new Haruna substation UHV test station [37].

Figure 2.27 shows Japan's New Haruna Substation UHV Test Station. Figure 2.28 shows Japan's UHV transmission lines.

China's UHV research and project construction. China has the most advanced UHV technology and widest project application in the world. In the mid-1980s, China began research into UHV technology and constructed UHV test bases or test stations. Since 2004, the SGCC has joined hands with many parties to vigorously study and develop UHV transmission technology and has done a lot of work in theoretical research, technological breakthroughs, planning and design, standard formulation, equipment development, project construction, and operation management [5–8]. After years of effort, in early 2009 China constructed and put into operation the 1000 kV Southeastern Shanxi-Nanyang-Jingmen UHV AC demonstration project that represented the world's highest voltage level at that time. In July 2010, China constructed and put into operation the ±800 kV, 6.4 GW, and 1970-km-long Xiangjiaba-Shanghai DC demonstration project with the highest voltage level, the largest transmission capacity, the longest transmission distance, and the most advanced technology. By end of 2017, China had constructed and put into operation 21 UHV projects including 8 AC projects and 13 DC projects. In addition, 4 UHV projects including 3 AC projects and 1 DC project are under construction. The length of approved UHV transmission lines and those under construction has

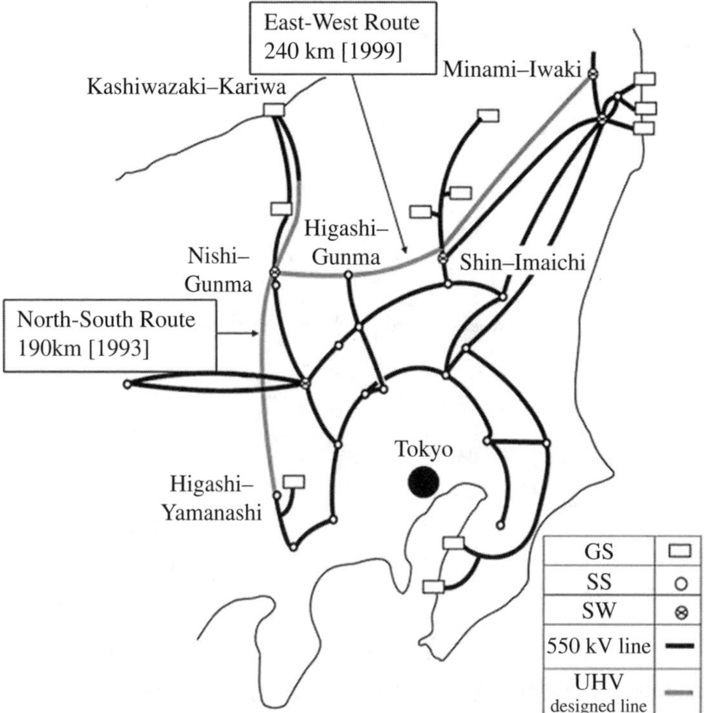

Figure 2.28 Japan's UHV transmission lines [37].

reached 37,000 km and the transformation (inversion) capacity totals 370 million kVA (kW). The UHV projects have become the energy channels of China's West-East and North-South electricity transmission. Thanks to these large UHV grids, by the end of 2017, China's installed hydropower capacity, wind power capacity and solar power capacity had reached 340, 160, and 130 GW, respectively, representing a 3.3-, 464-, and 6800-fold increase compared to that of 2000, respectively, all of which rank first in the world. By July 2016, the electricity transmitted by SGCC's UHV projects had totalled 530 billion kWh, of which 199.8 billion kWh was clean energy power generation, reducing coal consumption by 240 million tons, reducing sulfur dioxide emissions and other pollutants by 2.4 million tons, and reducing carbon dioxide emissions by 170 million tons in East and Central China. In 2015, the Jinsu, Fufeng, and Binjin UHV DC projects transmitted 21.6 GW power to East China at full power in flooded summer season, with a daily transmission of 520 GWh and an annual transmission of 102 billion kWh, playing an important role in meeting the demands for electricity at the load center in East China. At present, the longest UHV AC project under construction is the Yuheng-Weifang 1000 kV UHV AC project with a transmission capacity of 10 GW and a transmission distance of 1049 km (see Figure 2.29). At present, the UHV DC project with the highest voltage under construction is Zhundong-South Anhui ±1100 kV UHV DC project with a transmission capacity of 12 GW and a transmission distance of 3324 km (see

Figure 2.29 Yuheng-Weifang 1000 kV UHV AC project (under construction) (transmission capacity: 10 GW, distance: 1049 km). Picture source: SGCC.

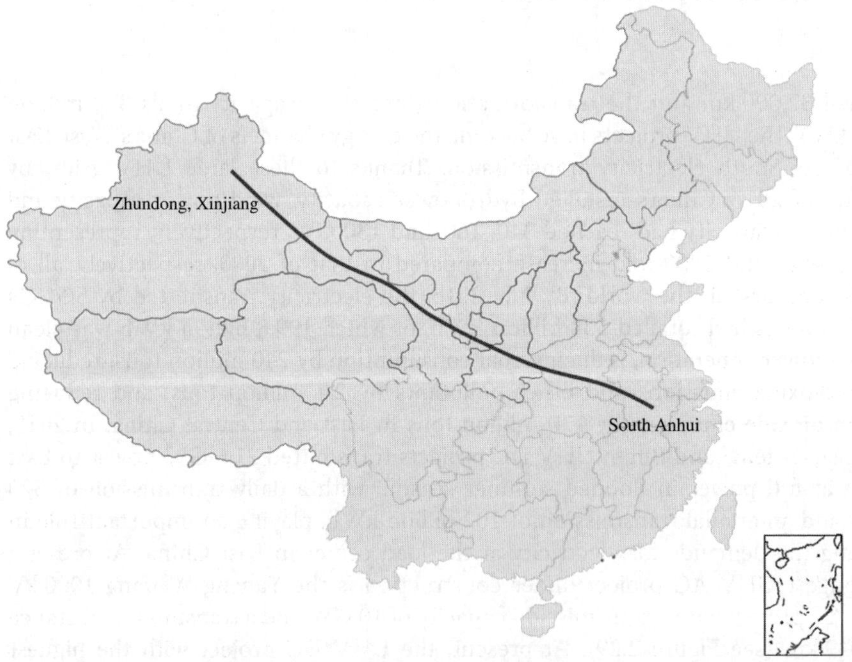

Figure 2.30 Zhundong-South Anhui ±800 kV UHV DC project (under construction) (transmission capacity: 12 GW, distance: 3324 km). Picture source: SGCC.

Table 2.3 SGCC's UHV projects put into operation and under construction by 2015.

	Number	Project	Voltage (kV)	Line Length (km)	Transmission Power (10 MW)
Projects put into operation (4 AC lines and 5 DC UHV projects)	1	Southeastern Shanxi-Nanyang-Jingmen	1000	640	500
	2	Huainan-North Zhejiang-Shanghai	1000	2×648.7	675
	3	North Zhejiang-Fuzhou	1000	2×603	680
	4	Xilin Gol League-Shandong	1000	2×730	900
	5	Xiangjiaba-Shanghai	± 800	1907	640
	6	Jinping-South Jiangsu	± 800	2059	720
	7	South Hami-Zhengzhou	± 800	2210	800
	8	Xiluodu-West Zhejiang	± 800	1669	800
	9	Ningdong-Zhejiang	± 800	1720	800
Projects under construction (4 AC lines and 6 DC UHV projects)	1	Huainan-Nanjing-Shanghai	1000	2×780	1000
	2	West Inner Mongolia-South Tianjin	1000	2×608	1000
	3	Yuheng-Weifang	1000	2×1049	1000
	4	Xilin Gol League-Shengli	1000	2×240	900
	5	Jiuquan-Hunan	± 800	2383	800
	6	North Shanxi-Jiangsu	± 800	1119	800
	7	Xilin Gol League -Taizhou	± 800	1620	800
	8	Shanghaimiao-Shandong	± 800	1238	800
	9	Zhundong-South Anhui	± 1100	3338	1200
	10	Jarud-Qingzhou	± 800	1234	1000

Figure 2.30). Table 2.3 and Figure 2.31 show SGCC's UHV projects that have been put into operation and under construction. Figure 2.32 shows China's UHV AC and DC transmission equipment.

Brazil's and India's UHV transmission projects. Brazil, India, and other countries are actively developing UHV power grids and in these countries a number of UHV power transmission projects are being constructed. Brazil is constructing two UHV DC transmission projects for Belo Monte hydropower station that will be completed and put into operation in 2017 and 2018, respectively. In the long run, UHV transmission technology is required for the interconnection of power grids in the whole of Asia, the Grand Inga hydropower project, and the supply of solar power generated in North Africa to Europe. Figure 2.33 and Figure 2.34 show Brazil's and India's UHV DC transmission projects.

2.7.1.3 Clean Energy

Development of clean energy is the fundamental task of building Global Energy Interconnection (GEI). For a long time, electricity has mainly been generated by using coal,

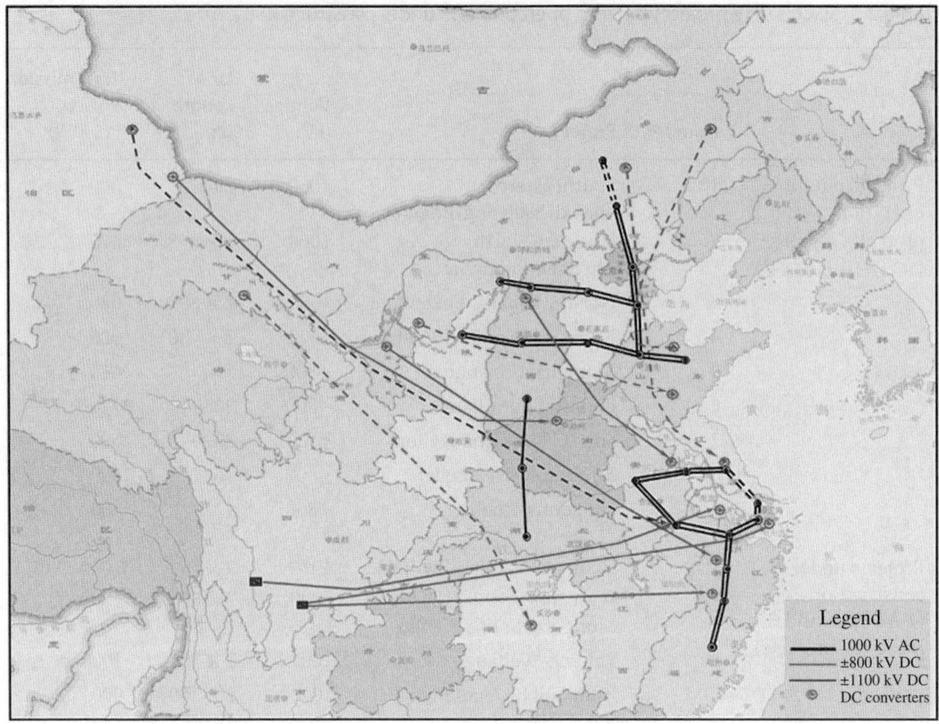

Figure 2.31 Overview of SGCC's UHV projects. Picture source: SGCC.

natural gas, nuclear, and hydroelectric. The dependence on fossil energy has led to problems such as increasingly severe resource shortages, environmental pollution, and climate change. Since the Industrial Revolution, the global average surface temperature has increased by 1°C. If the situation is not controlled, by the end of 21st century, the global temperature will rise by more than 3°C, resulting in serious ecological disasters. To cope with the challenges, the fundamental way out is to change the development mode of relying on fossil energy, speeding up the implementation of the "two replacements"; namely, in energy development replacing fossil energy with clean energy alternatives such as solar energy, wind energy, and hydro-energy; and in energy consumption replacing direct consumption of coal and oil with electricity. In the future, the GEI will mainly transmit clean energy power. The world is endowed with abundant clean energy resources, only 5/10,000 of which can meet global energy needs. However, clean energy resources are unevenly distributed in the world. The technically exploitable wind energy resources in the Arctic exceed 300 trillion kWh, accounting for 20% of the world's total. The technically exploitable solar energy resources around the equatorial region exceed 1000 trillion kWh, accounting for 30% of the world's total. The technically exploitable hydro-energy resources in the world are 16 trillion kWh, most of which is concentrated in Asia, Africa, and South America. Eighty-five percent of the hydro-energy, wind energy, and solar energy resources in Asia, Europe, and Africa are concentrated in the energy belt running from North Africa through Central Asia to Russian Far East, forming a 45° angle with the equator. The load is mainly concentrated in

UHV AC Transformer

UHV AC GIS

UHV DC Converter Transformer

UHV DC Converter Valves

Figure 2.32 China's UHV AC and DC transmission equipment. Picture source: SGCC.

East Asia, South Asia, Europe, South Africa, and other regions. Figure 2.35 shows the characteristics of distribution of clean energy resources in Asia, Europe, and Africa. In addition, wind power generation and PV power generation are highly random, intermittent, and fluctuating. Only after they are integrated into large power grids, can they realize great development, which means we must construct the GEI to efficiently develop clean energy and distribute and utilize it in a wider range.

Constant breakthroughs have been made in the field of wind turbine technology and wind power prediction accuracy has increased significantly, which is manifested in the change of main technical indicators, including increased capacity (the maximum single-machine capacity of wind power has reached 10 MW), improved adaptability (the minimum average annual wind speed of low-speed wind turbines is reduced to about 5.2 m/s), improved efficiency (power generation efficiency of intelligent wind turbines improves by 15–20%), and improved prediction level (the wind power prediction system is applied to a variety of terrains and climates with accuracy higher than 85%). The change in trend of single-machine capacity of world wind turbines is shown in Figure 2.36.

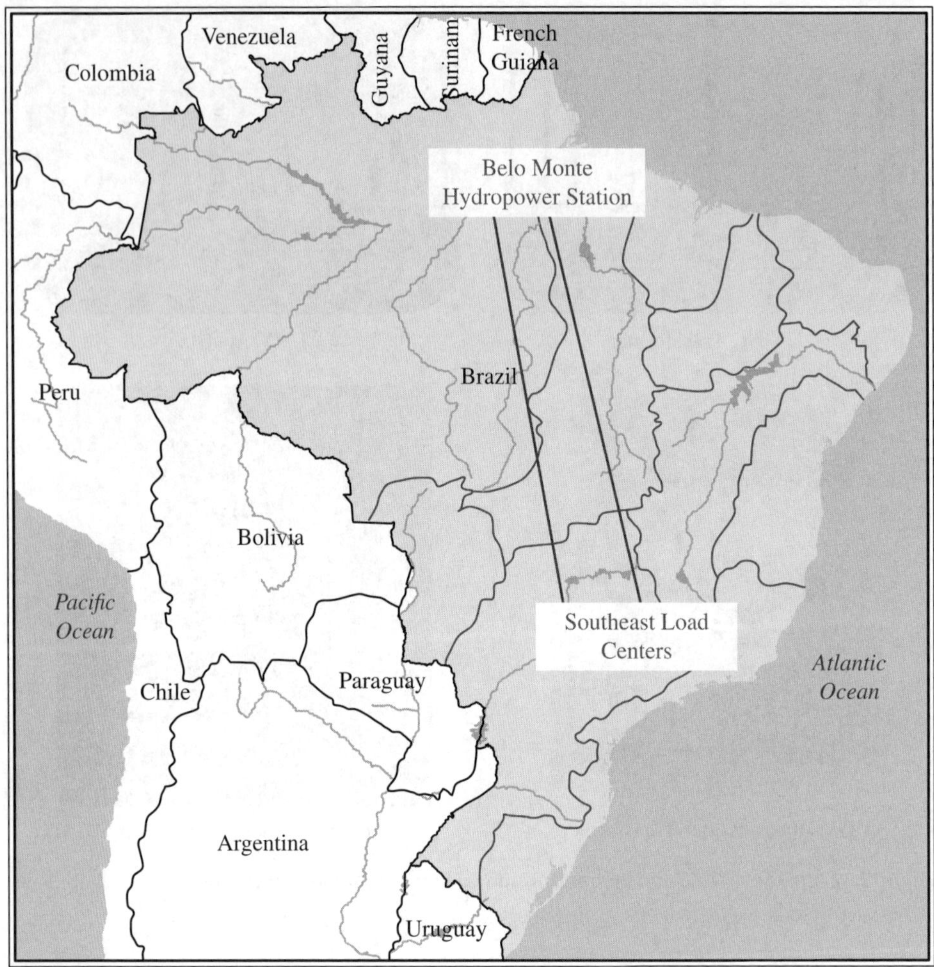

Figure 2.33 Brazil's UHV DC transmission projects. Picture source: SGCC.

Currently, onshore wind power and PV power generation cost are about 5 US cents and 9 US cents per kWh, respectively. Recently, the bid price of some PV power projects in United Arab Emirates and Chile has been as low as 3 US cents per kWh. It is expected that by 2025, the average cost of wind power and PV power generation will fall to 3 US cents per kWh and the competitiveness of wind power and PV power will exceed that of fossil energy. Figure 2.37 shows the per kWh cost for different power supplies. Figure 2.38 shows China's wind power bases with a planned capacity of above 10 GW.

2.7.1.4 GEI

GEI has broad development prospects. By the mid-twenty-first century, it will usher in an important development period that can be divided into three stages; domestic interconnection, intra-continental interconnection, and intercontinental interconnection.

Figure 2.34 India's UHV DC transmission projects.

By 2050, the GEI will be basically complete. Figure 2.39 shows the development stages of the GEI and Figure 2.40 shows an illustration of it.

In the decade to come, China will accelerate the R&D and application of new technologies like UHV, flexible DC grid and wind/PV/energy storage and transmission. According to planning by SGCC, China's Energy Interconnection will be primarily built by 2025, when a power development pattern will be formed to transmit electricity from the West to the East and from the North to the South, with electricity from all sources such as coal-fired, hydro, wind, and solar energy complementing each other, and interconnections with neighboring countries (see Figure 2.41).

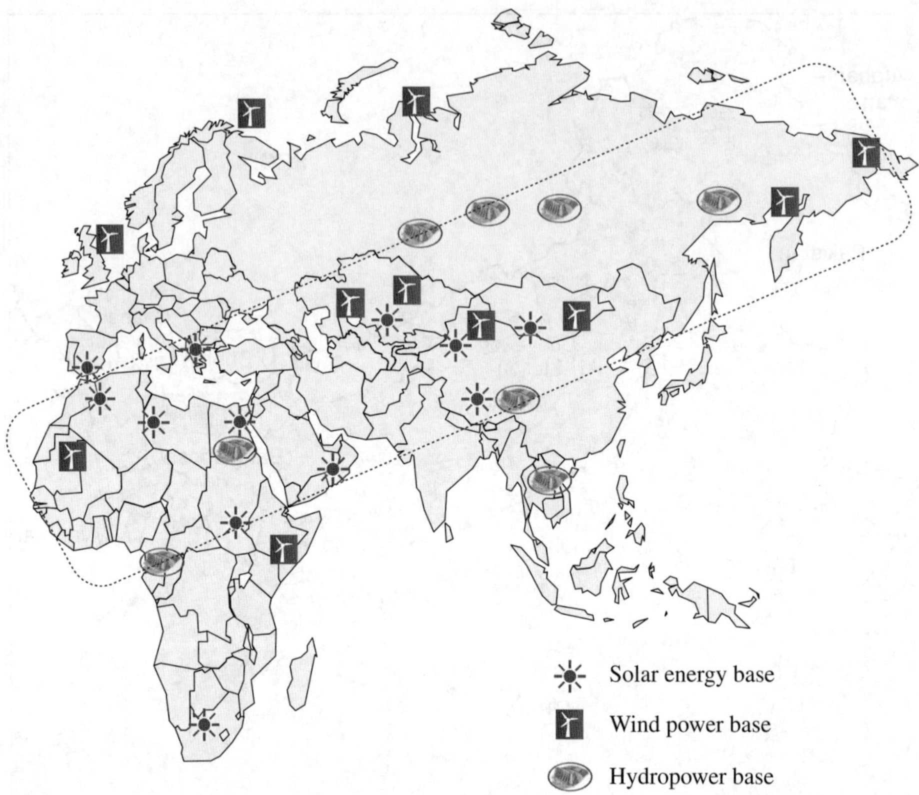

Figure 2.35 Characteristics of the distribution of clean energy resources in Asia, Europe, and Africa. Picture source: SGCC.

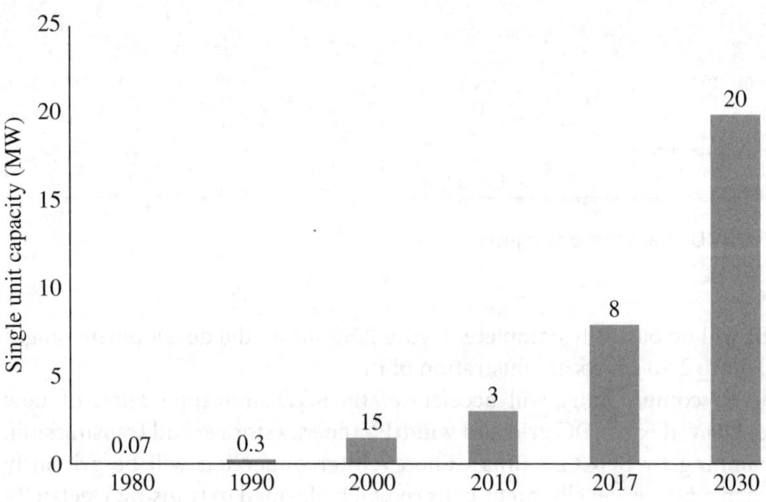

Figure 2.36 Change in trend of the single-unit capacity of wind turbines. *Source*: European Environmental Agency.

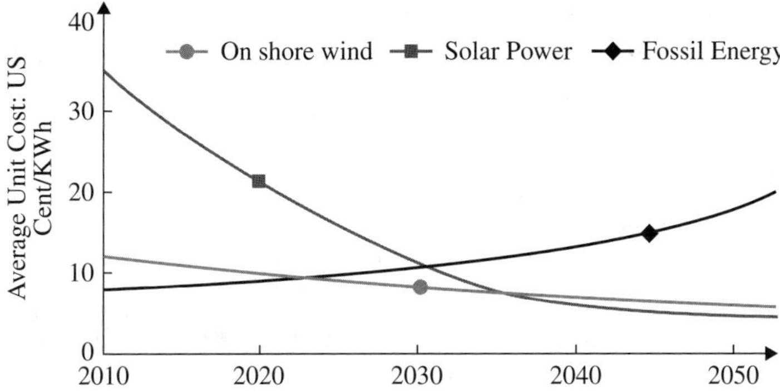

Figure 2.37 Per kWh cost for different power supplies. Picture source: GEIDCO, Bloomberg New Energy Finance.

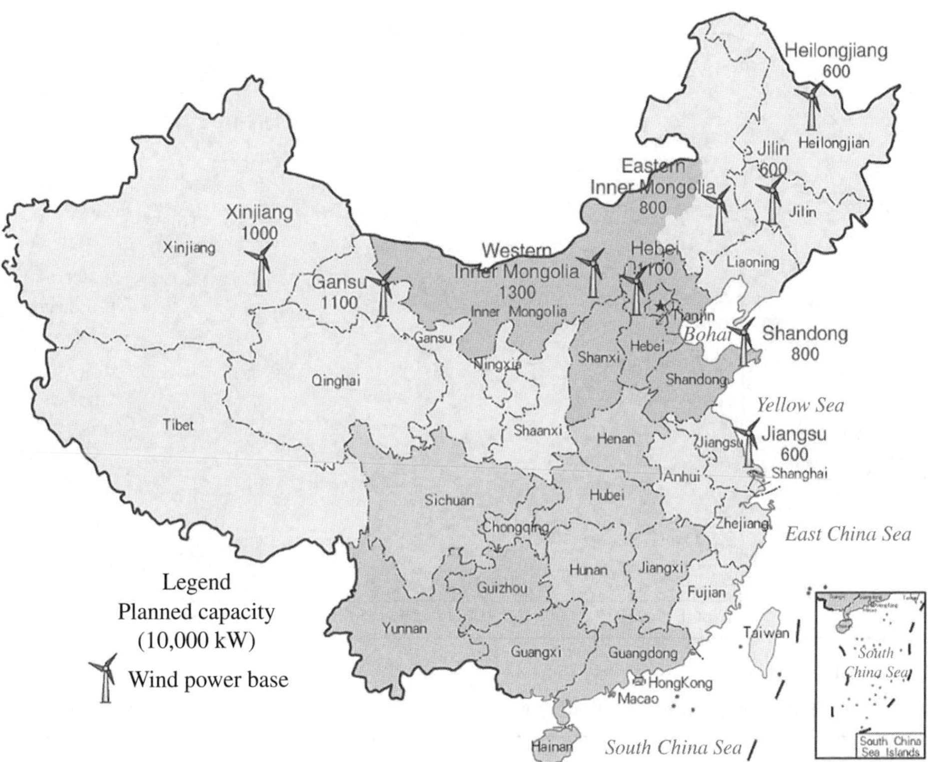

Figure 2.38 China's wind power bases with a planned capacity of above 10 GW. Data source: China Electricity Council.

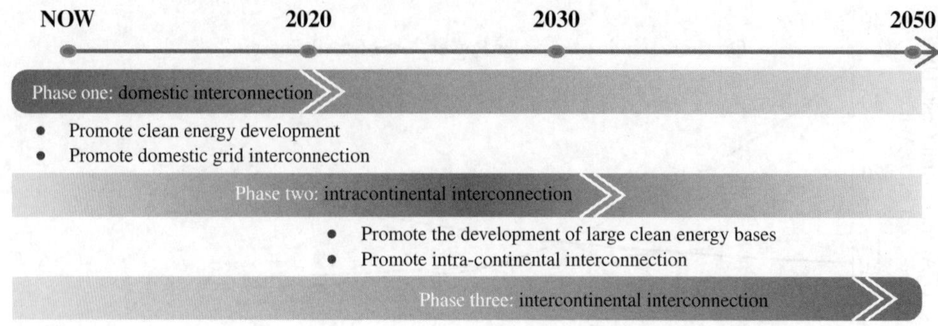

Figure 2.39 Development stages of GEI. Data source: SGCC.

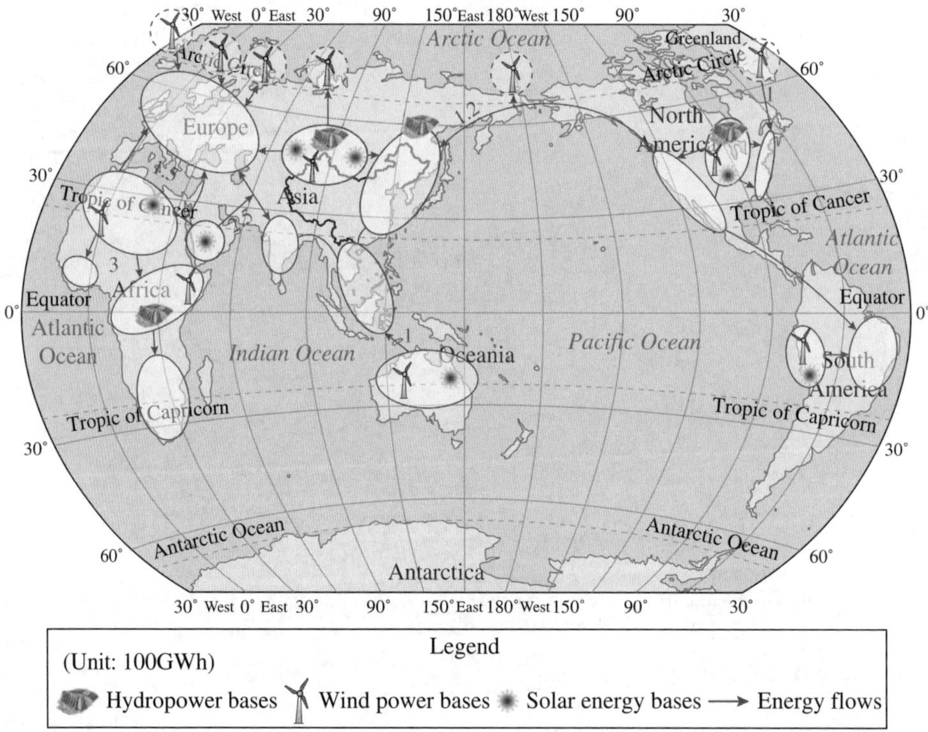

Figure 2.40 Illustration of GEI. Picture source: GEIDCO.

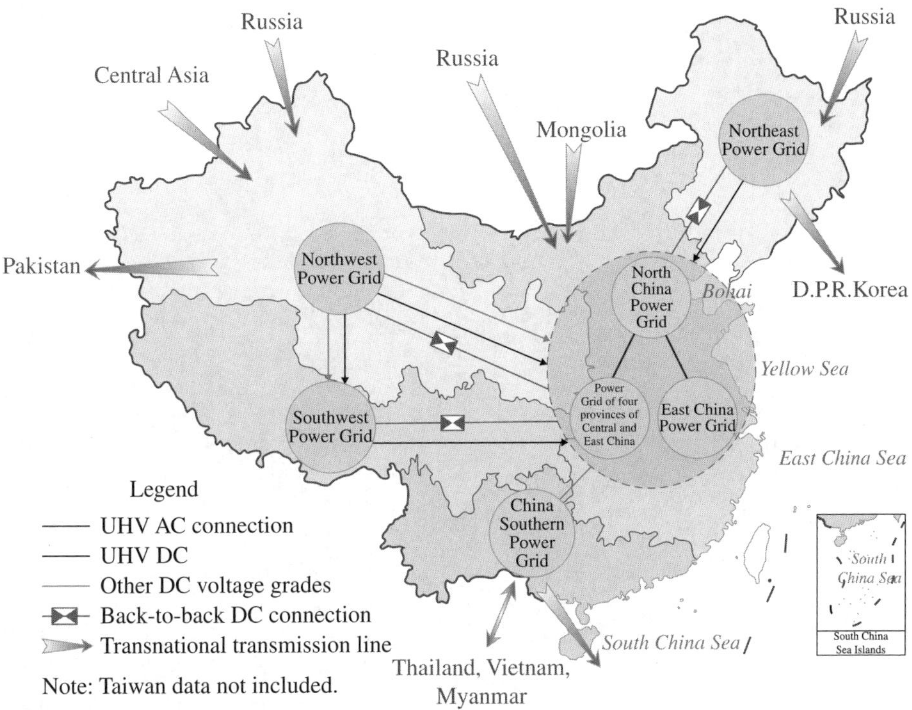

Figure 2.41 General layout of China's future power flow. Picture source: SGCC.

The Global Energy Interconnection Development and Cooperation Organization (GEIDCO) has conducted an in-depth study on potential grid interconnection in the world. According to research by GEIDCO, the development of clean energy in North China, Mongolia, and Russia will be accelerated and electricity will be transmitted to load centers in East China, South Korea, and Japan, realizing the Northeast Asia Grid Interconnection (see Figure 2.42). In Asia, a grid pattern of "1+5" will be formed, connecting the power grids of China, Northeast Asia, Southeast Asia, Central Asia, South Asia, and West Asia (see Figure 2.43)

By 2050, clean energy account for more than 80% of the world's total energy consumption and carbon dioxide emissions will only be 50% of that in 1990; the goal of controlling the global temperature rise to within 2°C will be realized. Construction of the GEI can produce grid interconnection benefits and reduce energy cost through time difference, seasonal difference, and electricity price difference (see Figure 2.44 and Figure 2.45); promote South-South and North-South cooperation; convert the resource advantages into economic ones to narrow the gap between regions; reduce international disputes; and make sustainable energy available for all.

2.7.2 Potential Applications of VFTs in GEI Systems

As a new type of interconnection equipment, VFT is being applied in North America. By 2010, a total of five VFTs had been put into operation in electric power systems throughout the world. The application of VFT technology will play an important role

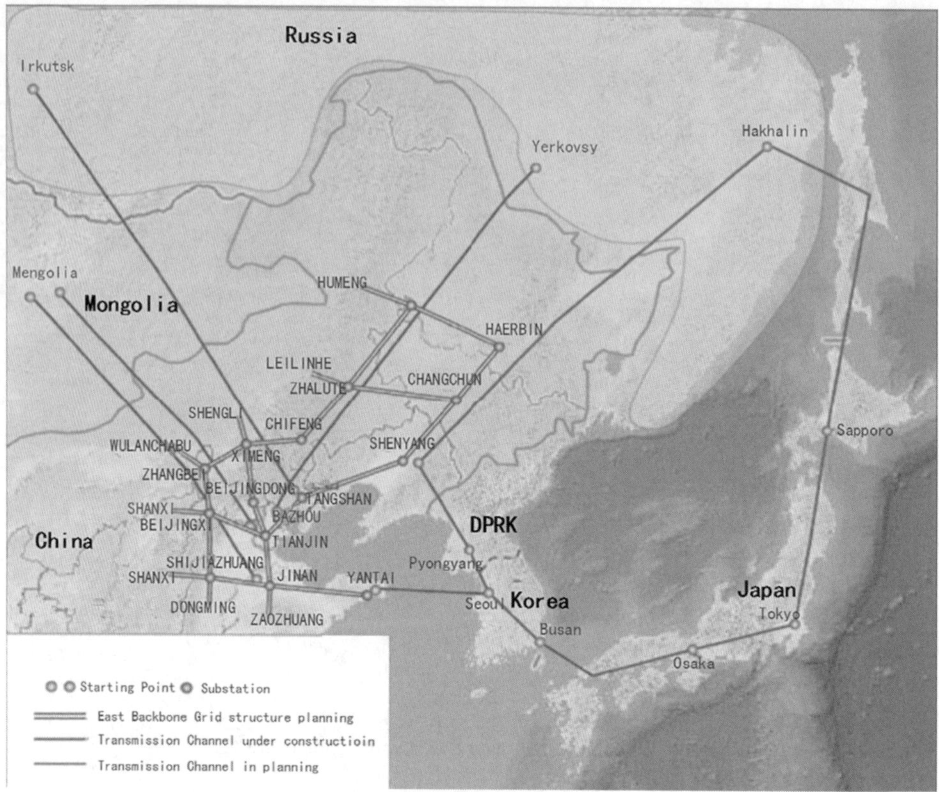

Figure 2.42 Schematic diagram of interconnected power grids in Northeast Asia. Picture source: GEIDCO.

in further increasing the power exchange capacity and system operation flexibility of interconnected power grids. Particularly, the new VFT technology can be used for the interconnection of weak power grids and large power systems, wind farms, and main power grids as well as the marginal interconnection of adjacent large power grids.

At present, aimed at promoting West-East electricity transmission, facilitating mutual supply between South China and North China, and strengthening large-scale optimal allocation of energy resources, China is actively constructing national interconnected system covering a larger scope and with stronger cross-regional power transmission capability and more reasonable grid structure. It is expected that during the "13th Five-Year Plan" period and the "14th Five-Year Plan" period, China will build up four synchronous power grids; North China-Central China-East China ("Three Chinas"), Northeast, Northwest, and South China and realize the asynchronous interconnection of Three Chinas-Northeast, Three Chinas-Northwest, and Three Chinas-South China, forming the new national interconnection pattern characterized by reasonable structure and large-scale optimal allocation of energy resources. As a new type of asynchronous interconnection device with certain economic and technical advantages, VFT will play an active role in China's large power grid interconnection and smart grid development process.

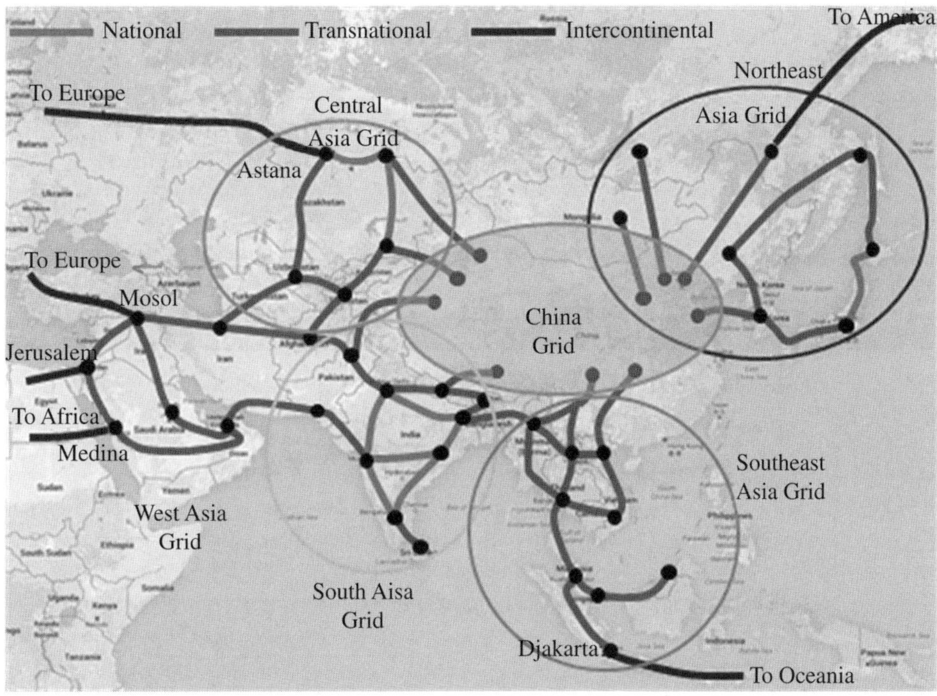

Figure 2.43 Schematic diagram of interconnected power grids in Asia. Picture source: GEIDCO.

2.7.2.1 Using VFTs to Loop-off Electromagnetic-looped Networks

With the formation of the 500, 750, and even 1000 kV main grid frames, the original low-voltage electromagnetic loop network requires open-loop operation; otherwise, when the tripping operation occurs to the high-voltage parallel connection circuit, it will lead to problems such as serious overload of low-voltage circuit and system voltage instability caused by the potential large-scale power transfer. The current countermeasure is the open-loop operation of the low-voltage power grid, which, however, will lead to the decline of the reliability and economic efficiency of power supply in some areas. An effective solution to this problem is to connect a VFT of a certain capacity in series in the low-voltage shunt circuit, which can not only prevent the risk of large power transfer, but also ensure the reliability and economic efficiency of power supply. Figure 2.46 shows a wiring diagram of the use of VFTs to realize loop opening of the electromagnetic loop network.

2.7.2.2 Using VFTs to Realize the Marginal Interconnection of Asynchronous Grids

Currently, China has six synchronous power grids and even by 2020 China will have four synchronous power grids. Generally speaking, each synchronous grid has relatively weak power supply reliability in its peripheral areas. VFT can improve the power supply reliability of some boundary power grids if it is used to connect adjacent asynchronous power grids. Take Central China Grid and Northwest Grid as an example. At present, at the Bikou Hydropower Plant in Gansu, Northwest China, sometimes the hydropower generating units are connected to Guangyuan Grid in Sichuan, Central China through

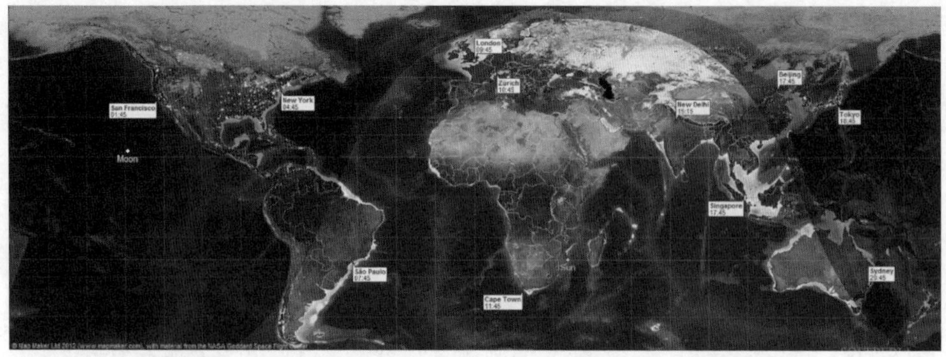

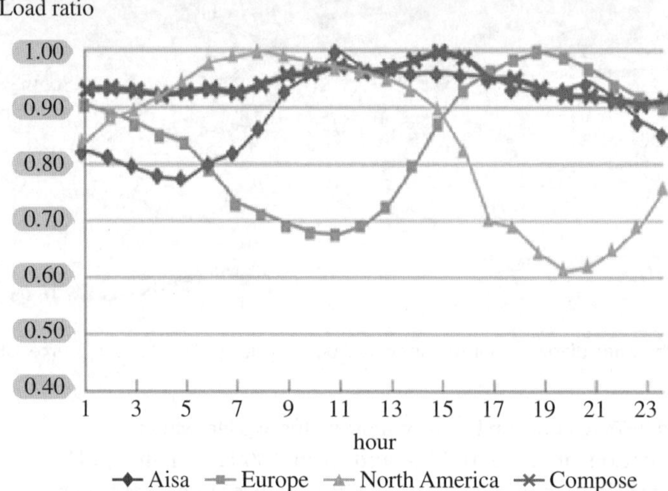

Figure 2.44 Illustration of complementary relationships of load curves in Europe, North America, and Asia. Picture source: GEIDCO.

the circuit breaker; sometimes the hydropower generating units are connected to the Longnan Grid in Gansu through the 220 kV transmission line; and sometimes the hydropower generating units are connected to the Shaanxi Grid through the 220 kV transmission line. If two or more VFTs are installed at Bikou Hydropower Plant, it can realize the interconnection of power grids in Guangyuan, Sichuan, and power grids in Longnan, Gansu, and power grids in southern Shaanxi, effectively improving the power supply reliability of power grids. Figure 2.47 shows the wiring method for using VFTs to realize the marginal interconnection of asynchronous grids. In this kind of wiring method, if a serious fault of losing power supply occurs to the system on either side, the sound system on the other side will provide the faulty system with a black-start power supply through the the VFT.

2.7.2.3 Using VFTs to Suppress System Low-Frequency Oscillation

Low-frequency oscillation is a common problem that emerges with large power grid interconnection. With the expansion of the system scale, system inertia increases,

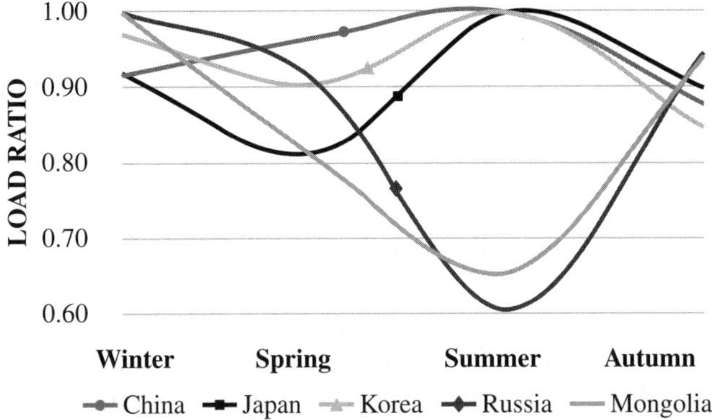

Figure 2.45 Complementary characteristics of load in Northeast Asian countries.

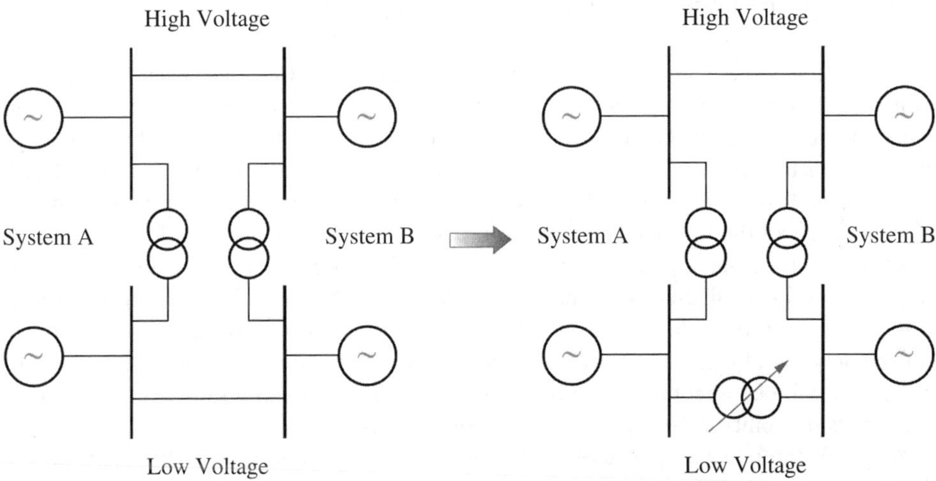

Figure 2.46 Wiring diagram of the use of VFTs to realize loop opening of the electromagnetic loop network.

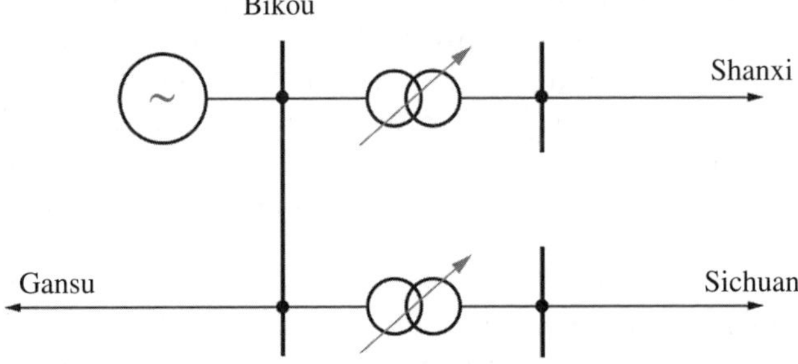

Figure 2.47 Schematic diagram of use of VFTs to realize marginal interconnection of asynchronous grids.

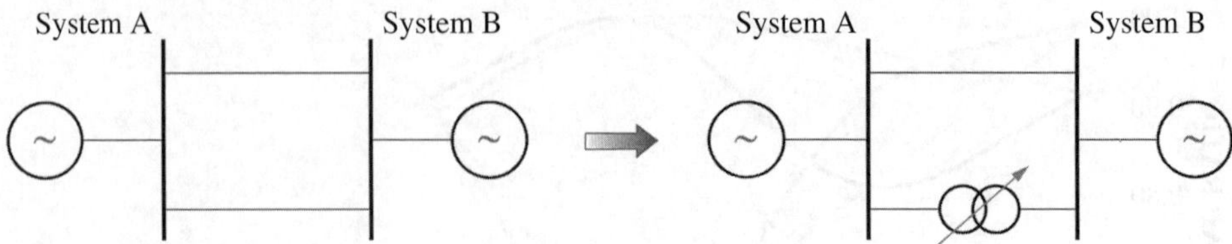

Figure 2.48 Line diagram showing use of VFT to suppress system low-frequency oscillation.

the electrical distance between units increases, and the probability of occurrence of low-frequency oscillation increases. For this kind of system, a VFT of a certain capacity can be connected in series in a branch current in the section of the interconnected system or a parallel branch in which a series-connected VFT can be increased to dynamically control the power flow of the section, increase the system damping, and prevent and suppress low-frequency oscillation. Figure 2.48 shows the wiring diagram of using VFT to suppress the system low-frequency oscillation.

2.7.2.4 Using VFTs to Improve Operation Characteristics of an Unstable Power Supply

Both wind power and solar power generation have problems such as randomness and unstable power supply. Generally, the generated power has to be integrated into large power grids for accommodation and better utilization. If local grids are connected with these unstable power supplies, this might affect the local power supply and even endanger the safety of the whole electric system. If VFT is used to isolate the local grids with these unstable power supplies, it can effectively reduce the impact of the unstable power supplies on the local small systems, maximally ensure the reliable power transmission and supply, and provide more controllable grid connection methods for wind power, solar energy, and other types of power supplies. Meanwhile, when local load is connected, it will be beneficial to improving the system stability and improving the power transmission ability, safety, and stability of power grids.

As VFTs can effectively isolate the impact of the change in the small system on the large system, with the deepening research on VFTs, we have found that for power grids connected with large-scale wind power VFTs can significantly improve system stability and play an active role in the new energy development. In Reference [33], there is an in-depth study on the system stability of a large wind farm that transmits power to the load center through two 765 kV transmission lines. The results show that if a wind farm is connected to the local weak power system through the VFT, it can significantly improve the system transient stability and power generation capability of the wind farm. Figure 2.49 shows the system wiring diagram for using a VFT to improve the transmission capacity of a wind farm. On the contrary, if the VFT is not used, the system will have to be equipped with a large number of shunt capacitors and/or series compensation of a certain capacity.

For China, a lot of wind farms are far away from load centers. As the local power grid has a small load and limited ability to accommodate wind power, if it is directly connected with the wind farm, there is a great risk of instability. If a VFT is used to connect wind farms with local loads, it will be beneficial to the accommodation of some wind power on the spot, enhancing the sending end power grid and improving the transmission capacity of the wind farm.

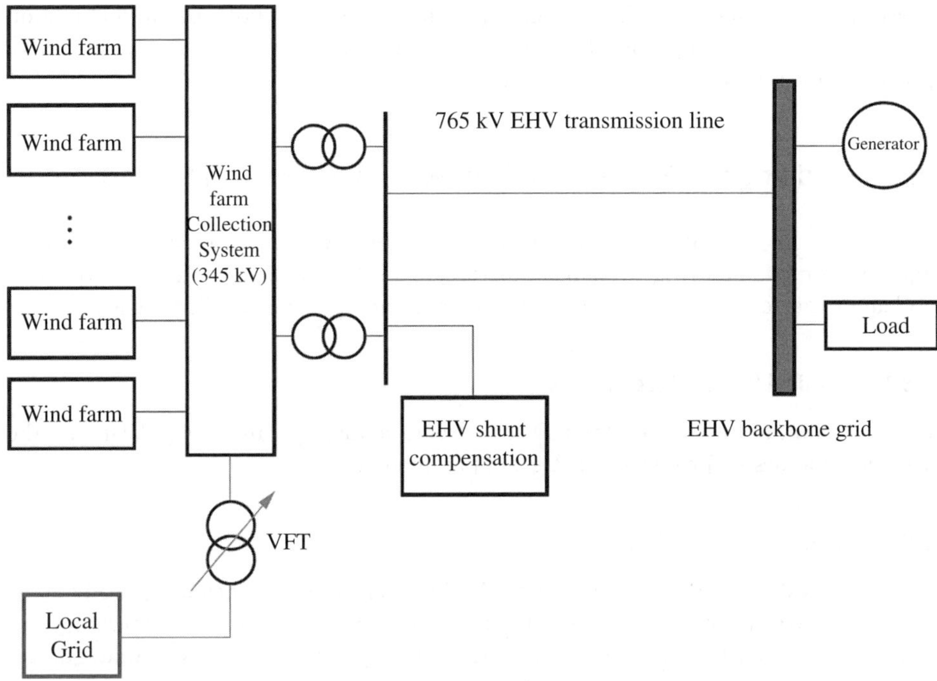

Figure 2.49 System wiring diagram showing the use of a VFT to improve the transmission capacity of a wind farm [33].

2.7.2.5 Using VFT to Connect Weak Grids to the Main Grid

China is a country with a vast territory. Some local power grids are at the terminals of the electrical power. These grids often have fragile structure, low stability level, and poor reliability of power supply. If they are directly connected to the main grid, it might affect the safety, stability, and reliability of the main grid. If a VFT is used to connect these power grids with the main grid, it can connect the terminal weak grids with the main electric power system. Meanwhile, it will not bring new system stability risks. Figure 2.50 shows the use of a VFT to connect weak grids with the main grid.

2.7.2.6 Using VFTs to Optimize System Power Flow

With the development of power grids, there are usually many loops in the electric power system. However, for historical reasons, in the early stages the line conductor had a small section and a large loss in the large power flow. In addition, some lines might run

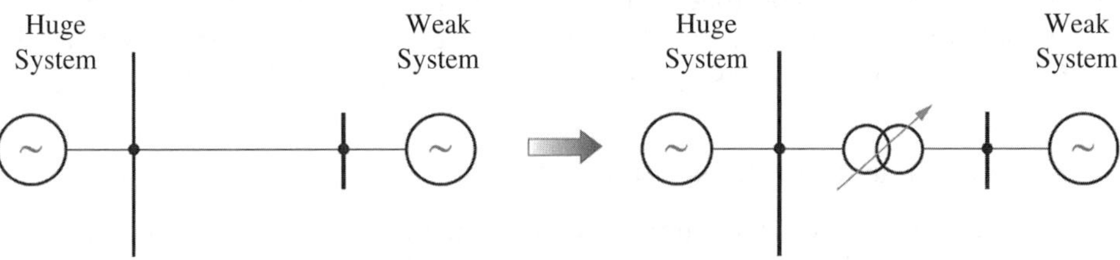

Figure 2.50 Use of a VFT to connect weak grids with the main grid.

overloaded without control. In these circumstances, the power flow regulation function of VFT can be used to optimize the system power flow distribution, reduce system loss, prevent some lines from running overloaded, and so on.

2.8   Studying the Prominent Problems of VFTs to be Solved

As a new technology of asynchronous interconnection, VFT involves structure design, equipment manufacturing, system simulation, protection, control, and so on. To study and apply this technology, we need to make breakthroughs in the following aspects.

2.8.1   Physical Parameters of VFTs

It is important to understand the physical structure, main parameters of VFT, and the selection and design principles of relevant parameters.

2.8.2   Basic Theory of VFTs

We need to establish various equations of VFTs in various states including steady-state, transient, and electromagnetic transient to form system equations under different working conditions, master the operation characteristic equation of this device under different working conditions, and provide theoretical support for the analysis and study of the characteristics of VFTs.

2.8.3   Simulation Tools for VFTs

In modern large system control, research into simulation tools is very important. At present, there are no VFT-related models available in the electromagnetic transient simulation program, electromechanical transient simulation program, and power flow calculation program we have adopted. As a result, we need to combine the characteristics of the different kinds of software (such as PSASP, EMTP, PSCAD/EMTDC, etc.) to establish VFT modules including the control module based on the characteristics and provide support for the large system performance simulation of VFTs.

2.8.4   Control Protection of VFTs

The VFT has a very complex control system. Not only the rotor speed, but also the phase-angle difference between the rotor winding magnetic field and stator winding magnetic field has to be controlled. Meanwhile, the low-frequency oscillation has to be controlled. It is a typical multi-input and multi-control objective complex system. The performance of this system has a great impact on the functions of a VFT.

2.8.5   Development and Manufacturing of VFTs

As the VFT is a new type of interconnection equipment, we need to develop corresponding physical devices based on actual engineering development, and realize the control and other functions of VFTs.

2.8.6 System Application of VFTs

The purpose of studying and developing a new technology is to promote and apply it and form productive forces. The VFT has a broad application space in China. Based on deepening the research, we need to promote and apply this technology in a timely manner.

2.8.7 Technical Economy of VFTs

A comparative study of VFTs, flexible DC transmission systems, and back-to-back DC transmission systems should be carried out from technical and economic perspectives, so as to conduct an evaluation and comparison for applying VFTs.

2.9 Summary

1. The VFT is a new kind of electric power system equipment integrating electromagnetic conversion and mechanical motion control. With strong asynchronous interconnection and other functions, it is applicable to different system conditions.
2. The successful development and operation of the VFT has verified its superior performance. It has been successfully applied in North America. This technology has broad development prospects in China.
3. Studying and applying VFTs is a systematic and innovative work. We need to study the basic principles, research tools, control simulation, and design analysis of VFTs in a comprehensive way.
4. A VFT can dynamically regulate the phase shift of the rotor winding magnetic field and the stator winding magnetic field on a large scale. Theoretically, it can improve system stability. The qualitative analysis of the dual system connected through the VFT further proves that is can improve the stability of the interconnected system.

References

1 E. Larsen, R. Piwko, D. McLaren, D. McNabb, M. Granger, M. Dusseault, et al. Variable-frequency transformer – A new alternative for asynchronous power transfer, presented at *Canada Power, Toronto, Ontario, Canada*, September 28–30, 2004.
2 P. Doyon, D. McLaren, M. White, Y. Li, P. Truman, E. Larsen, et al. Development of a 100 MW VFT, presented at *Canada Power, Toronto, Ontario, Canada*, September 28–30, 2004.
3 J.M. Gagnon, D. Galibois, M. Granger, D. McNabb, D. Nadeau, J. Primeau, et al. First VFT application and commissioning, presented at *Canada Power, Toronto, Ontario, Canada*, September 28–30, 2004.
4 H. Dayu. First "face to face" type of interconnecting device for asynchronous power grids and its remarkable significance. *Electric Power*, 2004, 37(10): 4–7 [in Chinese].

5 D. Mc Nabb, D. Nadeau, A. Nantel, E. Pratico, E. Larsen, G. Sybille, et al. Transient and dynamic modeling of the new Langlois VFT asynchronous tie and validation with commissioning tests, presented at the *6th International Conference on Power System Transients (IPST'05), Montreal, Québec, Canada,* June 20–23, 2005.

6 C. Gesong, Z. Xiaoxin. Digital simulation of variable frequency transformers for asynchronous interconnection in power systems. *IEEE/PES Transmission and Distribution Conference & Exhibition: Asia and Pacific Proceedings, Dalian, China,* 2005.

7 A. Merkhouf, S. Upadhyay, P. Doyon. Variable frequency transformer-an overview. *Power Engineering Society General Meeting,* 2006.

8 R.J. Piwko, E.V. Larsen. VFT – FACTS technology for asynchronous power transfer. *Transmission and Distribution Conference and Exhibition,* 2006.

9 J.M. Gagnon, A 100 MW variable frequency transformer (VFT) on the Hydro-Québec network: A new technology for connecting asynchronous networks. *The 2006 CIGRE General Session Proceedings, Paris, France,* 2006.

10 R. Piwko, E. Larsen, C. Wegner. VFTeg: The new technology for asynchronous networking transmission power. *Southern Power System Technology,* 2006, 2(1): 29–34.

11 M. Spurlock, R. O'Keefe, D. Kidd, E. Larsen, J. Roedel, R. Bodo, and P. Marken. AEP's selection of GE energy's variable frequency transformer (VFT) for their grid interconnection project between the United States and Mexico, presented at the *North American T&D Conference, Montreal, Canada,* 2006.

12 H. Dayu. A new type of "face to face" interconnecting device for asynchronous power grids based on VFT. *Electrical Equipment,* 2007, 8(7): 116–118 [in Chinese].

13 D. Nadeau. A 100-MW variable frequency transformer (VFT) on the Hydro-Québec TransÉnergie network: The Behavior during disturbance. *Power Engineering Society General Meeting,* 2007.

14 J. Paserba. Recent power electronics/FACTS installations to improve power system dynamic performance. *Power Engineering Society General Meeting,* 2007.

15 P. Hassink, V. Beauregard, R. O'Keefe, et al. Second & future applications of stability enhancement in ERCOT with asynchronous interconnections. *Power Engineering Society General Meeting,* 2007.

16 B. Bagen, D. Jacobson, G. Lane, et al. Evaluation of the performance of back-to-back HVDC converter and VFT for power flow control in a weak interconnection. *Power Engineering Society General Meeting,* 2007.

17 J.J. Marczewski. VFT applications between grid control areas. *Power Engineering Society General Meeting,* 2007.

18 E.R. Pratico, C. Wegner, E.V. Larsen, et al. VFT operational overview: The Laredo Project. *Power Engineering Society General Meeting,* 2007.

19 J.J. Marczewski. VFT interconnection study process with ISOs/RTOs and grid managers/operators. Power Engineering Society General Meeting, 2007.

20 P. Truman, N. Stranges. A direct current torque motor for application on a variable frequency transformer. *Power Engineering Society General Meeting,* 2007.

21 A. Merkhouf, S. Uphadayay, P. Doyon. VFT electromagnetic design concept. *Power Engineering Society General Meeting,* 2007. IEEE 24–28 June, 2007, 1–6.

22 A. Merkhouf, P. Doyon, S. Upadhyay, et al. Variable frequency transformer – Concept and electromagnetic design evaluation. *Energy Conversion, IEEE Transactions on,* 2008, 23(4): 989–996.

23 P. Marken, J. Roedel, D. Nadeau, et al. VFT maintenance and operating performance. *Power and Energy Society General Meeting 9 – Conversion and Delivery of Electrical Energy in the 21st Century*, July, 2008.

24 H. Elahi, S. Venkataraman, E. Larsen, K. Schreder, J. Marczewski. The Linden VFT Merchant Transmission Project, *CIGRE 2008, Paris*.

25 C. Yin, C. Gesong, Y. Rongxiang. Mathematical model and simulation analysis of variable frequency transformers. *Power System Technology*. 2008, 32(17): 73–77.

26 P. Hassink, P.E. Marken, R. Ond. *Simulation Analysis of Improving Power System Dynamic Performance*. IEEE: Laredo, TX, 978-1-4244-1904-3/08, 2008.

27 J. Adams, H. Chao, C. Custer, et al. *Planning for uncertainty: NYISO planning process and smart grid. Power and Energy Society General Meeting – Conversion and Delivery of Electrical Energy in the 21st Century*, 2008.

28 P.E. Marken, J.J. Marczewski, R. D'Aquila, et al. VFT – A smart transmission technology that is compatible with the existing and future grid Power Systems. *2009 PES'09, March*, 2009.

29 Y. Rongxiang, C. Yin, C. Gesong, S. Yong. Simulation model and characteristics of variable frequency transformers used for grid interconnection. *Power & Energy Society General Meeting, 2009. PES '09*. IEEE, 26–30, July, 2009.

30 E. Paul. P.E. Marken. Variable frequency transformer – A simple and reliable transmission technology, *EPRI HVDC Conference, New Orleans*, 2010.

31 C. Gesong. *Modeling and Control of Variable Frequency Transformer for in Power Systems*. D. Phil Thesis. Beijing: China Electric Power Research Institute, 2010.

32 E.R. Pratico, C. Wegner, P.E. Marken, J.J. Marczewski. First multi-channel VFT application – The Linden Project, power electronic/FACTS installations to improve power system dynamic performance, at the *2010 IEEE PES T&D Conference, New Orleans*, 2010.

33 P.E. Marken, K. Clark, E. V. Larsen. *VFTs and Bulk Power Transmission*. Report, GE, 2009.

34 C. Gesong, Z. Xiaoxin, S. Ruihua. Design of self-adaptive damping controller to low-frequency power oscillation in interconnected power systems based on VFTC, *Proceedings of the CSEE*, 2011. p. 2.

35 Z. Xiaoxin, G. Jianbo, L. Jimin, et al. TCSC in Power System. Beijing: Science Press, 2009.

36 C. Gesong, Y. Yonghua, T. Yong. Studies on resonance phenomena caused by energization of transformers, *PowerCon 2002, Kunmin, China*, 2002.

37 Y. Yamagata, M. Ono, K. Sasamori, K. Uehara, *Important Technologies Applied for UHV AC Substations in Japan*, Published online 21 January 2011, Wiley Online Library (wileyonlinelibrary.com). DOI: 10.1002/etep.568.

3

Basic Equations and Simulation Models of VFTs

3.1 Overview

The electrical power system is a huge and complex system designed, constructed, operated, and controlled by human beings. The planning, design, and operation of the electrical power system, the development, manufacturing, and control of important electrical power equipment such electrical motors should be based on analysis of the electrical power network, which includes power flow analysis, stability analysis, electromagnetic transient analysis, and short-circuit current calculation. The VFT is a new type of grid interconnection equipment and its operation relates to the safety of interconnected power grids. Therefore, in project implementation we must carry out project decision-making, device development, manufacturing, and optimizing the control strategies based on sufficient theoretical research, system simulation, and practical tests. In order to provide the necessary technical means for research on this new device [1], researchers must be able to establish basic equations and mathematical models of the VFT and realize its functions in an electrical power system analysis and simulation software.

Objectively, the operation status of the power system, such as its current, voltage, power flow, and so on, is always changing. However, in order to facilitate the analysis of the characteristics of the electrical power system, we categorize the state of the power system into steady-state and transient state so that we can focus on the main aspect of the issue based on the research subject.

1. *Steady-state* refers to the normal and relatively static operation state of the electrical power system. In this state the power flow distribution, voltage level, and current amplitudes of the system remain relatively unchanged. In simulation research, we use relatively simplified calculation models and methods to simplify simulation process and accelerate the calculation speed, which is applicable to calculations of the large electrical power system. Meanwhile, the calculation results build the initial stage for various transient states, which is usually the basis of the calculation of the transient state.
2. The *transient state* of the electrical power system is a process of changing from one state to another one. The transient state is divided into the wave process, electromagnetic transient process, and electromechanical transient process.
   a. The *wave process* is mainly related to the running operations or the overvoltage in case of lightning strike, which involves the propagation of the current and the

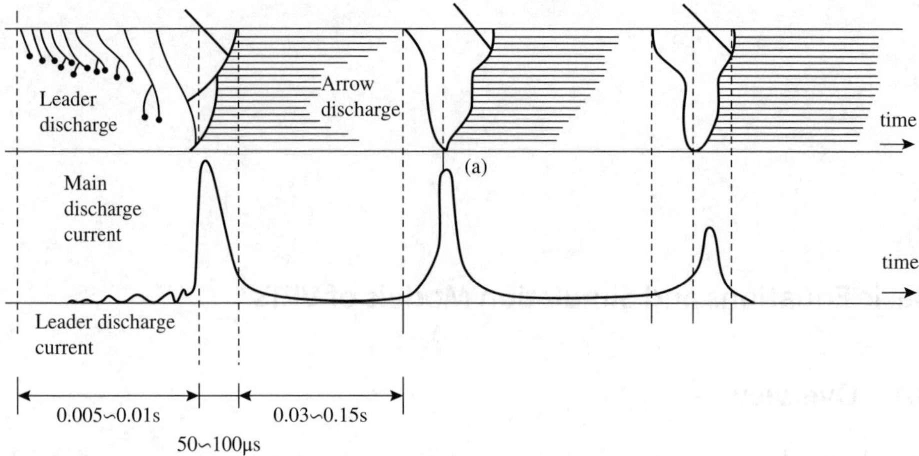

Figure 3.1 Typical wave processing in a power system.

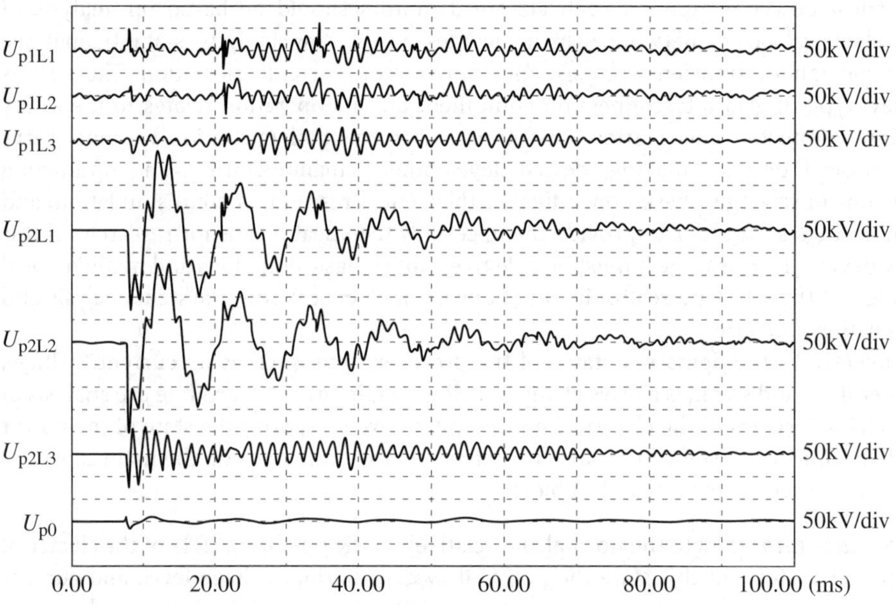

Figure 3.2 Typical electromagnetic transient of a power system.

voltage wave. The wave process is very short, generally ranging from a few to hundreds of microseconds, and its results are mainly used for research and design of the overvoltage level and insulation coordination of the electrical power system equipment and transmission lines (Figure 3.1).

b. The *electromagnetic transient process* is mainly related to short-circuit, self-excitation, control system actions, and equipment operations, involving current, voltage, control parameters, and sometimes involving power angle changing with the time (see Figure 3.2). This kind of process is longer than the

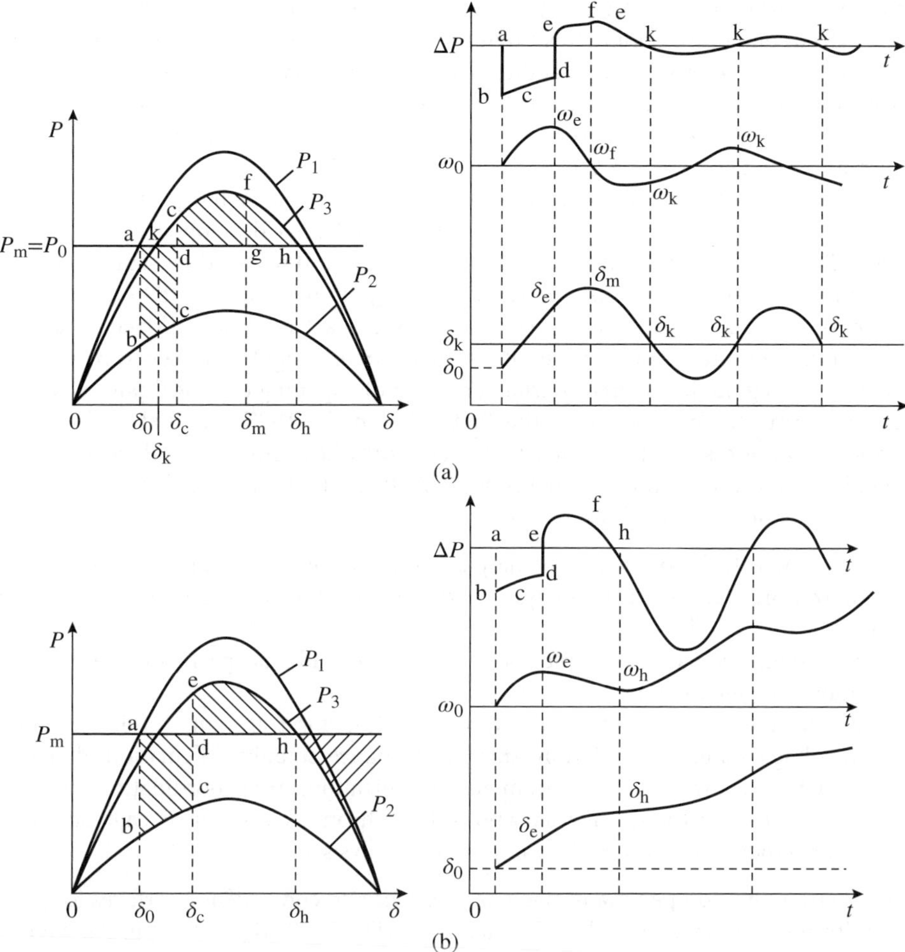

Figure 3.3 Typical electromechanical transient of a power system. (a) Stable condition. (b) Unstable condition.

wave process, usually ranging from a few milliseconds to hundreds of milliseconds and its results are mainly used for the selection, design of power system equipment parameters, research on control systems and control strategies, and so on.

c. The *electromechanical transient process* is mainly related to the system oscillation, instability, asynchronous operation, and so on, which involves the change of the power, power angle, and frequency of power system. This process usually ranges from a few cycles to a few seconds (see Figure 3.3) and its results are mainly used for the research on the system stability characteristics, overall configuration, and operation control.

In practice, the wave process, electromagnetic transient process, and electromechanical transient process are not separated from each other. Instead they are often related;

for example, when lightning damages the equipment insulation and results in a serious short-circuit failure leading to system instability. This is a system failure event that includes these three processes. In addition, in order to carry out the device design and model selection, we generally have to conduct short-circuit current calculations and the results are mainly used to determine the ability of the electrical power equipment to withstand impact current and short-circuit current [2–7].

A lot of software can be used for the power flow and stability calculations in an electrical power system; for example, PSASP and PSD developed by the China Electric Power Research Institute, the Simpow Program developed by ABB, Netmark Program developed by Siemens, and the PSSE Program developed by PTI. PSASP is a user-friendly large electrical power system software package characterized by a long history and strong functions whose intellectual property rights are owned by CEPRI. As a highly integrated, resource sharing, and friendly opening electrical power system R&D platform, this program won the first prize in the 1985 National Science and Technology Progress Awards. Based on the support of the power grid database, fixed model base, and user-defined model base, PSASP can be used for a variety of power system calculations and analysis including:

1. Power flow calculation, net loss analysis, optimal power flow and reactive power optimization, static security analysis, harmonic analysis, and static equivalence in steady-state analysis;
2. Short-circuit calculation, complex fault calculation, and relay protection setting calculations in fault analysis;
3. Transient stability calculation, the direct method for transient stability calculation, voltage stability calculation, small disturbance stability calculation, dynamic equivalence, control system parameter optimization, and coordination in electromechanical transient analysis, and subsynchronous resonance calculation in electromagnetic-electromechanical transient analysis.

With a friendly and open man–machine interface and convenient interfaces for popular software analysis tools such as Excel, AutoCAD, MATLAB, and so on, PSASP can make full use of the resources and functions of these software packages and is an important tool for the power system planning and design engineer to determine economical and technically feasible of the given schemes; an effective means for dispatchers to determine the system operation mode, analyze system accidents, and take anti-accident measures; and a helpful tool for researchers to study new equipment and apply new elements into the system. In particular, the user-defined (UD) and user programming interface (UPI) function of PSASP provides researchers with a useful tool to carry out the research on the electrical power system new equipment and new technologies. Based on its flexibility and perfect functions, PSASP can be used for the power flow, stability, and short-circuit calculation of VFT [8].

There is also a lot of software for the transient calculation in electrical power systems; for example, EMTP founded by Professor H. W. Dommel, EMTPE developed by the China Electric Power Research Institute, PSCAD/EMTDC developed by Manitoba HVDC Research Center (Canada), and TRANS developed by UBC. EMTPE is developed based on EMTP, integrating main power electronic equipment and components and improving the algorithm. The core of PSCAD is EMTDC, which is a kind of electromagnetic transient analysis program with friendly interface. With their flexibility and perfect functions, these two kinds of software can be used for the study and simulation of the electromagnetic transient of a VFT [9, 10].

Taking all these factors and the needs of VFT research and application into consideration, and based on the author's experience in power system simulation research and software development; in this book, we mainly carry out VFT model development based on PSASP, EMTPE, and PSCAD/EMTD. In this chapter, we first study and deduce basic equations of VFT in the calculation of steady-state power flow, electromechanical transient, electromagnetic transient, and short-circuit current, and use existing PSASP, EMTPE, and PSCAD/EMTDC to establish corresponding digital simulation models and form the full digital simulation models used for the study of the operation characteristics of VFT in the electrical power system, which provides the necessary research tools and means for the development of VFT and its system application.

3.2 The Steady-State Equation and the Power Flow Calculation Model of VFTs

Power flow calculation is an important part in the study of the power system steady-state operation. Based on the given operating conditions and system configuration, it determines the operating state of each section of the whole power system, including the voltage of each bus, transmission power of each element, system power loss, and so on. The early power system was composed of power supply, transformer, line, and load that are linear elements; the system power flow calculation was simple. With a large number of controllable nonlinear devices and switching devices such as DC transmission, power electronic devices, flexible AC transmission systems (FACTS), and phase shifters put into operation, the power flow calculation has become more complicated. For a VFT, a new piece of controllable equipment in the electrical power system, we need to establish element models and power flow algorithms based on the existing analysis programs of the electrical power system [11, 12].

3.2.1 Steady-State Frequency Equation

DC transmission and VFT are two different types of asynchronous interconnection equipment. Interconnection of asynchronous power grids with the DC transmission is mainly achieved through the rectification and inversion of the commutation valve to complete the AC-DC-AC conversions and achieve the controllable transmission of the line power. For the asynchronous interconnection achieved through the VFT, in essence it is to superimpose an electrical frequency component corresponding to the rotating speed of the rotor of rotary transformer on the frequency of the system on one side to make the superimposed synthetic frequency equal to the frequency of the system on the other side. Therefore, in order to realize the steady power transmission between two asynchronous systems through a VFT, the magnetic field of the VFT rotor winding current is required to be synchronous with the magnetic field of the VFT stator winding current formed in the magnetic gap space; otherwise it will cause cyclical changes in the power transmission and even system instability.

The rotating speed of the space magnetic field formed by the stator winding current is decided by the stator winding current frequency while the rotating speed of the magnetic field formed by the rotor winding current is the synthesized result of the rotor current frequency and the rotor (electrical) speed. So, in the steady-state, the following steady-state frequency equation exists

$$f_{rm} = f_s - f_r \tag{3.1}$$

Where,

f_{rm} = rotor frequency of VFT (converted into electrical frequency), Hz;
f_s = stator side power grid frequency, Hz;
f_r = rotor side power grid frequency, Hz.

It can be seen from Equation (3.1) that when the frequency of the systems on both sides of VFT is exactly the same, if there are parallel AC transmission channels, in the steady-state the VFT rotor is almost static and only makes proper acceleration and deceleration adjustments in the power regulation process. For the systems on both sides of 50 or 60 Hz, as relevant regulations have strict requirements for system frequency, the change in the rated frequency should not exceed 0.2 Hz. So in most cases the rotating speed of the VFT is quite small.

What needs to be explained is that the VFT winding generally consists of 4 or 2 × N (N ≥ 2) poles, namely, Npole pairs. f_{rm} and the actual rotor speed f_{rotor} have the following relationship

$$f_{rm} = Nf_{rotor} \tag{3.2}$$

It can be seen from Equation (3.2) that the greater the number of the pole pairs, the smaller the actual rotor speed.

3.2.2 Steady-State Power Flow Equation

In the steady-state, without taking into consideration the impact of the VFT circuit resistance, a VFT can be simplified to be the series circuit of an equivalent reactance and an ideal phase shifter and a parallel magnetizing reactance branch. Figure 3.4 shows the VFT connection system in which both the leakage reactance of VFT rotor winding and leakage reactance of VFT stator winding are calculated on the rotor side. If the loss is taken into consideration, corresponding resistance should be increased in the corresponding parallel and series branches. The phase shift angle of the ideal phase shifter can be smoothly regulated between 0 and 360° to realize the flexible control of power flow between interconnected electrical power system. Assuming voltage is maintained as a constant, in order to achieve the constant power transmission between two asynchronous systems through VFT, not only must it satisfy the frequency Equation (3.1),

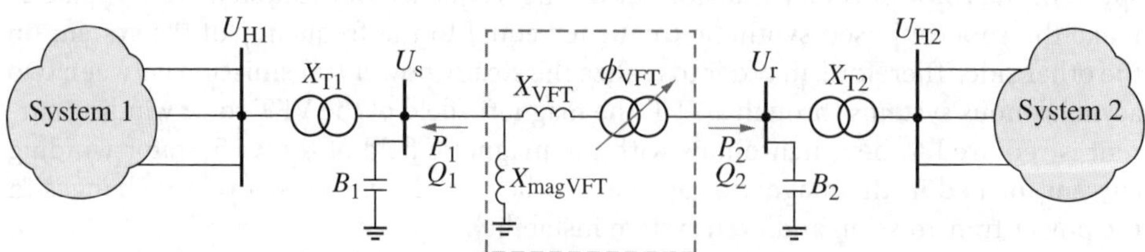

Figure 3.4 Schematic diagram of a VFT connection system.

but also the phase difference between the magnetic field formed by the VFT rotor winding current in the magnetic gap space and the magnetic field formed by the VFT stator winding current in the magnetic gap space should be constant [13–15].

The parameters in Figure 3.4 are defined as follows:

$\dot{U}_{H1}$ is the voltage of high-voltage bus in System 1;

$\dot{U}_{H2}$ is the voltage of high-voltage bus in System 2;

$\dot{U}_s$ is the voltage of the grid connected to the stator side of VFT;

$\dot{U}_r$ is the voltage of the grid connected to the rotor side of VFT;

P_1 and Q_1 are the active power and reactive power output by the VFT from the system on the stator side, respectively;

P_2 and Q_2 are the active power and reactive power output by the VFT from the system on the rotor side, respectively;

B_1 and B_2 are the susceptance of the capacitor banks input on the stator and rotor side of the VFT, respectively;

X_{VFT} and X_{magVFT} are the equivalent short-circuit reactance and excitation reactance of VFT, respectively;

Φ_{VFT} is the equivalent phase shift angle of VFT.

Ignore the impact of the excitation reactance and loss of VFT. The phasor of voltage and current are shown in Figure 3.5.

The parameters in Figure 3.5 are defined as follows:

$\dot{U}'_r$ is the voltage phasor which has the same amplitude as $\dot{U}_r$, but rotate counterclockwise θ_{rm} angle;

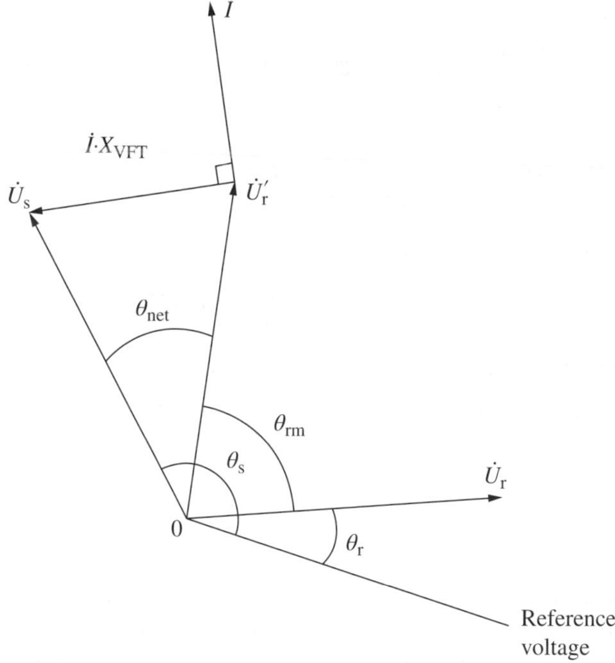

Figure 3.5 VFT voltage and current phasor diagram.

$\dot{I}$ is the phasor of the current flowing through VFT and its phase angle lags 90° behind the phasor difference between $\dot{U}_s$ and $\dot{U}'_r$;

θ_s is the phase angle of the bus voltage on the stator side;

θ_r is the phase angle of the bus voltage on the rotor side;

θ_{rm} is the phase shift generated by the rotation of the rotor (i.e., ideal phase shifter);

θ_{net} is equivalent to the phase difference corresponding to the equivalent impedance voltage drop and as $\theta_{net} = \theta_s - (\theta_r + \theta_{rm})$, in order to realize stable power transmission, θ_{net} must be maintained as a constant.

It can be seen from Figure 3.2 that if $\dot{U}'_r$ is taken as the new reference voltage, the rotor side voltage phasor can be expressed as

$$\dot{U}_r = U'_r \underline{/-\theta_{rm}} \tag{3.3}$$

Accordingly, the stator side voltage phasor can be expressed as

$$\dot{U}_s = U_s \underline{/\theta_{net}} = U_s \underline{/[\theta_s - (\theta_r + \theta_{rm})]} \tag{3.4}$$

Expand Equation (3.4) to obtain

$$\dot{U}_s = U_s \cos[\theta_s - (\theta_r + \theta_{rm})] + jU_s \sin[\theta_s - (\theta_r + \theta_{rm})] \tag{3.5}$$

The current flowing through the branch of VFT, namely, the current flowing through equivalent reactance $X(= X_{VFT})$ can be expressed as

$$\dot{I} = \frac{\dot{U}_s - \dot{U}'_r}{jX} = \frac{\dot{U}_s - U_r}{jX} \tag{3.6}$$

Substitute Equation (3.5) into (3.6) to obtain

$$\dot{I} = \frac{U_s \cos[\theta_s - (\theta_r + \theta_{rm})] + jU_s \sin[\theta_s - (\theta_r + \theta_{rm})] - U_r}{jX} \tag{3.7}$$

Simplify it to obtain

$$\dot{I} = \frac{U_s \sin[\theta_s - (\theta_r + \theta_{rm})]}{X} - j\frac{U_s \cos[\theta_s - (\theta_r + \theta_{rm})] - U_r}{X} \tag{3.8}$$

The output complex power on the rotor side of VFT is

$$\tilde{S}_r = \dot{U}_r \overset{*}{I} = \frac{U_s U_r \sin[\theta_s - (\theta_r + \theta_{rm})]}{X} + j\frac{U_s U_r \cos[\theta_s - (\theta_r + \theta_{rm})] - U_r^2}{X} \tag{3.9}$$

The input complex power on the stator side of VFT is

$$\tilde{S}_s = \dot{U}_s \overset{*}{I} = \frac{U_s U_r \sin[\theta_s - (\theta_r + \theta_{rm})]}{X} + j\frac{U_s^2 - U_s U_r \cos[\theta_s - (\theta_r + \theta_{rm})]}{X}$$

The active power is

$$P = \frac{U_s U_r \sin[\theta_s - (\theta_r + \theta_{rm})]}{X} \tag{3.10}$$

Equation (3.10) is the power transmission formula of VFT. When $\theta_{net} = \theta_s - (\theta_r + \theta_{rm}) = 0$, power transmitted by VFT is 0. The rotation of the rotor of VFT changes the relative position of the magnetic field of the stator winding current and the magnetic field of the rotor winding current. Changing θ_{rm} and θ_{net} can control the transmission power of VFT. Meanwhile, it can be seen from Equation (3.10) that when $\theta_{net} = \theta_s - (\theta_r + \theta_{rm}) = 90°$, the transmission power of VFT reaches its limit, namely,

$$P_{max} = \frac{U_s U_r}{X} \tag{3.11}$$

Within this power limit, the flexible control of the transmission power of the VFT can be realized.

3.2.3 Power Flow Calculation Model

As VFT is a new type of power system device, in the existing power flow calculation programs there is no available component model for it. However, corresponding model functions of the phase-shifting transformer can be used for the calculation of power flow of VFT. See the specific power flow model in Figure 3.6 in which the VFT can be expressed with the serious connection of a 1:1 ideal transformer and a phase shifter the phase angle of which can be smoothly regulated [1].

The phase angle of the phase-angle regulator is the phase difference between the voltage of the stator side winding and the voltage of the rotor side winding and can be regulated based on the needs of power flow. B_1 and B_2 are the switched capacitors used for reactive power compensation and can be switched based on the grid voltage level and the reactive power absorbed by the VFT.

3.2.4 Using PSASP to Realize the VFT Power Flow Calculation Model

PSASP, an open simulation platform of electrical power systems providing a powerful UD modeling method, can conveniently realize the power flow calculation of user-defined elements, which is very useful for the study of a new power system device. In recent years, many universities and research institutes in China have done a lot of research using the UD function of PSASP and have achieved remarkable results.

For the so-called UD modeling method, it means that without understanding the internal structure of programs and programming design users can, based on their calculation and analysis needs, use the familiar concepts and methods that are easy to master and design various models so that they can flexibly simulate any system element, automatic device, and control function in principle. Through the UD modeling method, on

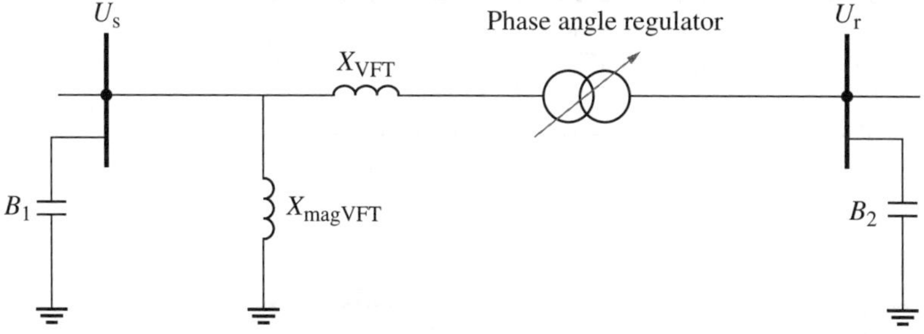

Figure 3.6 Power flow model of VFT. Where: U_s = stator side voltage; U_r = rotor side voltage; X_{VFT} = total leakage reactance of the primary side and secondary side winding of VFT; and X_{magVFT} = excitation reactance of transformer.

the one hand, users can build models to meet their special needs; on the other hand, some common models built with PSASP can be stored in the "system model base" as a component of PSASP and directly provided for users. These functions include using a certain bus voltage, generator reactive power, and load reactive power to control the voltage of a certain bus and the reactive power of a certain branch, using the reactance of a certain branch to control the voltage of a certain bus and line power and building various relay models and DC models with the control system. Users can conveniently modify these models provided by the program or use them after increasing the model parameters.

The main components of UD modeling include basic functional blocks and input and output information. The basic functional blocks are the smallest part of the UD model. Through the combination, assembly, and integration of some functional blocks, users can build a UD model. Each basic functional block can be used to complete the operation of solving the output variable (Y) based on the input variable (X). Figure 3.7 shows the structure of basic functional blocks [8].

More than 50 functional blocks are set up in PSASP and they can be further expanded according to actual needs. By nature these functional blocks can basically be categorized into calculus operation, algebraic operation, basic function operation, logic control operation, linear and nonlinear function operation, and other operations that can be commonly used by the power flow, transient stability, and other related calculation programs.

Figure 3.8 shows the schematic diagram of a UD model and power system connection. Each UD model is connected to the studied electrical power system through its input information X (X_1, X_2, …) and output information Y (Y_1, Y_2, …).

The input and output information provided by PSASP must be sufficient and necessary. In the power flow program, the known calculation variables and relevant unknown variables are, respectively, used as input and output information. In transient stability program, the input information includes bus, generator, load, DC line, and variables

Figure 3.7 Structure of basic functional blocks.

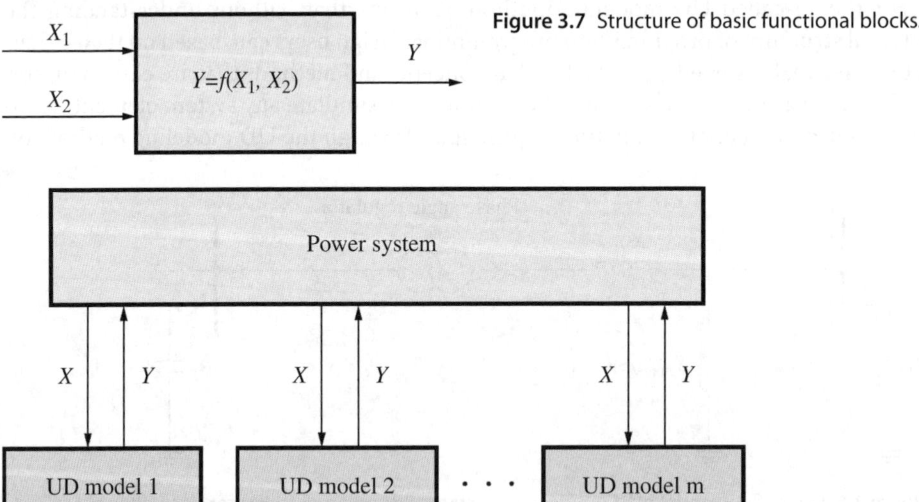

Figure 3.8 Schematic diagram of the UD Model and power system connection.

related to positive, negative, and zero sequence network while output information includes variables related to generator and DC line, generating unit tripping, load shedding, network operation, and so on. With the input and output information, the connection of any UD model with the electrical power system can be achieved.

Based on the functional characteristics of a VFT and the aforementioned Equations (3.1)–(3.10), the VFT UD power flow calculation model shown in Figure 3.9 is established to conveniently realize the electrical power system power flow calculation including the VFT elements. In the model, the input variable is the voltage of the bus of the systems on both sides of a VFT; the output variable is the current output by the VFT to the systems on both sides; VFT impedance is the input parameter; and the transmission power command is the control variable. For the power flow calculation

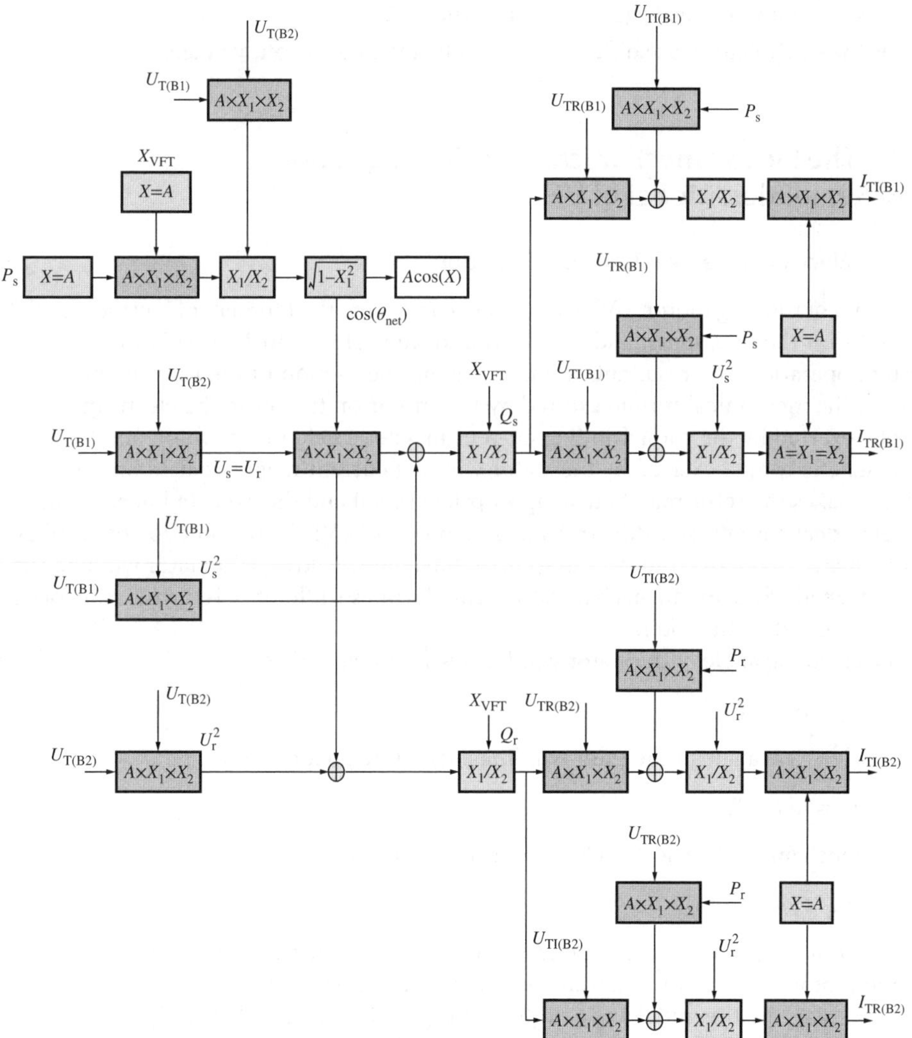

Figure 3.9 UD power flow calculation mode in PSASP.

model that can realize more complex controls, the input and output variables can be further expanded based on it.

The parameters in Figure 3.9 are defined as follows:

$U_{T(B1)}$ and $U_{T(B2)}$ are the bus voltage on both sides (input variable), respectively;

$U_{TI(B1)}$ and $U_{TI(B2)}$ are the imaginary part of bus voltage on both sides (input variable);

$U_{TR(B1)}$ and $U_{TR(B2)}$ are the real part of bus voltage on both sides (input variable);

$I_{TI(B1)}$ and $I_{TI(B2)}$ are the imaginary part of current injected into the bus current on both sides (output variable);

$I_{TR(B1)}$ and $I_{TR(B2)}$ are the real part of current injected into the bus current on both sides (output variable);

P_s and P_r are the stator side active power and the rotor side active power;

X_{VFT} is the equivalent leakage reactance of the VFT;

θ_{net} is the equivalent phase difference of winding the equivalent voltage.

3.3 The Electromechanical Transient Equation and Simulation Model of VFTs

3.3.1 Electromechanical Transient Equation

As a kind of rotating motor, a VFT is an inertial system. The moment of inertia of a VFT depends on the size, mass, and distribution of the rotor shaft in a VFT. In the actual system operation, the resultant moment driving the motion of the rotor is the vector sum of the mechanical torque exerted by DC motor on the rotor, the electromagnetic torque exerted by the rotor winding current magnetic field on the rotor, and the electromagnetic torque exerted by the stator winding current magnetic field on the rotor, which makes the rotor maintain an appropriate speed and the expected phase shift. In the electrical power system, the transmission power of a VFT is the function of the phase difference between the rotor winding current magnetic field and the stator winding current magnetic field; in the mechanical system, the phase difference is the function of the torque exerted on the rotor.

The electrical angle of the stator winding current magnetic field is:

$$\theta_s = \omega_s t + \theta_{s0}$$

The electrical angle of the rotor winding current magnetic field is

$$\theta_r = \omega_r t + \theta_{r0}$$

The mechanical electrical angle of the rotor rotation is

$$\theta_{rm} = \omega_{rm} t + \theta_{rm0}$$

where $\omega_s, \omega_r, \omega_{rm}$ are the corresponding electrical angular frequency of the stator side system, rotor side system, and transformer shaft rotation speed;

$\theta_{s0}, \theta_{r0}, \theta_{rm0}$ are the initial angle of the corresponding electrical angle $\theta_s, \theta_r, \theta_{rm}$ of the stator winding current magnetic field, rotor winding current magnetic field, and rotor mechanical rotation.

The angle difference between the stator winding current magnetic field and the rotor winding current magnetic field is:

$$\begin{aligned}
\theta_{net} &= \theta_s - (\theta_r + \theta_{rm}) \\
&= \omega_s t + \theta_{s0} - (\omega_r t + \theta_{r0} + \omega_{rm} t + \theta_{rm0}) \\
&= [\omega_s - (\omega_r + \omega_{rm})]t + [\theta_{s0} - (\theta_{r0} + \theta_{rm0})]
\end{aligned} \tag{3.12}$$

When the system is in the steady-state, θ_{net} should be a constant value, so

$$\omega_s - (\omega_r + \omega_{rm}) = 2\pi \times [f_s - (f_r + f_{rm})] = 0$$

namely

$$f_s = f_r + f_{rm}$$

Where:

f_s = frequency of the grid connected to the stator side winding;

f_r = frequency of the grid connected to the rotor side winding;

f_{rm} = product of the rotor rotation mechanical frequency and the number of pole pairs; the positive direction of f_{rm} is the rotation direction of the rotor side winding current magnetic field and the opposite direction is the negative direction of f_{rm}. The value of f_{rm} depends on the frequency difference between the systems on both sides and it can be positive or negative.

Accordingly, the rotor motion equation of the rotary transformer is

$$\begin{cases}
\dfrac{d\theta_{rm}}{dt} = \omega_{rm} = 2\pi f_{rm} \\[2mm]
\dfrac{d\omega_{rm}}{dt} = \dfrac{1}{T_{Jr}}(T_d + T_r - T_s)
\end{cases} \tag{3.13}$$

where T_d = mechanical torque exerted on the rotor;

T_r = electromagnetic torque exerted by the rotor winding on the rotor;

T_s = electromagnetic torque exerted by the stator winding on the rotor;

T_{Jr} = moment of inertia of the rotor.

In Equation (3.13), the stray loss such as the rotor winding is ignored. The relationship between various torques and related power is as follows:

$$\begin{aligned}
T_s &= \frac{P_s}{\omega_s} \\[2mm]
T_r &= \frac{P_r}{\omega_r} \\[2mm]
T_d &= \frac{P_d}{\omega_{rm}}
\end{aligned} \tag{3.14}$$

where P_s = electromagnetic power of stator winding, W:

P_r = electromagnetic power of rotor winding, W:

P_d = output power of driving motor, W.

These are the system dynamic equations that need to be taken into consideration in the study of the electromechanical transient characteristics of VFT.

3.3.2 Electromechanical Transient Model

As the VFT is a new type of power system equipment, in our commonly used PSASP, PSD, PSSE, and other programs, there is no readily available model that can directly be used. Through comprehensive comparison of different transient stability analysis software, in this book we will first use PSCAD/EMTDC to build a relatively complete and detailed VFT model to study the dynamic characteristics of the VFT and use it as a criterion for reference and comparison. Then we will use the UP function of PSASP to build the electromechanical transient simulation module including a VFT and compare it with the PSCAD/EMTDC simulation results and the actual measurement results provided by GE.

In this book in PSASP, a VFT is regarded as a two-port element; the element input parameter is the voltage of the two buses connected to it while the output parameter is the current injected into the buses. In other words, based on the steady-state power flow calculation, the UD mode with the inertia characteristics of the rotary transformer and the driving characteristics of the DC motor taken into consideration is used to control the phase shift of VFT stator and rotor winding so as to achieve the VFT power regulation and control function in a dynamic simulation process. Figure 3.10 shows the schematic diagram of a VFT electromechanical transient model [8].

In this model, the VFT rotor motion equation is Equation (3.13) while the difference between the VFT stator voltage phasor and the VFT rotor voltage phasor is Equation (3.12). Shown in Table 3.1 are the typical parameters of a VFT.

3.3.3 Using PSASP to Realize the Electromechanical Transient Model of VFT

PSASP is an open system analysis platform that provides a good user interface and UPI. PSASP develops the user program interface (UPI) function (PSASP/UPI) to provide users with a more free and open environment. To be specific, through programming

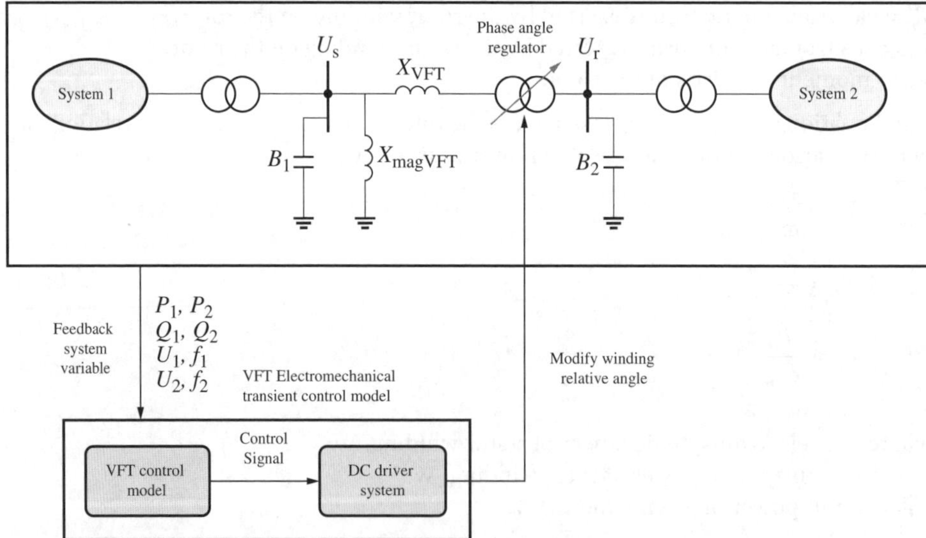

Figure 3.10 Schematic diagram of the VFT electromechanical transient model.

Table 3.1 Typical parameters of VFT.

Name of Parameter	Typical Value (with the rated value of VFT as reference)
VFT leakage reactance X_{VFT}	18%
VFT excitation reactance X_{magVFT}	10 pu
Reactance of the transformer winding on both sides of VFT X_T	10%
Conductance of the capacitor banks B_1 and B_2	20~80%
Moment of inertia of VFT $J(= 2H)$	50s

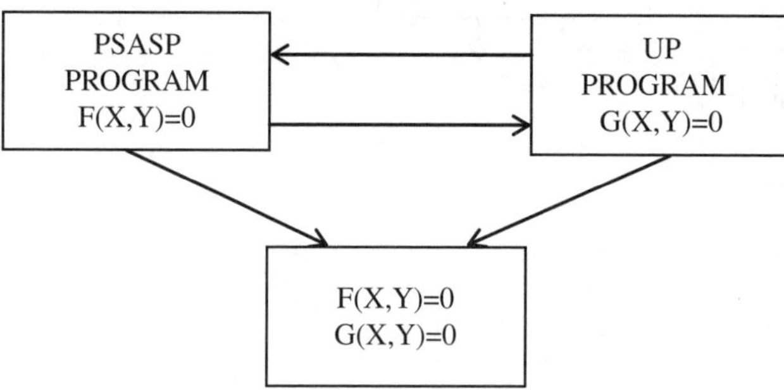

Figure 3.11 Operation mode of PSASP and UP.

(with no language tool limit, such as FORTRAN, C++, etc.) the PSASP resources are used to achieve the function expansion of PSASP to make the execution module of PSASP and user program operate alternately and jointly to complete a calculation task. Figure 3.11 shows its operation mode [8].

The data transfer between them is realized through data file. The input data X, namely the data obtained through UP from PSASP, is stored in the input information file **. F1 (with ** being the user program name); the output data Y, namely the output data of the UP calculation results to be calculated and processed by PSASP, is stored in the output information file **. F2 (with ** being the user program name). Therefore, the process of PSASP and UP is shown in Figure 3.12.

PSASP can be operated together with multiple user programs or in its UD model. In addition, it provides the function of inter-combination between user programs, UD models as well as user programs, and UD models, which provides a powerful means for the study of system characteristics of the new equipment in the electrical power system.

As a two-port device, a VFT can be seen as a dynamic element connecting the two buses of the electrical power system. The dynamic equation of the VFT can be established through the derivation of the VFT port voltage, current, power, power angle, and angular speed of rotary transformer. In PSASP, the UP model can be used to realize the mathematical model of VFT in which the input signal of the UP module X is the system bus voltage while the output signal X is the current injected from the VFT into the system bus. Figure 3.13 shows the flow chart of the whole simulation of the UP subprogram.

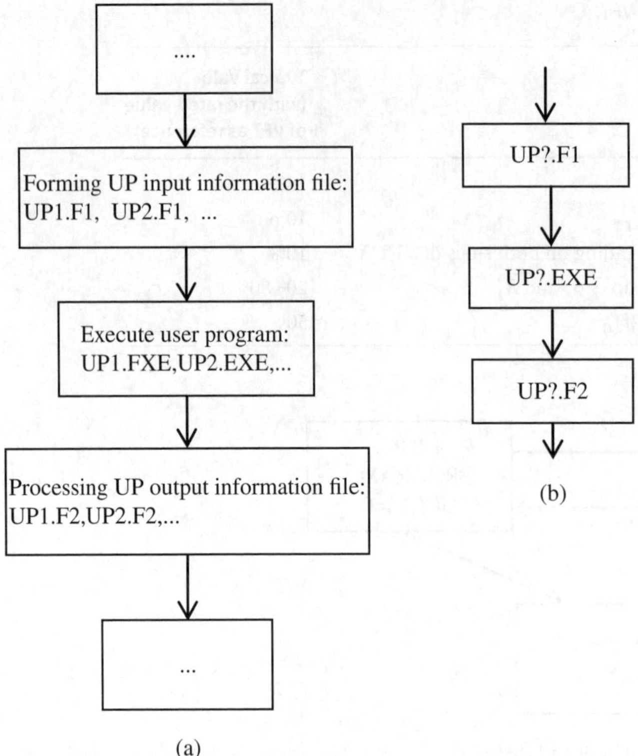

Figure 3.12 Schematic diagram of the process of (a) PSASP and (b) UP.

Now we are going to derive the relational formulas of the main electrical parameters of the VFT and the system interface. The switching complex power of the VFT and the bus of the systems on both sides is;

$$\tilde{S}_r = P_r + jQ_r = \frac{U_s U_r \sin[\theta_s - (\theta_r + \theta_{rm})]}{X}$$
$$+ \frac{j\{U_s U_r \cos[\theta_s - (\theta_r + \theta_{rm})] - U_r^2\}}{X} \tag{3.15}$$

$$\tilde{S}_s = P_s + jQ_s = \frac{U_s U_r \sin[\theta_s - (\theta_r + \theta_{rm})]}{X}$$
$$+ j\frac{U_s^2 - U_s U_r \cos[\theta_s - (\theta_r + \theta_{rm})]}{X} \tag{3.16}$$

Where,

P_s and Q_s = stator side active power and reactive power;
U_s and U_r = stator side and rotor side voltage;
X = VFT circuit reactance (including the leakage reactance of transformers on both sides).

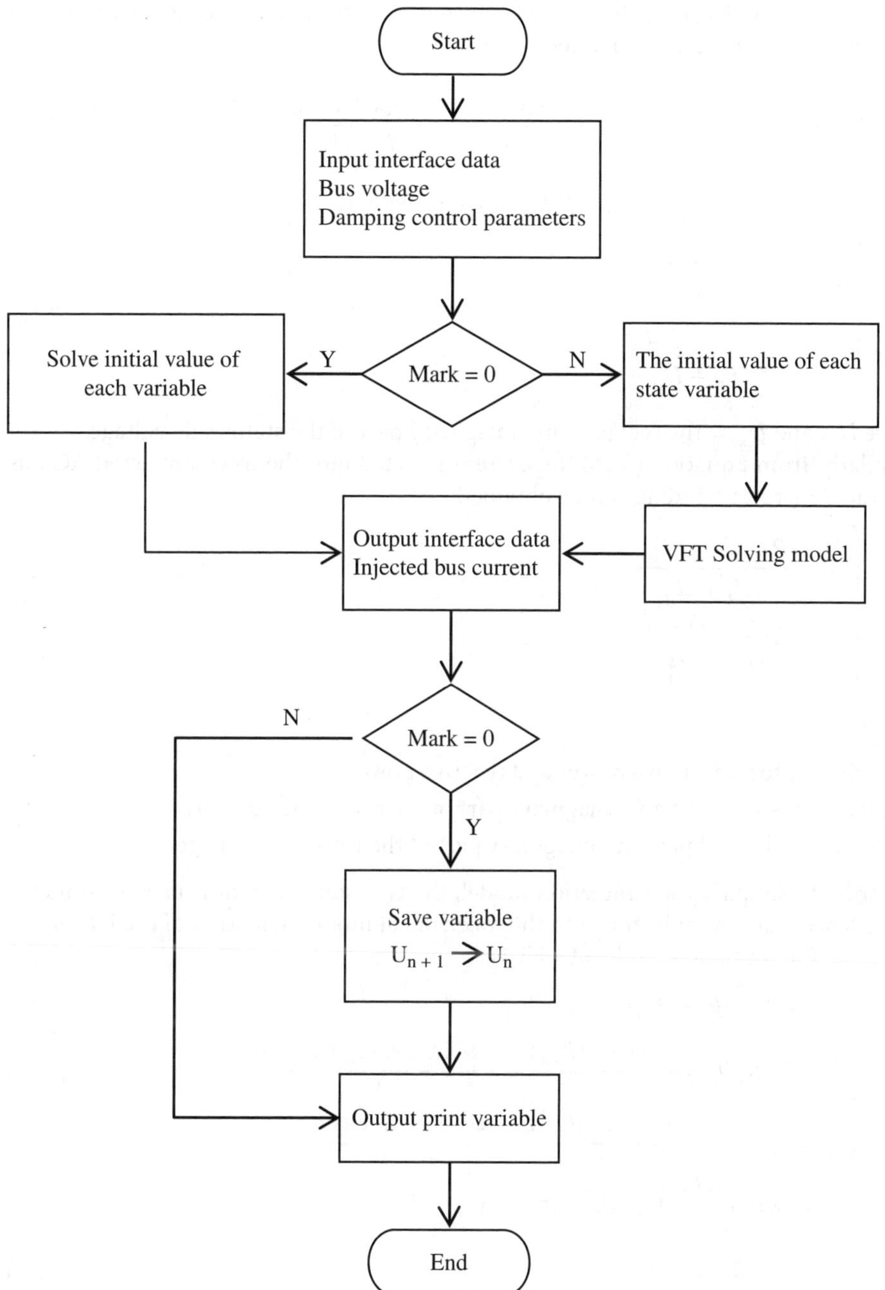

Figure 3.13 Flow chart of the VFT UP subprogram.

From Equation (3.15), I_s, the current injected into the stator side AC bus from the VFT stator winding, can be obtained

$$\dot{I}_s = I_{sR} + jI_{sI} = -\frac{\overset{*}{S}_s}{\overset{*}{U}_s} = -\frac{P_s - jQ_s}{U_{sR} - jU_{sI}} = -\frac{P_s U_{sR} + Q_s U_{sI}}{U_{sR}^2 + U_{sI}^2} - j\frac{P_s U_{sI} - Q_s U_{sR}}{U_{sR}^2 + U_{sI}^2} \quad (3.17)$$

The real part and imaginary part of in the injected current I_{sR} and I_{sI} are

$$\begin{cases} I_{sR} = -\dfrac{P_s U_{sR} + Q_s U_{sI}}{U_{sR}^2 + U_{sI}^2} \\[3mm] I_{sI} = -\dfrac{P_s U_{sI} - Q_s U_{sR}}{U_{sR}^2 + U_{sI}^2} \end{cases} \quad (3.18)$$

Where U_{sR} and U_{sI} = the real part and imaginary part of the stator side voltage.

Similarly, from Equation (3.16) the current injected into the interconnected AC bus from the VFT rotor winding can be obtained

$$\begin{cases} I_{rR} = \dfrac{P_r U_{rR} + Q_r U_{rI}}{U_{rR}^2 + U_{rI}^2} \\[3mm] I_{rI} = \dfrac{P_r U_{rI} - Q_r U_{rR}}{U_{rR}^2 + U_{rI}^2} \end{cases} \quad (3.19)$$

Where,

P_r and Q_r = rotor side active power and reactive power;

I_{rR} and I_{rI} = the real part and imaginary part of the rotor side current;

U_{rR} and U_{rI} = the real part and imaginary part of the rotor side voltage.

In order to simplify the simulation model, the two-step element is used to simulate the VFT drive unit rectification and the mechanical motion equation of the DC motor and rotor. The simplified model of VFT is

$$\begin{cases} \theta_{net} = \theta_s - (\theta_r + \theta_{rm}) \\[2mm] \tilde{S}_r = P_r + jQ_r = \dfrac{U_s U_r \sin(\theta_{net})}{X} + \dfrac{j[U_s U_r \cos(\theta_{net}) - U_r^2]}{X} \\[2mm] \tilde{S}_s = P_s + jQ_s = \dfrac{U_s U_r \sin(\theta_{net})}{X} + j\dfrac{U_s^2 - U_s U_r \cos(\theta_{net})}{X} \end{cases} \quad (3.20)$$

$$\theta_s = \arctan\left(\frac{U_{sI}}{U_{sR}}\right), \quad \theta_r = \arctan\left(\frac{U_{rI}}{U_{rR}}\right) \quad (3.21)$$

$$\theta_{rm} = \frac{K_1}{1 + bs + as^2}\theta_{PI} \quad (3.22)$$

$$\theta_{PI} = \left(K_2 + \frac{1}{T_2 s}\right)(P_{sRef} - P_s + \Delta P_{sR}) \quad (3.23)$$

where K_1, K_2, T_2, a, and b are all coefficients.

The VFT electromechanical transient model in the simulation of PSASP shown in Figure 3.14 is obtained.

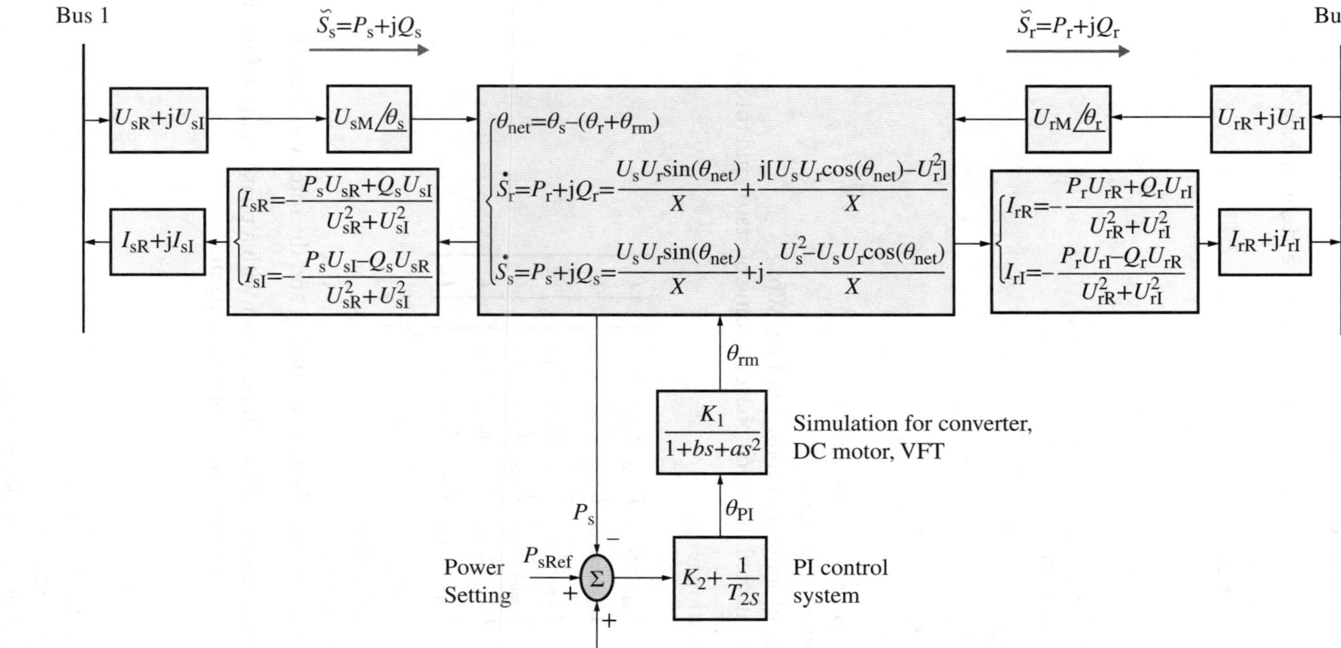

Figure 3.14 VFT electromechanical transient simulation model in PSASP.

3.4 The Electromagnetic Transient Equation and Simulation Model of VFTs

3.4.1 Electromagnetic Transient Equation

The electromagnetic transient equation of VFT can be derived based on the basic equations of the motor [9–16]:

Stator side voltage equation

$$u_{sa} = p\psi_{sa} + r_{sa}i_{sa}$$
$$u_{sb} = p\psi_{sb} + r_{sb}i_{sb}$$
$$u_{sc} = p\psi_{sc} + r_{sc}i_{sc} \tag{3.24}$$

Rotor side voltage equation

$$u_{ra} = p\psi_{ra} + r_{ra}i_{ra}$$
$$u_{rb} = p\psi_{rb} + r_{rb}i_{rb}$$
$$u_{rc} = p\psi_{rc} + r_{rc}i_{rc} \tag{3.25}$$

where p = derivative operator of time, $p = \mathrm{d}/\mathrm{dt}$;
u = phase voltage of stator and rotor, V;
ψ = phase flux linkage of stator and rotor winding, Wb;
i = excitation current injected into each phase winding of stator and rotor, A.

The flux linkage equation is

$$\begin{bmatrix} \psi_{sa} \\ \psi_{sb} \\ \psi_{sc} \\ \psi_{ra} \\ \psi_{ra} \\ \psi_{ra} \end{bmatrix} = \begin{bmatrix} L_{saa} & L_{sab} & L_{sac} & L_{sraa} & L_{srab} & L_{srac} \\ L_{sba} & L_{sbb} & L_{sbc} & L_{srba} & L_{srbb} & L_{srbc} \\ L_{sca} & L_{scb} & L_{scc} & L_{srca} & L_{srcb} & L_{srcc} \\ L_{sraa} & L_{srab} & L_{srac} & L_{raa} & L_{rab} & L_{rac} \\ L_{srba} & L_{srbb} & L_{srbc} & L_{rba} & L_{rbb} & L_{rbc} \\ L_{srca} & L_{srcb} & L_{srcc} & L_{rca} & L_{rcb} & L_{rcc} \end{bmatrix} \cdot \begin{bmatrix} i_{sa} \\ i_{sb} \\ i_{sc} \\ i_{rc} \\ i_{rb} \\ i_{rc} \end{bmatrix} \tag{3.26}$$

where L_s = stator side inductance, H;
L_r = rotor side inductance, H;
L_{sr} = mutual inductance between the stator winding and the rotor winding, H.

The VFT in function is equivalent to a transformer with a phase-shifting function, so a VFT satisfies the magnetic motive force equation, namely,

$$N_s i_s = -N_r i_r \tag{3.27}$$

where,

N_s = number of stator winding turns;
N_r = number of rotor winding turns;
i_s = current flowing out of the stator winding, A;
i_r = current flowing out of the rotor winding, A.

Based on the flux linkage equation of the transformer, we have

$$U_s = 4.44 f_s N_s \Phi_a \tag{3.28}$$

$$U_r = 4.44 f_r N_r \Phi_a \tag{3.29}$$

So Equation (3.29) can be changed into

$$U_r = U_s \frac{f_r N_r}{f_s N_s} \tag{3.30}$$

where,

f_s = stator winding current frequency, Hz;
f_r = rotor winding current frequency, Hz;
N_s = number of stator winding turns;
N_r = number of rotor winding turns;
Φ_a = air-gap flux, Wb $\cdot$ S.

When the system operates stably, namely when the VFT rotor is in the steady-state and the stray loss such as the rotor winding is ignored, the active power balance equation is

$$P_s = P_d - P_r \tag{3.31}$$

Equation (3.31) can be rewritten as

$$P_d = U_s I_s + U_r I_r \tag{3.32}$$

From Equation (3.27) we can obtain

$$I_r = -I_s \frac{N_s}{N_r} \tag{3.33}$$

Substitute Equations (3.30) and (3.33) into Equation (3.32)

$$P_d = U_s I_s - U_s \frac{f_r N_r}{f_s N_s} I_s \frac{N_s}{N_r}$$

$$= U_s I_s \left(1 - \frac{f_r}{f_s} \right) \tag{3.34}$$

As

$$T_d = \frac{P_d}{2\pi f_{rm}} \tag{3.35}$$

that

$$T_d = \frac{U_s I_s \left(1 - \frac{f_r}{f_s} \right)}{2\pi f_{rm}} = \frac{U_s I_s}{2\pi f_s} = \frac{P_s}{2\pi f_s} \tag{3.36}$$

It can be seen from Equation (3.36) that in stable operation the VFT transmission power is proportional to the DC motor driving torque that can be regulated through the regulation of the output power of the DC motor. Meanwhile, it can be seen from Equation (3.35) that when the frequency difference between the systems on both sides of the VFT is very small, the small driving power can produce large driving torque, which is more conducive to the control of the VFT transmission power.

3.4.2 Electromagnetic Transient Model

EMTPE and PSCAD/EMTDC are professional power system electromagnetic transient simulation software packages that, with powerful power system electromagnetic transient simulation functions, can simulate the electrical power system equipment well, including complex electrical equipment and power electronic devices. In addition, PSCAD/EMTDC has a user-friendly interface, which is convenient for power electronic rectification and control links in the simulation of a VFT. As a result, it is convenient and effective to use EMTPE and PSCAD/EMTDC to simulate a VFT.

In EMTPE, the VFT can be seen as a doubly fed induction generator (DFIG) as the winding on both sides of which is connected to an AC system. In the simulation, the Type 4 General Motor in EMTPE can be used to express it; in the internal calculation of EMTPE its electrical power system can be expressed with the power system equivalent circuit shown in Figure 3.15, while its mechanical inertia system can be expressed with the mechanical system equivalent circuit shown in Figure 3.16 [9, 16].

Where,

$I_{d,q}$, $I'_{d,q}$ = d and q component of the primary side current and secondary side current;
$U_{d,q}$, $U'_{d,q}$ = d and q component of the primary side voltage and secondary side voltage;
$R_{d,q}$, $R'_{d,q}$, = primary side and secondary side d and q resistance;
$L_{kd,q}$, $L'_{kd,q}$ = primary side and secondary side d and q leakage reactance;
$L_{md,q}$ = excitation reactance.

In PSCAD/EMTDC, although there is no model or example of a VFT, the double fed motor model and DC motor model can still be used to simulate VFT. The double fed motor model is used to simulate the rotary transformer of a VFT while the DC motor model is used to simulate the a VFT driving motor with the ontology model shown in

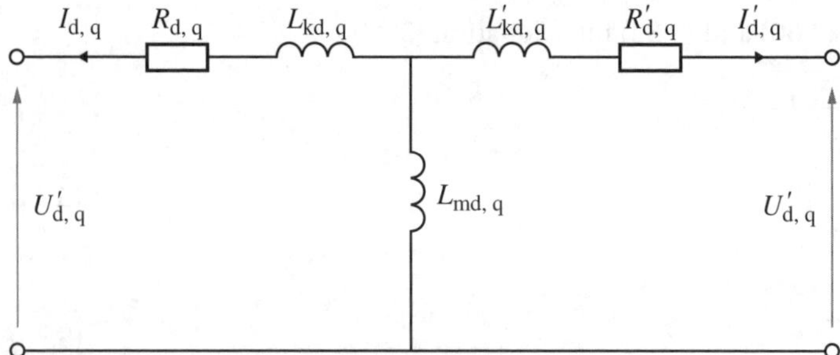

Figure 3.15 Schematic diagram of power system equivalent circuit of VFT in EMTPE.

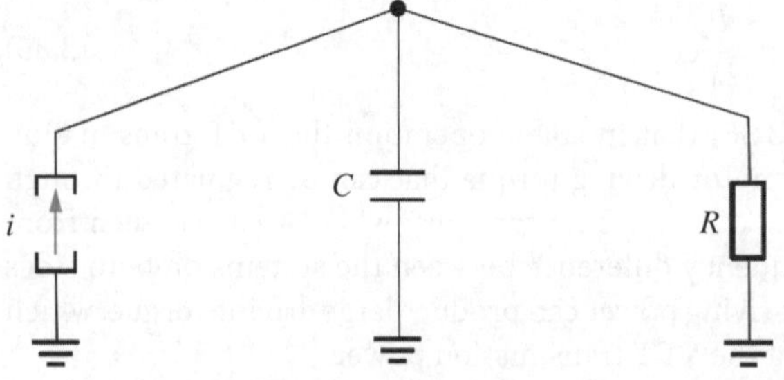

Figure 3.16 Mechanical system equivalent circuit of VFT in EMTPE: i = torque; C = inertia; and R = loss.

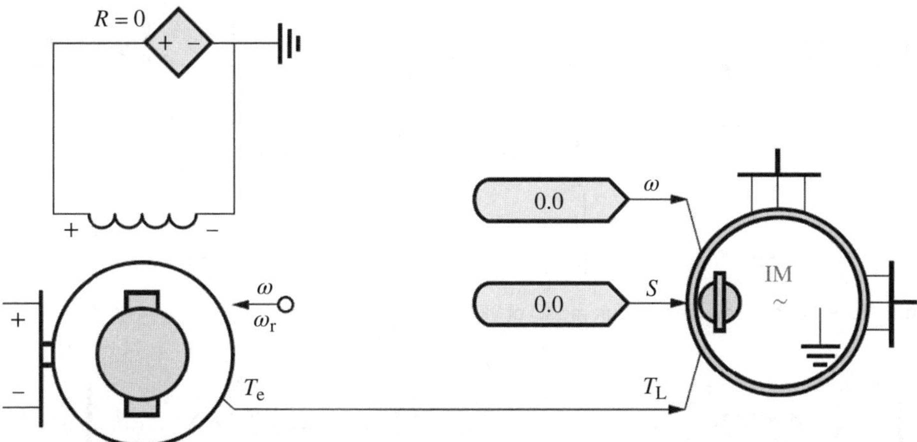

Figure 3.17 Ontology model of VFT in PSCAD: ω (ω_r) = rotor angular speed; S = mechanical rotational speed; T_L = input mechanical torque; and T_e = output electromagnetic torque of motor.

Figure 3.17 used as its electrical circuit. In Figure 3.17, the output torque of the DC motor is used as the input torque of the induction motor, and the output speed of the asynchronous motor is used as the speed input of the DC motor for the purpose of equivalent realization of the characteristic that the DC motor and the rotary transformer rotor are on the same shaft. On the other hand, the size and direction of the torque of the DC motor are controlled by the output voltage of the rectifying circuit [10].

3.5 Short-Circuit Impedance and Calculation Model of VFTs

3.5.1 Short-Circuit Impedance

In short-circuit current calculations, generally only the positive sequence impedance value and zero sequence impedance value of each device are needed. The VFT is used to connect points A and B in the system. In the short-circuit analysis, the positive sequence impedance between Point A and Point B is the sum of leakage resistance of VFT and the leakage resistance of the step-down transformers on both sides, generally being about 0.38 pu. The short-circuit capacity of the new bus might be 150–250% of the rated capacity of the VFT, mainly depending on the short-circuit level of the transmission grid on the other side of the VFT.

At the same time, the step-up transformer similar to that used in the generator is used for the step-down transformer connected with the VFT. On the high-voltage side, the Y-shaped wiring and the neutral point grounding mode are used while on the low-voltage side triangle-shaped wiring is used. As a result, a zero sequence reactance to earth with amplitude of about 0.1 pu is added to the bus on each side.

3.5.2 Short-Circuit Calculation Model

The short-circuit calculation models of VFT, namely the positive sequence circuit model and the zero sequence circuit model, are shown in Figures 3.18 and 3.19, respectively.

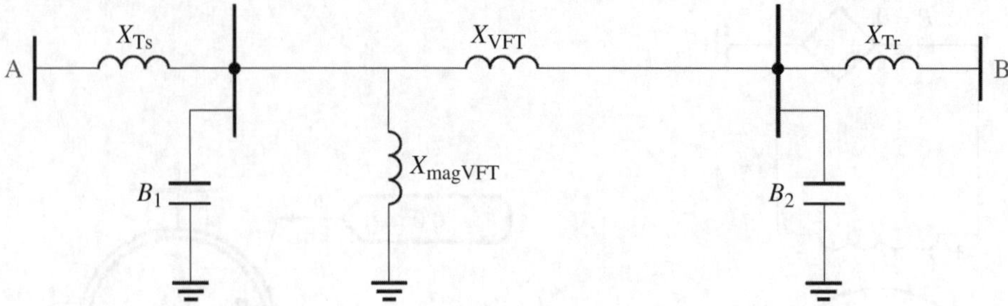

Figure 3.18 Positive sequence circuit model of VFT.

Figure 3.19 Zero sequence circuit model of VFT.

3.6 VFT Simulation Model Availability Verification

3.6.1 VFT Power Flow Calculation Model Verification

We have used the Northeast China-North China-Central China interconnected system to conduct the VFT power flow calculation verification with a command power of 100 MW (namely, 0.1 pu, with 1000 MW as the benchmark). Figure 3.20 shows the convergence process of the VFT-involved power flow iterative calculation. It can be seen from the number of UD power flow iterations shown in Figure 3.20 that, after three to four iterations, the electrical variables such as the system voltage, reactive power, and active power are convergent. The results are at a reasonable level, which verifies the adaptability of the VFT power flow calculation model.

3.6.2 VFT Electromechanical Transient Model Verification

In order to test the adaptability of the VFT simulation model in PSASP, a simplified power system electromechanical model shown in Figure 3.21 is used for simulation. In Figure 3.21, Bus B1 and B4 are connected to a large power supply system and the system simulation parameters are as follows (with the base value $S_B = 100$ MVA).

Impedance: $X_{VFT} = 0.38$ pu, $X_{TB1B2} = 0.1$ pu, $X_{LB2B3} = 0.22$ pu, $X_{TB3B4} = 0.06$ pu;
Admittance: $B_{LB2} = 0.195$ pu, $B_{LB3} = 0.195$ pu;
Voltage: $U_{B1} = 1.04$ pu, $U_{B4} = 1.01$ pu;
Load: $LOAD_{B2} = 1.0 + j0.6$ pu, $LOAD_{B3} = 3.0 + j1.0$ pu;
Constant: $k_1 = -1.0$, $a = 0.003$, $b = 0.10$, $T_2 = 0.08$, $k_2 = 1.2$.

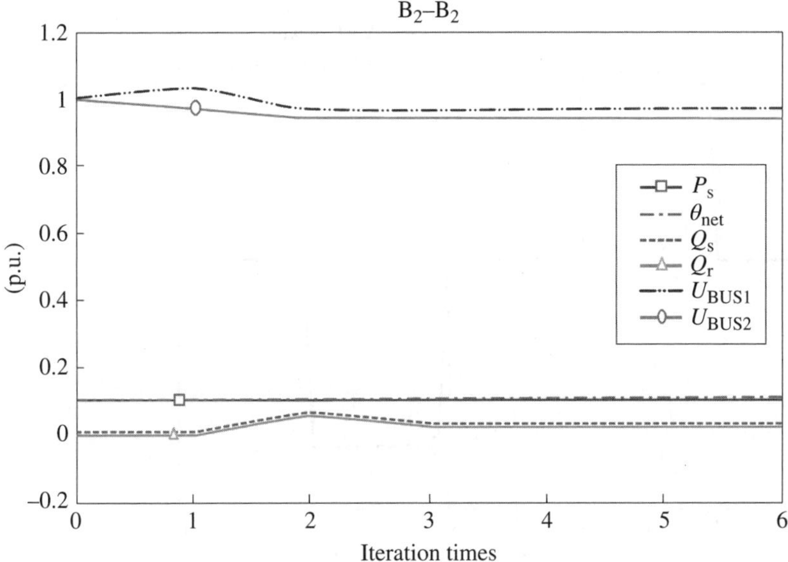

Figure 3.20 Number of UD power flow iterations of a VFT in the PSASP power flow calculation.

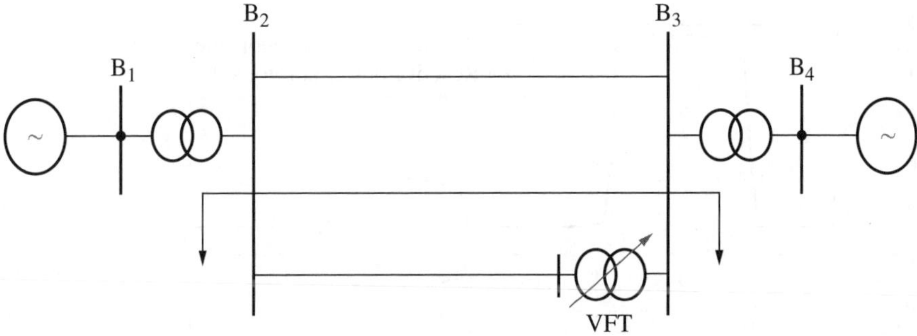

Figure 3.21 VFT electromechanical model simulation example.

As shown in Figure 3.20, in the power flow UD model, the active power transmitted by the VFT P_s is set to be 0.1 pu and the power flow calculation results are $U_{B2} = 0.96689$ pu and $U_{B3} = 0.93889$ pu.

The VFT UP model shown in Figure 3.14 is connected to simulate the electromechanical transient of the system shown in Figure 3.21. When the fixed power value of VFT changes by 0.3 pu, the step response of all variables of VFT are shown in Figures 3.22–3.26. It can be seen that the power response of VFT is fast, and basically reaches the steady-state 0.4 s after the step response. It can be seen from Figure 3.23 that the simulation results are basically consistent with the actual system test response characteristics provided by GE.

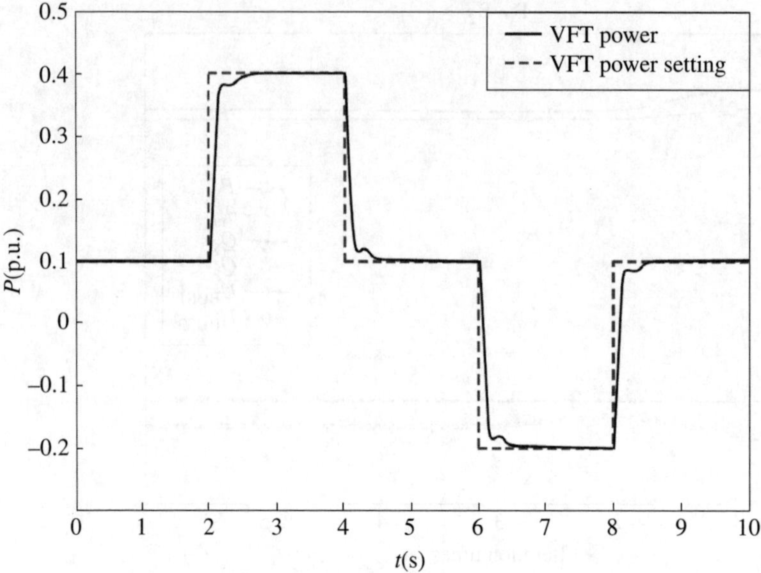

Figure 3.22 Step response of the transmission power of a VFT.

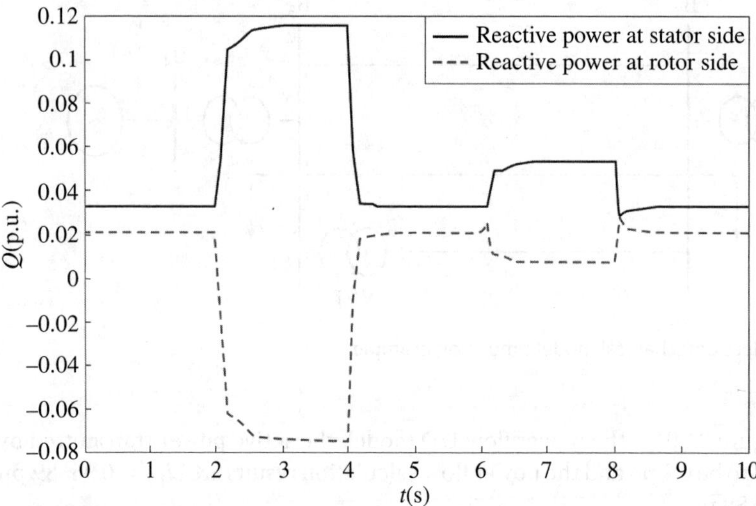

Figure 3.23 Step response of the reactive power of a VFT.

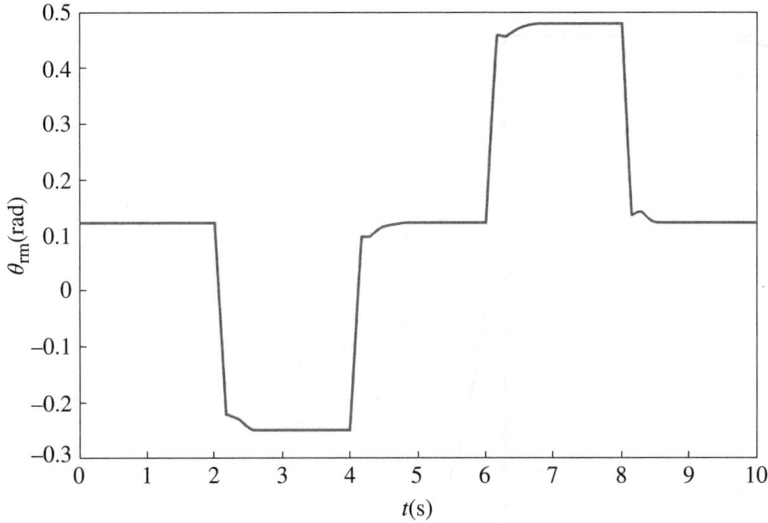

Figure 3.24 Step response of the rotor mechanical angle of a VFT.

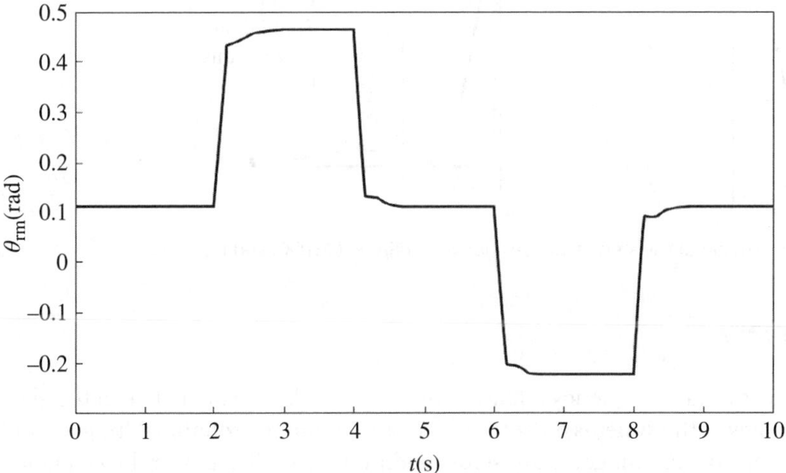

Figure 3.25 Step response of the angle difference between VFT rotor and stator windings.

3.6.3   VFT Electromagnetic Transient Model Verification

We use EMTPE and PSCAD/EMTDC to build the detailed VFT electromagnetic transient simulation model and corresponding system, and conduct step response research. Figure 3.27 shows the operational characteristics of a VFT calculated using EMTDC. The results show that the step response characteristics of VFT are basically consistent with the response characteristics in the actual system test (as shown in Figure 3.26), which also proves the adaptability of the VFT electromagnetic transient model. We are going to elaborate on further research and verification in Chapter 5.

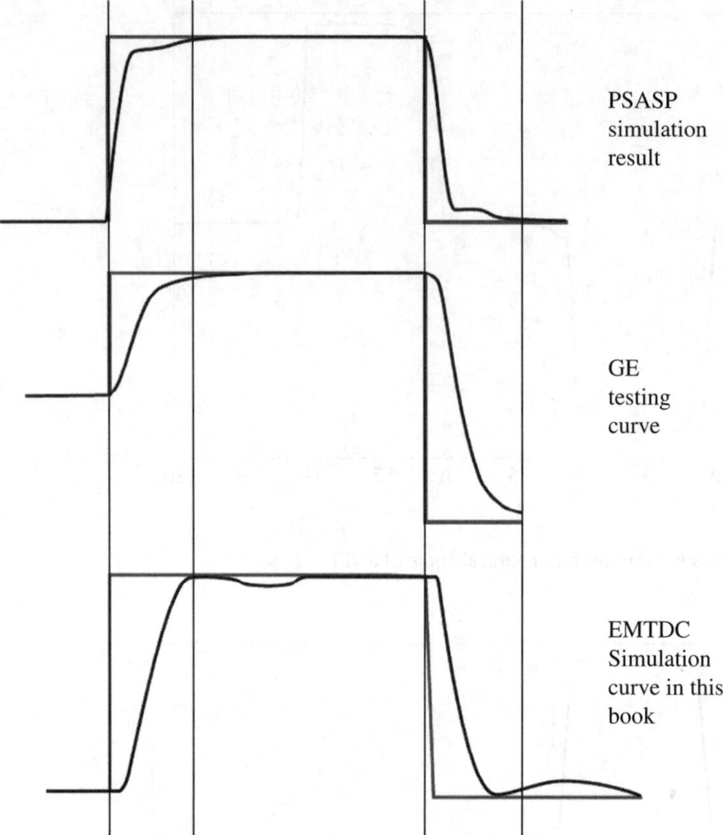

Figure 3.26 Comparison of the step responses between PSASP, EMTDC, and test [14].

3.7 Summary

1. The VFT steady-state frequency equation reveals the relationship between the electrical frequency of the systems on both sides of a VFT and the rotational frequency of the rotary transformer under steady-state conditions. The VFT power flow equation reflects the relationship between the transmission power and the equivalent phase shift. The VFT steady-state power flow model based on these static equations can be realized through the UD function provided by the PSASP program.

2. The dynamic equation of a VFT reveals the relationship between VFT transmission power, driving torque, and angular speed. The VFT electromechanical transient model based on these dynamic equations can be realized through the UPI function provided by the PSASP program.

3. The electromagnetic transient equation of a VFT reveals the temporal relationship between electrical variables such as VFT flux linkage, current, voltage, and power, and mechanical variables such as torque and damping. The VFT electromagnetic transient model including the electrical and mechanical equivalent circuits based on these electromagnetic transient equations can be realized through the general motor model in EMTPE and the double fed motor and DC motor model in EMTDC.

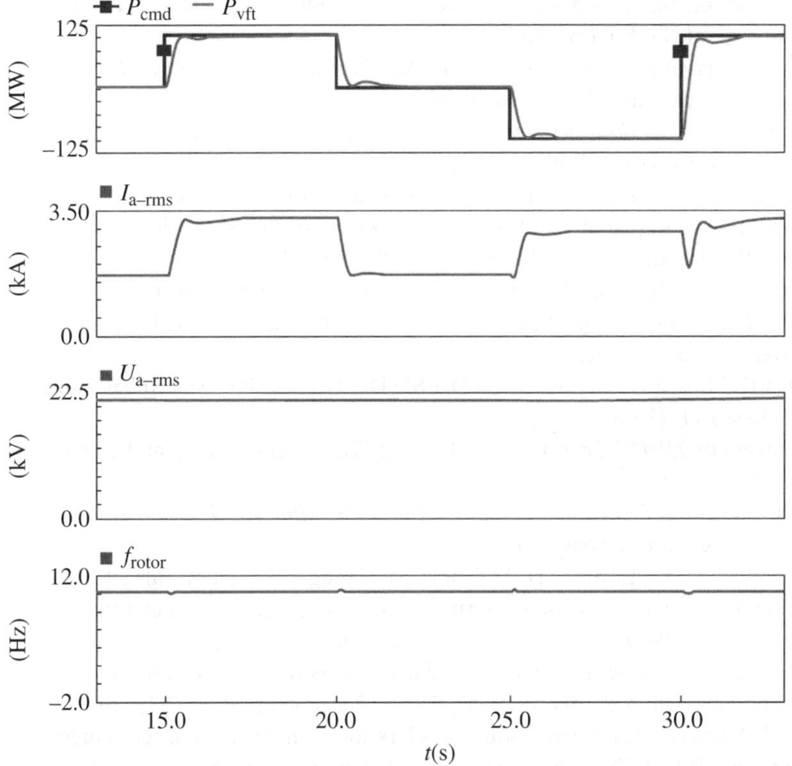

Figure 3.27 Operational characteristics of a VFT calculated using EMTDC: Where; P_{cmd} = command power; P_{vft} = VFT transmission power; $I_{a\text{-}rms}$ = VFT current; $U_{a\text{-}rms}$ = VFT stator side voltage; and f_{rotor} = electrical frequency corresponding to the rotor speed.

4. In this chapter, based on the structural characteristics of VFT, we proposed the positive sequence impedance and zero sequence impedance calculation model for the short-circuit current calculation of a VFT.

5. In this chapter, we used the power flow calculation, electromechanical transient, and electromagnetic transient models to conduct system simulation research and the results showed that the simulation model can meet the needs of power system simulation. The VFT power step response characteristics, proposed through the research on the double fed motor model built using PSASP and EMTDC, are basically consistent with the actual system test results, which verifies the rationality of the model.

References

1 C. Gesong. *Modeling and Control of VFT for in Power Systems*. D. Phil Thesis, Beijing: China Electric Power Research Institute, 2010.

2 C. Hang. *Steady-State Analysis of Power Systems*. Beijing: Water Conservancy and Electric Power Press, 1985.

3 L. Guangqi. *Transient Analysis of Power Systems*. Beijing: Water Conservancy and Electric Power Press, 1985.

4 Z. Xiaoxin, L. Hanxiang, W. Zhongxi. *Power System Analysis*. Beijing: Electric Power Research Institute of the Energy Ministry, 1989.

5 G. Jingde, W. Xiangyan, L. FaHai. *Analysis of AC Electrical Machinery and its System*. Beijing: Tsinghua University Press, 1993.

6 P. Kundur. *Power System Stability and Analysis*. Z. Xiaoxin and S. Yonghua (trans.). Beijing: China Electric Power Press, 2001, pp. 548–549 [in Chinese].

7 X. Shizhang. *Electrical Machinery*. Beijing: Machinery Industry Press, 1988.

8 China Electric Power Research Institute. User's *Manual of Power System Analysis Program (PSASP)*. Beijing: China Electric Power Research Institute.

9 China Electric Power Research Institute. Power electronics and electromagnetic transient simulation software package EMTP/EMTPE Instructions. Beijing: China Electric Power Research Institute.

10 Manitoba HVDC Research Center. PSCAD/EMTDC User Guide. Manitoba: Manitoba HVDC Research Center.

11 Zhejiang University. *HVDC Transmission*. Beijing: Water conservancy and electric power press, 1991.

12 X. Xiaorong, J. Qirong. *Principle and Application of Flexible AC Transmission Systems*. Beijing: Tsinghua university press, 2006.

13 E. Larsen, R. Piwko, D. McLaren, D. McNabb, M. Granger, M. Dusseault, et al. Variable frequency transformer – A new alternative for asynchronous power transfer, presented at *Canada Power, Toronto, Ontario, Canada*, 2004.

14 R. Piwko, E. Larsen, C. Wegner. VFT – A asynchronous interconnection power transfer technology. *China Southern Power System Technology*, 2006, 2(1): 29–34.

15 C. Gesong, Z. Xiaoxin. Digital simulation of VFTs for asynchronous interconnection in power systems. *IEEE/PES Transmission and Distribution Conference & Exhibition: Asia and Pacific Proceedings, Dalian, China*, 2005.

16 H.W. Dommel. *Electromagnetic Transient Calculation Theory of Power Systems*, L. Yonghuang, L. Jiming, Z. Zhanghua (trans.): Beijing: Water Conservancy and Electric Power Press, 1991.

4

VFT Control System Research and Modeling

4.1 Overview

The electrical power system is very complicated, while the power generation, transmission and distribution are completed in an instant. The VFT is an important device connecting transmission and distribution systems. In order to ensure the safe and reliable power transmission between the two systems, the rotary motion of rotary transformer as well as the phase change of magnetic field, composited by rotor winding in a magnetic gap space, must meet the requirements of system control. It should be said that this is a complex system involving electricity, magnetism, motion, and mutual conversion, and its control process is also very complex and stringent. The design of a VFT control system is the key to the promotion and application of this new technology.

In the development of electrical power systems, especially in the past 20 years, with the development of large power grids and the promotion and application of the FACTS device, the research into electrical power system control theories has constantly deepened. A lot of advanced control theories and design methods have formed in practice that can be generally classified into the following three categories [1–8]:

1. *Linear control theory*. This mainly includes linear PID control, power system stabilizer (PSS) control based on eigenvalue analysis and pole placement technique, and linear optimal control. In the linear control strategy, usually parameters are designed and optimized at certain motion points and in a certain running state. The damping characteristics near these designed operating points are very effective, but if the disturbance with a large scope of change to the operating points, the stability and effectiveness may not be satisfied, and the control parameters need to be verified and regulated.
2. *Nonlinear control theory*. Due to the inherent nonlinear characteristics of the electrical power system, with the development of the nonlinear control theory, the application of the nonlinear control in the electrical power system has developed rapidly. Nonlinear PID, differential geometric control, and direct feedback linearization control have been promoted and applied to some extent in the electrical power system. The nonlinear control reflects the nonlinear characteristics and uncertainty of the electrical power system that usually has good adaptability and robustness, but the nonlinear control-based nonlinear controller has a complicated structure

that requires massive calculations. In engineering applications, there are higher requirements for physical theory and the engineering experience of engineering technical personnel.

3. *Intelligent control theory.* Intelligent control uses the fuzzy logic theory (Fuzzy), genetic algorithms (GA), neural networks (NN), and so on. These methods are intelligent, robust, adaptive, and fault-tolerant, but intelligent control also faces greater challenges in practical applications. For example, the fuzzy control of complex systems has difficulty with knowledge acquisition and descriptions; GA requires massive computation; and for NN, the training is difficult and even samples exceeding the training scope usually cannot ensure the effectiveness of its control.

These different control theories and methods have their own advantages. They are suitable for different applications or different target needs and can also be integrated for utilization. In terms of a VFT, it is a motion system and electromagnetic power switching device with a large inertia, similar to the hydro-generator. Judging from this situation, using the mature linear PID control can meet the self-control requirements of this device. Meanwhile as a VFT is a series flexible AC transmission system device, in studying how it automatically regulates the control parameters to adapt to the changing situation of the system in its suppression of system low-frequency oscillation, it is necessary to find a parameter regulation method with a self-adaptive function. Based on these considerations, in this book, the ontology control of the VFT will be dominated by the conventional PID control, and for the use of the VFT to suppress the power system low-frequency oscillation, the Prony method will be used for identification of the system oscillation mode, which can adaptively adjust corresponding control system parameters. However, what needs to be explained is that it is necessary and helpful to study and apply other types of control theories and methods in order to further improve the control characteristics of VFTs and it is also our next research focus.

At the same time, the control object of a VFT includes the rotary transformer, DC motor, capacitor banks, corresponding circuit breaker, and other devices while its control objectives include element level control such as the thyristor trigger angle of DC drive rectifying circuit, device-level control such as device switching, capacitor bank switching, as well as synchronizing close, and system-level applications such as system exchanging power, power flow optimization, reactive power balance, and suppression of low-frequency oscillation, forming the complete control system of the VFT in the electrical power system.

Based on these considerations, in this chapter based on the technical features and control requirements of VFTs, we are going to: propose the VFT three-level control system including the element level control, device-level control, and system-level control, provide the general block diagram of the VFT control system, build the VFT DC motor drive system including the rectifier trigger circuit, study and design the specific control block diagram, and provide the key parameters of main functions such as the VFT bottom level trigger angle control, middle level transmission power control, and high level system application control that include the rectifying circuit trigger angle, driving torque, VFT rotational speed, phase angle, grid integration, power, voltage and frequency modulation, low-voltage limit power, suppression of low-frequency oscillation, and power supply to the weak system of interconnected systems, so as to achieve various functions of VFTs and lay the foundations for the simulation of corresponding systems.

4.2   VFT Control Strategy and System Block Diagram

VFT operation control relates to the mechanical system, the electrical power system, electromagnetic conversion, rectifying circuit control, DC motor drive regulation, and many other factors. It is a complex control system with multi-objective, multi-input, multi-control condition, and multi-boundary constraints. It can be said that the control system is the nerve center and core of VFTs as well as the research focus of this book. According to the characteristics of the VFT and its role in the system, in this book we propose the three-level control ideas; namely, element-level control, device-level control, and system-level control [9].

4.2.1   Element-Level Control

This is the bottom level of control targeted at the thyristor trigger angle of the DC rectifying circuit. This forms the basis for realizing all control targets of VFTs. The element level control regulates the thyristor trigger angle of DC rectifying circuit, changes the magnitude and direction of the output voltage of rectifier circuit, and regulates the input current and direction of the DC motor in order to change the direction and size of the torque exerted by the DC motor on the rotor shaft, thus controlling the VFT rotor speed, the phase shift of the rotor winding current magnetic field, and stator winding current magnetic field to realize the controllable power transmission and flexible regulation between interconnected power systems. Figure 4.1 shows the element-level control block diagram and main input and output variables.

4.2.2   Device-Level Control

This is the middle level control with the VFT and related equipment as the control target, and mainly includes VFT transmission power regulation, rotor speed regulation, rotor winding current magnetic field, and stator winding current magnetic field phase shift control, and the startup, grid integration, and exit of a VFT, various types of relay protection actions, shunt capacitor bank switching, and so on. Figure 4.2 shows the device-level control block diagram and the main input and output variables. The main output variables include action commands of related circuit breakers, output command torque of VFT DC motor $T_{rq\text{-}cmd}$, and so on.

4.2.3   System-Level Control

This is the highest level of control to fit various power system application needs. Based on the power system operating conditions and arrangement of dispatching

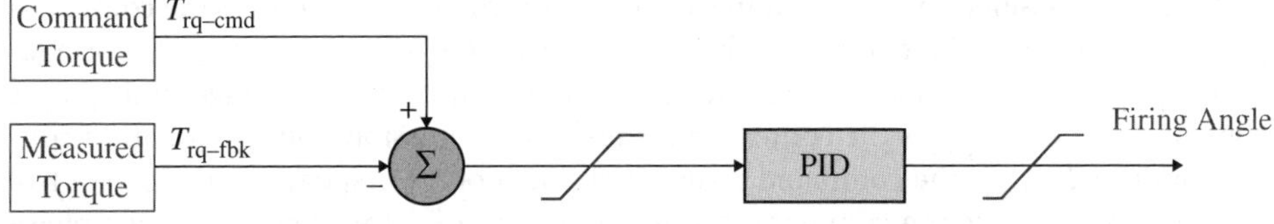

Figure 4.1 Schematic diagram of the VFT element-level control block diagram.

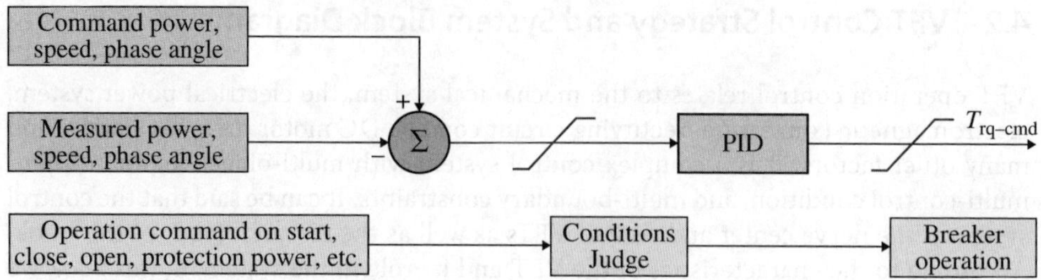

Figure 4.2 Schematic diagram of the VFT device-level control block diagram.

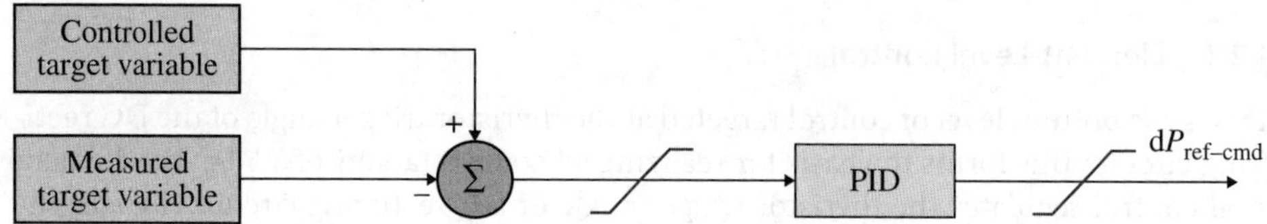

Figure 4.3 Schematic diagram of the VFT system-level control block diagram.

transactions, it controls the transmission power flow of the VFT, realizes functions such as the control of power regionally exchanged among interconnected grids, suppresses low-frequency oscillation in the power system, and undertakes system frequency regulation, system black start, and power supply to the weak system. The final output command is the command power of VFT or its regulation amount ($dP_{ref-cmd}$). Figure 4.3 shows the system-level control block diagram and main input and output variables. In Figure 4.3 the control target variables include VFT power, parallel branch transmission power, frequency of systems on both sides, system voltage, low-frequency oscillation power, system voltage, and so on.

It should be said that the element level control, device-level control, and system-level control of VFT are interrelated and supplementary to each other. Generally, the higher level control has to be realized through the lower level control. Figure 4.4 shows the total block diagram of the three-level control system. In Figure 4.4, all the parameters are defined as follows: P_{cmd} is the command power of VFT; X_{opt} is the system optimization control target; $X_{measure}$ is the measured value of system optimization control target; f_s is the stator side system frequency; f_r is the rotor side system frequency; U_s is the stator side bus voltage; U_r is the rotor side bus voltage; QF_{statu} is the status of circuit breaker; *Command* is the command that controls the switching of circuit breaker; $P_{measure}$ is measured value of transmission power of VFT; $S_{pd-measure}$ is the measured value of the speed of VFT; $T_{rq-measure}$ is the measured value of the torque of VFT; D_{P-opt} is the command power increment of the system optimization control module; D_{P-gov} is the command power increment of system frequency control module; D_{P-lpo} is the command power increment of the control module of suppressing system low-frequency oscillation; D_{P-vdpl} is the command power increment of the system low-voltage limit power control module; T_{rq-pwr} is the command torque of transmission power regulation control module; T_{rq-frq} is the command torque increment of rotor speed regulation control module; and T_{rq-ph} is the command torque increment of rotor winding voltage phase control module.

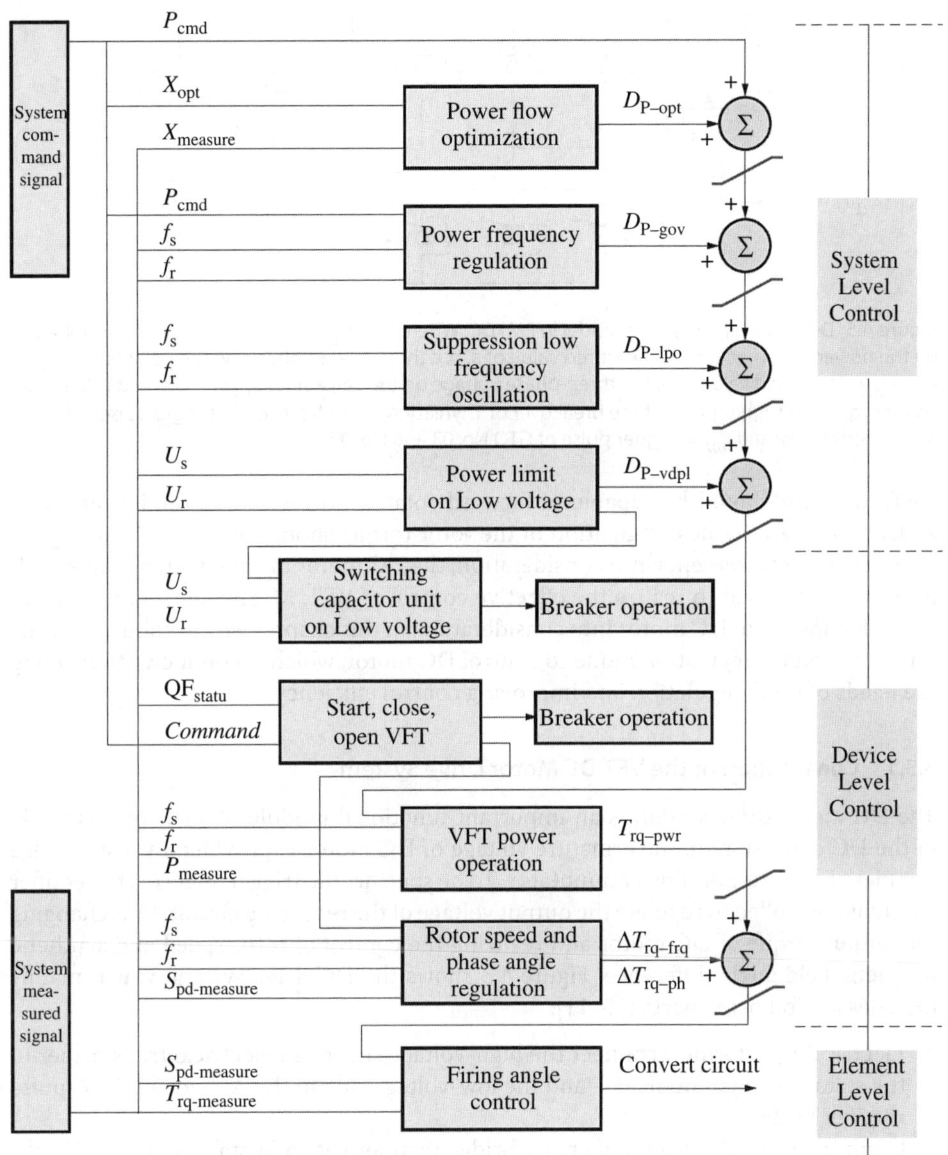

Figure 4.4 Total block diagram of the three-level VFT control system.

4.3 VFT Element-Level Control and DC Drive System Design

Based on the working principle of the VFT, the main control of it is, in essence, the control of rotor motion including the rotor speed control and the control of phase shift between rotor winding current magnetic field and stator winding current magnetic field. As a result, the accurate and efficient control of the rotor motion is the key to achieving the functions of a VFT. In addition, based on the power regulation requirements for

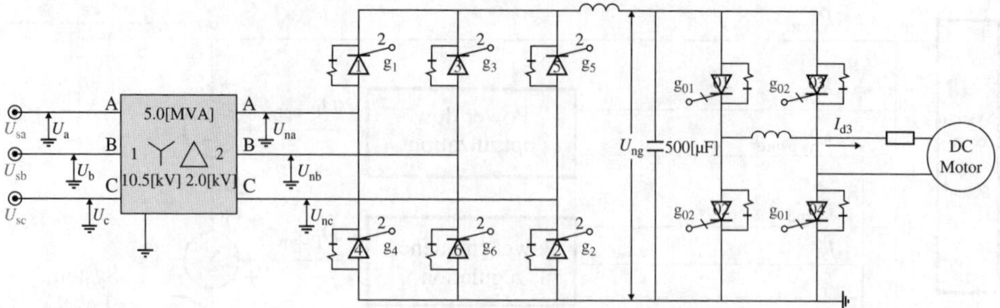

Figure 4.5 DC motor drive system model in EMTDC. Where: U_{sa}, U_{sb}, U_{sc} = a, b, c three-phase voltage on the power side; U_a, U_b, U_c = measured value of a, b, c three-phase voltage on the power side; U_{na}, U_{nb}, U_{nc} = measured value of a, b, c three-phase voltage on the valve side; U_{ng} = measured value of DC voltage; g_1–g_6 = trigger pulse of the breakover of Thyristor No. 1 to No. 6; g_{01} = trigger pulse of GTO No. 01 and No. 04; and g_{02} = trigger pulse of GTO No. 03 and No. 02

VFT, the rotor should be capable of forward rotation, reverse rotation, keeping still, and, accordingly, stepless regulation of the rotor torque should be realized. Taking the VFT control requirements into consideration, the DC motor installed on the same shaft of rotor can be used to realize the effective control of VFT. Meanwhile, taking various operation modes of DC motor into consideration, in this chapter we are going to use the armature voltage regulation mode to control DC motor, which is conducive to meeting the needs of rapid regulation and improving control efficiency.

4.3.1 Constitution of the VFT DC Motor Drive System

The DC motor drive system is an important functional module of VFT. In this book, in the DC drive system, the armature voltage of DC motor is provided by a six-bridge rectifier and corresponding commutator. To be specific, the trigger angle of the rectifier circuit is controlled to regulate the output voltage of the rectifying circuit, thus changing the output torque of DC motor and realizing the control of rotor speed and synthetic magnetic field angle difference. Figure 4.5 shows the DC drive system, which mainly includes the following parts [10, 11]:

1. Electrical transformer: connect the high-voltage side of an electrical transformer to the stator side system of VFT and the low-voltage side to the AC side of the 6-pulse rectifier bridge.
2. Controllable six-bridge rectifier: six-bridge rectifier with adjustable output DC voltage amplitude.
3. Auxiliary element: a flat wave reactor, filter capacitor.
4. Reversing switch: a bidirectional reversing device composed of a GTO bridge.
5. Thyristor trigger controller: a trigger pulse generator used to control the rectifier thyristor trigger conduction and commutator GTO conduction.
6. DC motor: generally the separate excitation is used and the output torque magnitude is adjusted by changing armature voltage.

The VFT DC drive system model includes a motor, rectifying circuit, and trigger control model. Its working principle is to obtain the power supply from the power grid through a step-down transformer and then through the rectification of a six-phase rectifier and the flat wave reactor, filter capacitor, and reversing device to obtain the DC voltage whose amplitude can be continuously regulated and polarity can be reversed. The rectifying output voltage of VFT will be exerted on the armature winding of the DC drive motor while the trigger pulse of the rectifier is provided by the trigger pulse control module. The simulation models of this rectification system in PSCAD/EMTDC – namely, the DC motor drive system model, triangular wave generator, commutation control circuit, trigger pulse generator – are shown in Figures 4.5–4.8 [12–16].

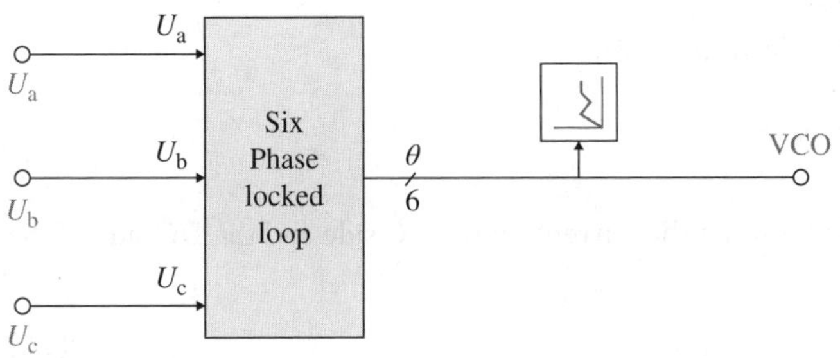

Figure 4.6 Triangular wave generator.

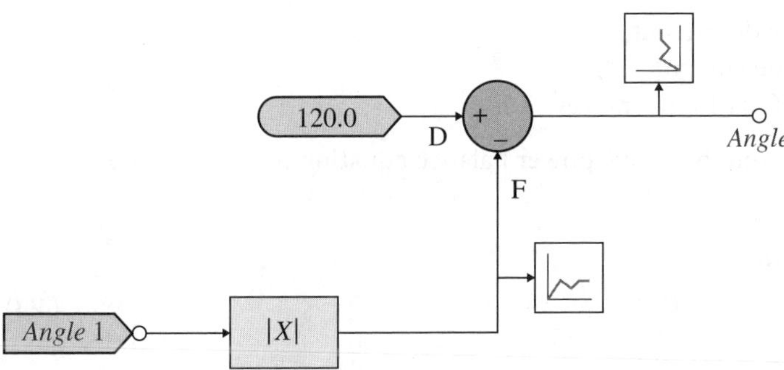

Figure 4.7 Commutation control circuit.

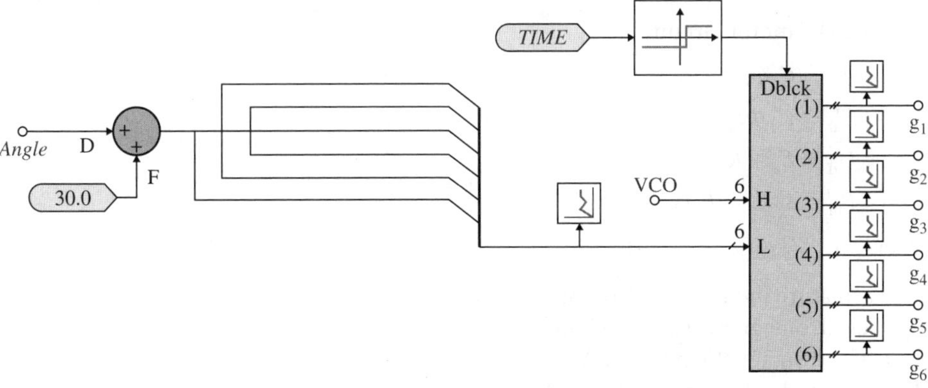

Figure 4.8 Trigger pulse generator.

4.3.2 Basic Equations of the DC Motor Drive System

Based on the basic equations of the rectifying circuit, the output voltage of the rectifying circuit is [17, 18]

$$U_{\rm d} = \frac{3}{\pi} \int_{\alpha-60°}^{\alpha} \sqrt{3}\, U_{\rm m} \cos(\theta + 30°)\mathrm{d}\theta - \frac{3X_{\rm r}}{\pi} I_{\rm d} = 1.35 U_{\rm LL} \cos\alpha - \frac{3X_{\rm r}}{\pi} I_{\rm d} \qquad (4.1)$$

where,

$U_{\rm d}$ = average value of DC voltage, kV;
$U_{\rm m}$ = peak value of AC side voltage, kV;
$U_{\rm LL}$ = effective value of AC side voltage, kV;
α = rectifying trigger angle, (°);
$X_{\rm r}$ = commutating reactance, Ω;
$I_{\rm d}$ = DC side current, kA.

Similarly, the relationship between the current on the AC side and the DC side of the converter is

$$I_{\rm L} = \frac{I_{\rm LM}}{\sqrt{2}} = 0.78 I_{\rm d} \qquad (4.2)$$

where,

$I_{\rm LM}$ = peak value of AC side current, kA;
$I_{\rm d}$ = amplitude of DC side current, kA;
$I_{\rm L}$ = effective value of AC side current, kA.

Ignore the circuit loss and the active power balance equation is

$$P_{\rm d} = U_{\rm d} I_{\rm d}$$
$$P_{\rm ac} = 3 U_{\rm LN} I_{\rm L1} \cos\alpha$$
$$P_{\rm d} = P_{\rm ac} \qquad (4.3)$$

where,

$P_{\rm ac}$ = AC side power, MW;
$P_{\rm d}$ = DC side power, MW.

In the DC motor circuit,

$$E_{\rm a} = C_{\rm e} \Phi n$$
$$T_{\rm d} = C_{\rm T} \Phi I_{\rm d}$$
$$U_{\rm d} = E_{\rm a} + R_{\rm a} I_{\rm d}$$
$$P_{\rm m} = \Omega T_{\rm d} = E_{\rm a} I_{\rm d} \qquad (4.4)$$

where,

$E_{\rm a}$ = armature induced electromotive force, kV;
$T_{\rm d}$ = electromagnetic torque, kN·m;
$P_{\rm m}$ = electromagnetic power, kW;
Φ = amplitude of magnetic flux, Wb;
$C_{\rm e}$ = electromotive force constant;

C_T = toque constant;
R_a = armature resistance, Ω;
n = rotor speed, r/min;
Ω = mechanical angular speed of rotor, rad/min.

It can be seen from Equation (4.1) and (4.4) that

$$
\begin{aligned}
T_d &= C_T \Phi (U_d - E_a)/R_a \\
&= C_T \Phi (U_d - C_e \Phi n)/R_a \\
&= C_T \Phi \left(1.35 U_{LL} \cos \alpha - \frac{3X_r}{\pi} I_d - C_e \Phi n \right) /R_a
\end{aligned}
\tag{4.5}
$$

It can be seen from Equation (4.5) that when the commutating reactance X_r is zero, namely equivalent to the infinite power, T_d is a function of α and n. As the normal speed of VFT n is very small, the armature reaction relative to the change of Φ can be ignored. When the frequency of the systems on both sides is identical, n is zero in the steady-state and at this point T_d is proportional to $\cos\alpha$.

Ignore the stray loss such as the rotor winding and based on Equations (4.5) and (3.36)

$$
P_s = 2\pi f_s T_d = 2\pi f_s C_T \Phi \left(1.35 U_{LL} \cos \alpha - \frac{3X_r}{\pi} I_d - C_e \Phi n \right) /R_a
\tag{4.6}
$$

It can be seen from Equation (4.6) that when the frequency of the systems on both sides of VFT is constant, ignore the impact of the commutating reactance; when the rotor speed is constant, the transmission power of VFT P_s is proportional to $\cos\alpha$. In short, by changing the trigger angle α, the output torque of the DC motor can be changed so as to control the motion state of the rotor of VFT, control the magnitude and direction of the transmission power of VFT, and realize various control targets of VFT. However, when the commutating reactance X_r is taken into consideration, the relationship between P_s and $\cos\alpha$ is nonlinear.

It can also be seen from Equation (4.5) when the system transmission power is zero, T_d is 0 and for the speed n a given system frequency deviation corresponds to

$$
1.35 U_{LL} \cos \alpha = C_e \Phi n + \frac{3X_r}{\pi} I_d
$$

namely,

$$
\cos \alpha = \left(C_e \Phi n + \frac{3X_r}{\pi} I_d \right) /1.35 U_{LL}
\tag{4.7}
$$

Equation (4.7) can be used for the control before the grid-connected operation of VFT.

4.3.3 Trigger Control and Response Characteristics of the Rectifier Circuit

As mentioned before, the core task of the DC drive system is to control the rectifying trigger angle of the rectifying circuit, thus changing the armature voltage of the DC motor and regulating the output torque and rotor speed of the DC motor.

For systems shown in Figures 4.5–4.8, the phase shift of the rectifying trigger angle ranges from 0° to 120°. When the trigger angle is 0°, its output voltage is the largest; when the trigger angle is 120°, its output voltage is zero. The value of the trigger angle,

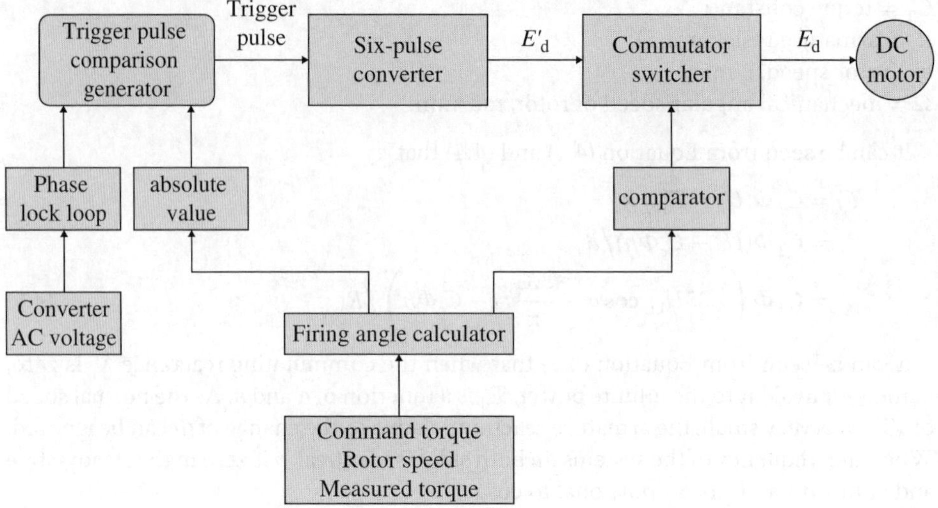

Figure 4.9 DC motor drive system control block diagram.

ranging from $-120°$ to $120°$, can be calculated based on the electromagnetic torque command issued by the higher level control system. The positive and negative signals are used to identify the direction of the current flowing into the armature winding of the motor; namely, the direction of the electromagnetic torque. These signals are used to control the GTO bridge conduction mode on the right-hand side in Figure 4.5. The trigger angle amplitude represents the trigger phase. Compare it with the triangular wave with an amplitude of $360°$ to generate corresponding trigger pulse so as to control the conduction angle of the thyristor elements and realize the control of the amplitude of the DC output voltage. See Figure 4.9 for the specific drive system control block diagram.

In Figure 4.10, the curves showing the relationships between trigger angle and electromagnetic torque, DC rectification voltage, DC current, and motor electromagnetic torque are given.

4.4　VFT Device-Level Control Design

4.4.1　Rotor Speed Control

The rotor speed control changes the electromagnetic torque of the DC motor and controls the speed of the rotary transformer to ensure $f_s = f_r + f_{rm}$, which is the basis of realizing the stable operation of the VFT and the premise of synchronizing its close grid-connected operation. It can be seen from Equation (3.13) that before the grid-connected operation of a VFT, the change of the rotor speed is decided by the output torque of the DC motor, and in this stage the motor torque is mainly regulated to make the electrical rotating frequency of the rotary transformer equal to the frequency difference between the systems on both sides of the VFT. After it is connected to the system, when current flows through the winding, the rotor speed is jointly affected by the output torque of the DC motor, the rotor winding electromagnetic torque, and the stator winding electromagnetic torque. This steady state is mainly to maintain the stability of the rotating speed. As in the actual control, the frequency of the systems on

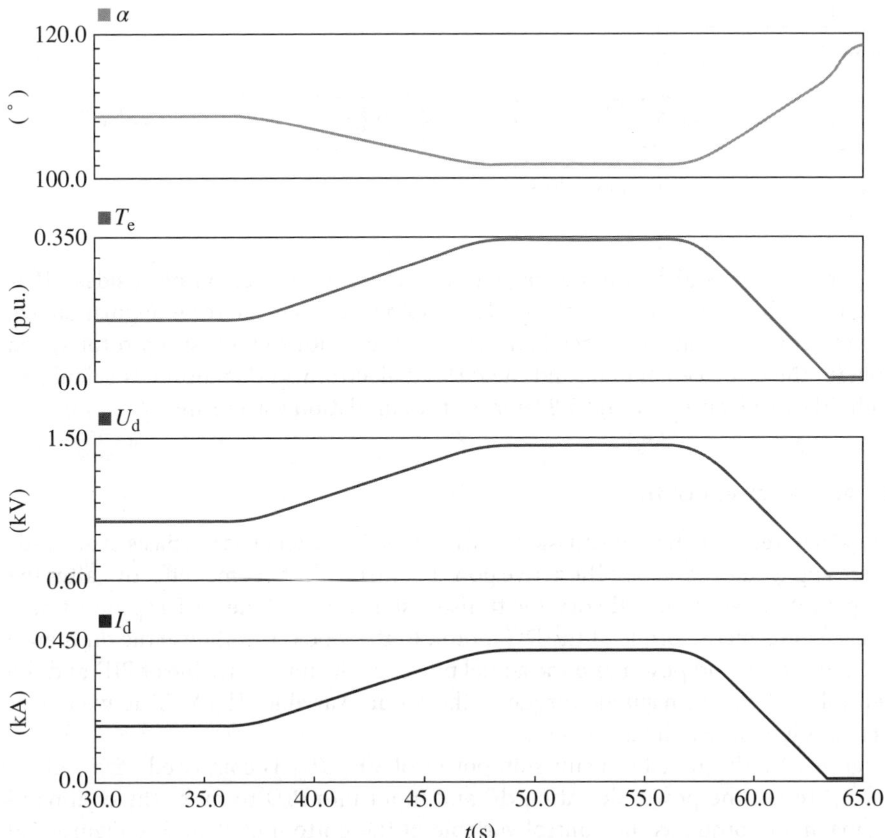

Figure 4.10 Curves showing the relationships between trigger angle and electromagnetic torque, DC rectification voltage, DC current: α = trigger angle; T_e = electromagnetic torque; U_d = DC rectification voltage; and I_d = DC current.

both sides of the VFT will be measured in order to reduce the impact of disturbance. Thevenin Equivalent Circuits can be used for measurement and calculation.

In this book, we first measure the frequency difference between the stator side system and the rotor side system, calculate the command electromagnetic torque needed by the realization of speed regulation through PID, and then use it as the input variable of the bottom control. See Figure 4.11 for the specific rotor speed control block diagram.

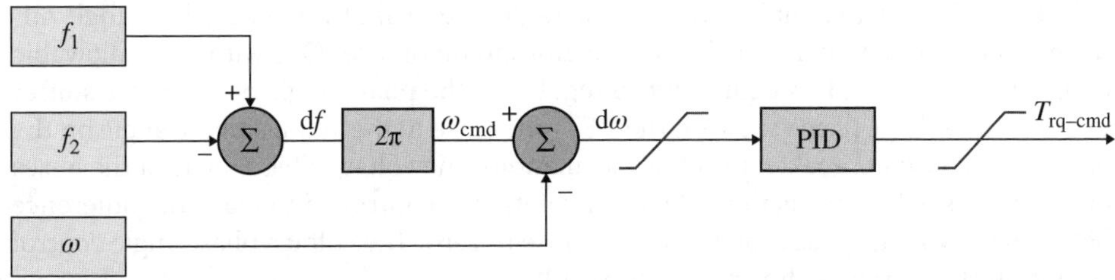

Figure 4.11 Rotor speed control block diagram.

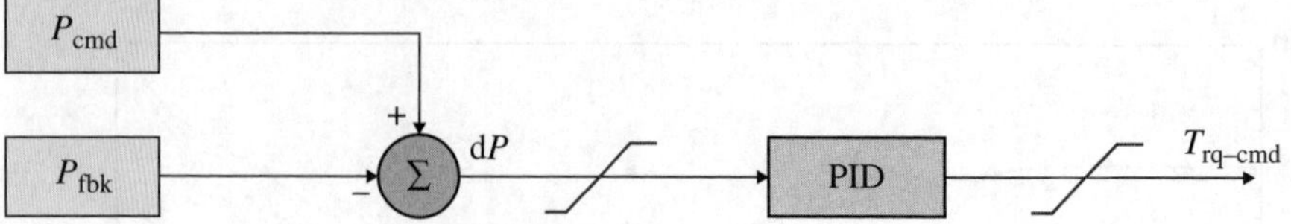

Figure 4.12 Active power control block diagram.

In Figure 4.11, f_1 and f_2 are the frequency of the power grids on both sides; df is the frequency difference between the grids on both sides; ω is the rotor angular speed; ω_{cmd} is the command angular speed and is compared with the measured rotor speed ω to obtain the displacement $d\omega$; and the desired electromagnetic torque is calculated through PID. See Figures 5.7 and 5.8 later for the simulation waveforms.

4.4.2 Active Power Control

This is used to regulate the transmission power of the VFT, which is the basis of all kinds of system application controls. The active power control only becomes effective after the VFT is put into operation. In theory, the transmission power of the VFT is proportional to the electromagnetic torque of the DC motor. In this book, we still use the difference between the command power and the actual power as the input variable of PID and the desired value of electromagnetic torque as the output variable. The VFT active power control circuit is shown in Figure 4.12.

In Figure 4.12, the actual transmission power of VFT P_{fbk} is compared with a given value P_{cmd} to get the power deviation dP and input into PID to make the command electromagnetic torque as the control variable of the bottom control. See Figures 5.9 and 5.10 for the simulation waveforms.

4.4.3 Voltage Phase Angle Control

Voltage phase angle control includes two functions. The first is to control the difference of voltage phase angles on both sides of synchronizing circuit breaker within the given range before the integration of the VFT; the second is to regulate the phase-angle difference between the VFT rotor winding voltage and stator winding voltage, equivalent to regulating the phase angle of phase shifter so as to control the VFT transmission power. In this book, we focus on the first function because the second function is realized through the power control module. In the first function, the control is mainly to provide conditions for VFT integration and reduce its impact on systems of both sides. Take Figure 5.4 as an example; its regulation target is to make the voltage phase angle difference on both sides of the VFT synchronizing circuit breaker QF3 within the allowable range. To be specific, it is equivalent to regulating the phase angle of the phase shifter to make the voltage phase angle on the VFT side close to the voltage phase angle on the other side of circuit breaker. In actual calculations, the voltage phase angle of the buses on both sides will be measured. The rotor rotation is controlled to make the difference between the voltage phases on both sides close to zero. The voltage phase angle control circuit block diagram is shown in Figure 4.13.

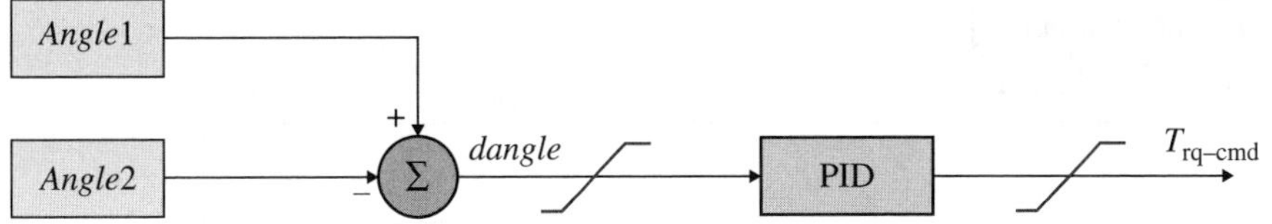

Figure 4.13 Voltage phase angle control circuit block diagram.

In Figure 4.13. Angle 1 and Angle 2 show the voltage phase of the two measured nodes and dangle is the voltage phase difference between the two nodes. The desired electromagnetic torque is calculated through PID and sent to the bottom control system to regulate the speed and position of VFT. Relevant simulation waveforms are shown in Figure 5.7 and Figure 5.8.

4.4.4 Synchronous Grid Connection Control

Synchronous grid connection control allows realization of the grid-connected operation of VFT when the grid connection control conditions are met. There is little difference from the grid connection of conventional power systems, however, VFT has its own advantages and can automatically regulate the phase angle to adapt to closing requirements. When the frequency difference and phase angle difference between the voltages on both sides of synchronizing circuit breaker of VFT meet the grid connection conditions, the synchronizing circuit breaker QF3 in Figure 5.4 can be capable of closing operation. Figure 4.14 shows the synchronizing close control logic block diagram: see Figures 5.7 and 5.8 for specific simulation waveforms.

4.4.5 Reactive Voltage Control

This controls the system reactive power balance and the bus voltage level by switching the parallel capacitor banks or limiting the transmission power of VFT. When current flows through VFT, the leakage reactance of VFT and the leakage reactance of the step-down transformer will absorb some reactive power, and might lead to voltage drop. At this point the reactive power absorbed by VFT can be balanced by switching the parallel capacitor banks. The VFT capacitor bank has three working modes:

1. *Self-balance mode.* Calculate the reactive power absorbed by VFT based on the current of VFT and switch the capacitor bank according to the predetermined order to make the reactive power of the capacitors as close as possible to the reactive power absorbed by the series leakage reactance.
2. *Voltage mode.* In this mode, the control target controls the bus voltage, namely to maintain the bus operation voltage level within the acceptable range through the graded switching of capacitor banks. When the bus voltage is lower than a certain amplitude, switch on the capacitor banks; and on the contrary, when the bus voltage is higher than a certain amplitude, switch off the capacitor banks.
3. *Manual mode.* The dispatching personnel judges the voltage control target and compensation level based on the system operation conditions to switch the capacitor banks manually.

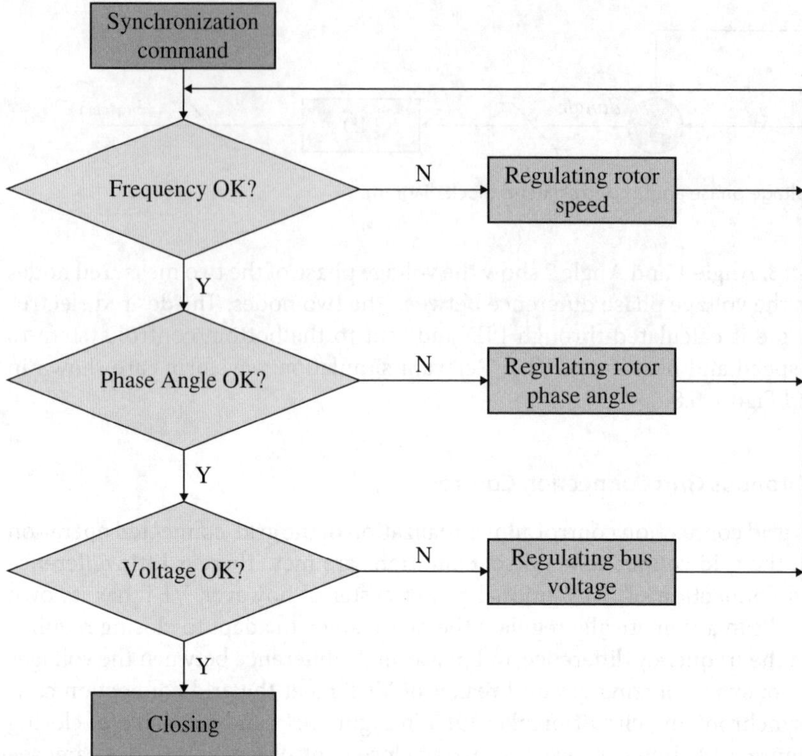

Figure 4.14 Synchronizing close control logic block diagram.

If, when all of the capacitors are put into operation, the voltage level is still lower than the allowable value, the transmission power of VFT needs to be further limited to maintain the system voltage. Figure 4.15 shows a block diagram of the reactive voltage regulation based on the voltage level control. See Figure 5.13 for relevant simulation waveforms.

4.5 VFT System-Level Control Design

4.5.1 Optimize System Power Flow

System power flow optimization uses the VFT to regulate the system power flow; namely, the transmission power and direction of power flow within the VFT design capacity and permitted range of dispatching power of the adjacent or parallel branches from exceeding the limit and reduce system loss. In the system power flow optimization calculation, the VFT can be regarded as a control variable in solving the optimal power flow equation and optimizing the system power flow with other control variables. The solution of optimal power flow will be used as the basis of VFT dispatching. In the actual system, the loss of the main transmission grids can be monitored and calculated in a real-time manner during VFT transmission power regulation process so as to provide an optimal VFT power dispatching and control strategy to realize the goal of

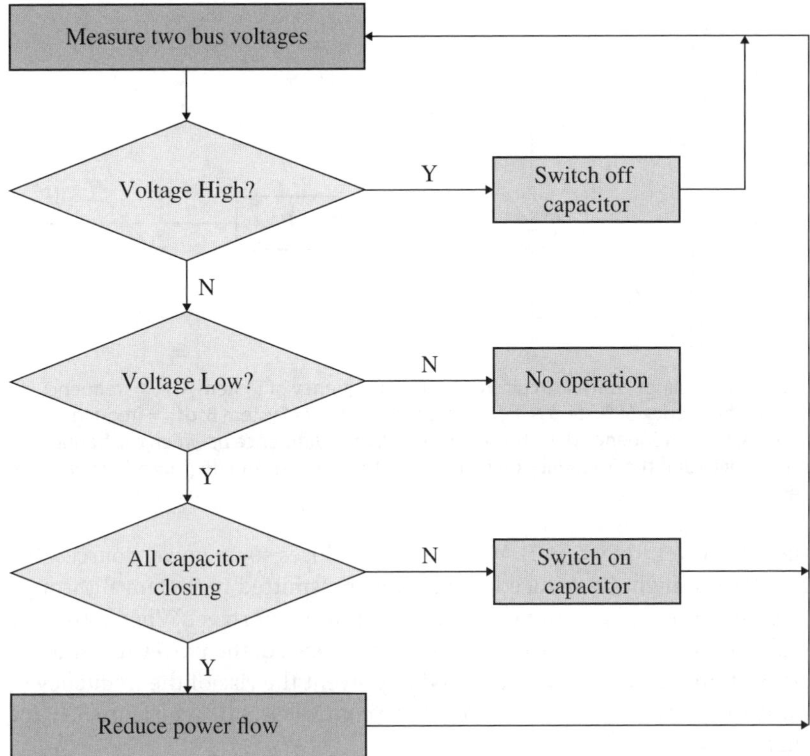

Figure 4.15 Block diagram of reactive voltage regulation based on voltage level control.

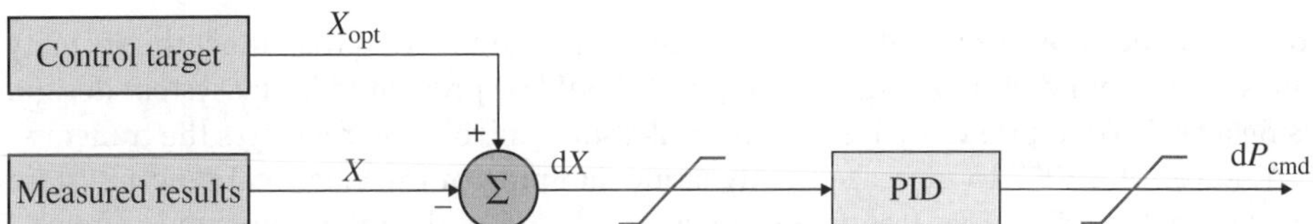

Figure 4.16 Schematic diagram of optimizing system power flow control. Where: X_{opt} = control target; X = measured result; dX = difference between control target and measured result; and dP_{cmd} = added value of command power.

reducing system loss. Figure 4.16 shows the schematic diagram of optimizing system power flow control: see Figure 5.12 for relevant simulation waveforms.

4.5.2 Regulate System Frequency

Regulating system frequency module uses the VFT to regulate the electrical frequency of the interconnected systems; namely, to control the power exchange between two systems within the VFT capacity range to maximally control the frequency of systems on both sides within a reasonable range, which is equivalent to regarding the system on one side as a frequency regulation power source or adjustable load. For example, System A and System B are interconnected with each other through VFT. In normal conditions,

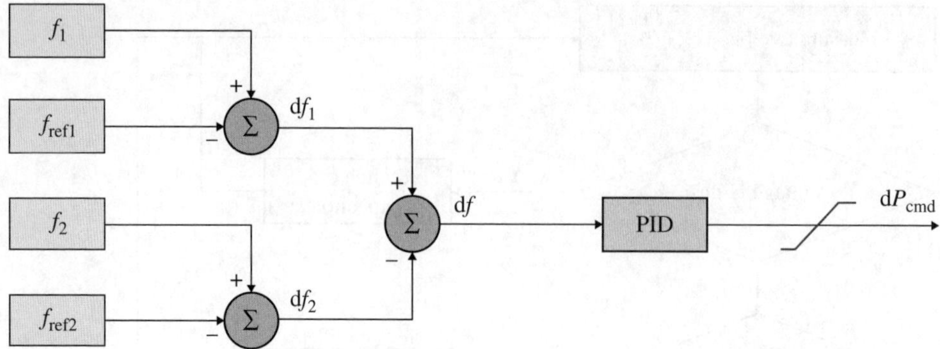

Figure 4.17 Frequency regulator control circuit. Where: f_1 = frequency of System A; f_2 = frequency of System B; f_{ref1} = rated frequency of System A; f_{ref2} = rated frequency of System B; df_1 = frequency deviation of System A; df_2 = frequency deviation of System B; df = difference between the frequency deviation of rotor winding and the frequency deviation of stator winding; and dP_{cmd} = added value of command power.

System A transmits power to System B. When System B loses some power sources, the frequency of System B might be reduced, the power transmitted to System B through VFT can be rapidly increased to compensate for the power shortage. When System B loses some load, the frequency of System B might be increased, the power transmitted by System A to System B can be rapidly reduced to prevent the rise of the frequency of System B. Figure 4.17 shows the frequency regulator control circuit. See Figure 5.17 for relevant simulation waveforms.

4.5.3 Suppressing Low-Frequency Oscillation

Low-frequency power oscillation is one of the important factors that affect the security and AC interconnection of large systems and should be prevented in the system design. Using a VFT to suppress low-frequency oscillation controls and regulates the transmission power of a VFT to increase the dynamic damping of the electrical power system and reduce the risk of low-frequency power oscillation of the interconnected system.

Obviously, if two systems are only interconnected through a VFT, the power oscillation within region can be damped through the VFT power control signal. However, if there are other parallel branches between the two systems, especially the AC branch with automatically balanced power flow, the section power flow including the VFT and its parallel AC circuit should be regarded as the control target. The transmission power of VFT is dynamically regulated to increase the damping of the system, thus effectively suppressing the low-frequency power oscillation in the synchronous interconnected system.

The damping controller used in the study is similar to the PSS in structure, which is composed of DC blocking, phase-lead, and phase-lag compensation links. Its input signal is the sum of the transmission power of studied sections while its output signal is the command power increment given to the VFT. Figure 4.18 shows the control circuit model of a VFT low-frequency oscillation controller.

In Figure 4.18, P_L is the active power of the cross section; T_w, T_1, T_2, T_3, and T_4 are time constants. The output value ΔP_{VFTRef} is obtained through the DC blocking,

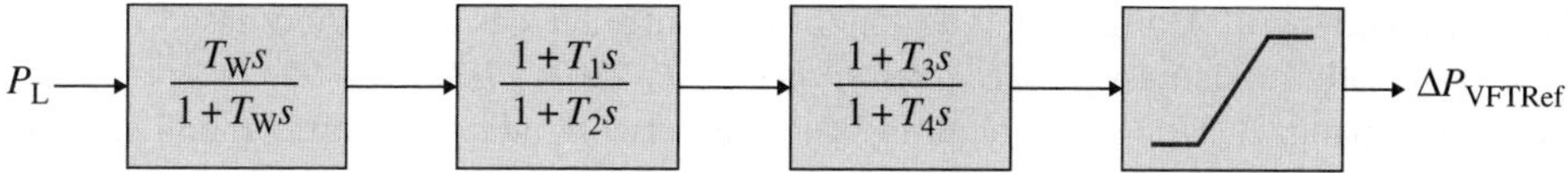

Figure 4.18 Control circuit of a low-frequency oscillation controller.

phase-lead, and phase-lag links. This output value is the reference power change value used for regulating command power amplitude. See Figures 5.29 and 5.30 for relevant simulation waveforms.

4.6 Summary

1. The three-level control system including element level, device-level, and system-level has been established and the integrated control protection system with clear objectives and cooperation, as well as specialization for each level, has been formulated. This is conducive to the realization of various functions of VFT and ensure the safe, reliable, and economical operation of the interconnected system.
2. The analysis of the DC rectifying-drive circuit equation reveals the mathematical relationship between rectifying circuit trigger angle and DC motor electromagnetic torque as well as output transformer power of VFT, which provides theoretical support for carrying out VFT control. The VFT rectifying circuit and trigger control system model is built by using power electronic components such as the thyristor, which realizes the bottom control of VFT and simulates and verifies the regulation characteristic of this bottom-level control system.
3. The established VFT device-level control system realizes the VFT transmission power regulation, rotor speed regulation, voltage phase regulation, and control functions of VFT such as startup, grid connection, exit, and capacitor bank switching, and are verified in the simulation.
4. The established VFT system-level control system realizes functions such as the system power flow optimization, system frequency regulation, and low-frequency oscillation suppression, providing conditions for system control.

References

1 Z. Xiaoxin, G. Jianbo, L. Jimin, et al. *TCSC in Power Systems*. Beijing: Science Press, 2009.
2 L.F. Jeng, S.P. Hung. A linear synchronous motor drive using robust fuzzy neural network control. *WCICA*, 2004: 4386–4390.
3 C.-F. Xue, X.-P. Zhang, K.R. Godfrey, Design of STATCOM damping control with multiple operating points: A multi-model LMI approach, 2006, *IEE Proc. – Generation, Transmission and Distribution*, 153(4): July, 375–382.
4 K.Y. Shieh, C. Ming. A fuzzy controller improving a linear model following controller for motor drives fuzzy systems. *IEEE Transactions on Electrical Applications*, 1994, 21(3): 194–202.

5 W. Chuanfeng, L. Donghai, J. Xuezhi. A PID controller design method based on probabilistic robustness. *Proceedings of the CSEE*, 2007, 27(32): 93–97.

6 R. Yongqiang, G. Hong, X. Zhanming. Neural network adaptive sliding mode control for direct-drive servo systems. *Transactions of the China Electrotechnical Society*, 2009, 24(3): 75–79.

7 P. Haiguo, W. Zhixin. Simulation of fuzzy-PID synthesis yawing control system of wind turbines. *Transactions of China Electrotechnical Society*, 2009, 24(3): 183–188.

8 N. Mithularmnthan, C.A. Canizares, I. Reeve, et al. Comparison of PSS, SVC, and STATCOM controllers for damping power system oscillations. *IEEE Trans on Power Systems*, 2003, 18(2): 786–792.

9 C. Gesong. *Modeling and Control of VFT for in Power Systems. D*. Phil Thesis, Beijing: China Electric Power Research Institute, 2010.

10 E. Larsen, R. Piwko, D. McLaren, D. McNabb, M. Granger, M. Dusseault, et al., Variable-frequency transformer – A new alternative for asynchronous power transfer, presented at *Canada Power, Toronto, Ontario, Canada*, 2004.

11 J.J. Marczewski. VFT applications between grid control areas. *Power Engineering Society General Meeting*, 2007.

12 H. Lipei. *Motor Control*. Beijing: Tsinghua University Press, 2003.

13 X. Shizhang. *Electrical Machinery*. Beijing: Machinery Industry Press, 1988.

14 Y. Gen, L. Yingli. *Motor and Motion Control System*. Beijing: Tsinghua University Press, 2006

15 P. Truman, N. Stranges. A direct current torque motor for application on a variable frequency transformer. *Power Engineering Society General Meeting*, 2007.

16 Manitoba HVDC Research Center. PSCAD/EMTDC User Guide, Manitoba: Manitoba HVDC Research Center.

17 Zhejiang University. *HVDC Transmission*. Beijing: Water Conservancy and Electric Power Press, 1991.

18 X. Xiaorong, J. Qirong. *Principle and Application of Flexible AC Transmission Systems*. Beijing: Tsinghua University Press, 2006.

5

Analysis of Operational Characteristics and Application of VFTs in the Electrical Power System

5.1 Overview

The rapid development in computing and the gradual improvement in simulation technology have both provided great convenience for research into electrical power systems, and have improved the efficiency of new technology research and development of electrical power systems. The development of new technologies and new devices in the modern electrical power system is mostly based on digital simulation instead of physical prototypes, which reduces the cost, saves time, and, more importantly, overcomes the limitations of the physical prototype thus enriching the research content. Using digital simulation to study and analyze the performance and characteristics of new equipment lays the foundation for development and application of new technologies and equipment. In this chapter, we use the VFT power flow calculation, electromechanical transient, and electromagnetic transient models established in the previous chapters, control systems and parameters for different application occasions, and general programs such as PSASP, EMTP, and PSCAD/EMTDC to conduct digital modeling of electrical power systems including VFTs on varying scales to simulate and calculate their operational characteristics in different system operating conditions and different fault conditions, to verify the rationality of the models and control systems, to simulate and analyze various functions of VFTs, and to further demonstrate the functions and advantages of VFTs in grid interconnection [1–4].

5.2 VFT Parameters and Research System Design

In this section, we first propose the main VFT parameters used in this book based on the application of VFT technology as well as China's developmental needs, and use the parameters of VFT equipment developed by GE as a reference. Meanwhile, in order to ensure the rationality, universality, and comparability of the simulation research results, we select three electrical power systems of varying scales for VFT simulation research. The simplified asynchronous interconnection system is used for VFT device-level control simulation research. The typical four-generator system, which is widely used for analysis in electrical power systems, is used for VFT system-level control simulation research. The Northeast China-North China-Central China interconnected system (2007 level year) is used for the research on the VFT power flow calculation, electromechanical transient stability, and low-frequency oscillation simulation calculation and analysis.

5.2.1 Basic Parameters of VFTs

Selection of equipment parameters and technical specifications is an important part of the electrical power system design. To study the new VFT equipment, we should first propose its main parameters based on the system requirements and equipment capability. Main parameters of VFTs include electrical parameters such as rated voltage, rated capacity, leakage reactance, excitation reactance, and mechanical parameters such as moment of inertia and damping. Meanwhile, we should also propose the main parameters of auxiliary equipment such as the capacitor bank, step-down transformer, rectifying circuit, and DC motor.

1. *Rated capacity of a VFT*. The capacity of a single VFT is restricted by the limitation of the insulation design of rotary transformer on the rated voltage (generally within 25 kV) as well as the maximum allowable current amplitude of the collector ring brush (generally being thousands of amperes). At present, the VFT manufactured by GE has a capacity of 105 MVA. Through discussion with relevant experts of GE and based on the existing technology, the design capacity of a single VFT can reach 200–300 MW. With the actual situation of power grids in China taken into consideration, the capacity of the single VFT studied in this book is generally considered to be 200 MW. In addition, multiple parallel VFTs are used based on actual needs to improve power transmission capacity.
2. *Rated voltage of a VFT*. This is the rated line voltage of a VFT. At a certain capacity, the higher the rated voltage is, the smaller the rated current and the more favorable it is for brush design, but this leads to higher requirements for the insulation design of rotating parts. At present, the rated voltage of the VFT manufactured by GE is 17 kV. Taking the voltage level of exiting hydro generators into consideration, in this book it is considered to be 20 kV so that it can properly reduce the brush current. The ratio of stator winding turns and rotor winding turns is considered to be 1:1.
3. *Leakage reactance of a VFT* (X_{VFT}): 18%.
4. *Excitation reactance of a VFT* (X_{magVFT}): 10 pu.
5. *Moment of inertia of a VFT* (J): 50 s.
6. *Capacitor bank*. Both VFTs and step-down transformers have leakage reactance, which absorbs reactive power when current flows through them. When necessary, the shunt capacitors can be used for compensation. Suppose the capacitors compensate all the reactive power loss of VFT circuit at the rated power, leaving a 20% margin; in this book the rated voltage of the capacitor bank is 20 kV and its capacity is 3×30 Mvar.
7. *Step-down transformer*. The transformation ratio is 230:20 or 525:20, depending on the application occasions; the rated capacity is 210 MVA; the short-circuit impedance is 10%; and there is a total of two groups.
8. *DC motor*. This provides sufficient torque for the VFT rotor, mainly related to the frequency difference between the systems on both sides of VFT and the transmission power of VFT. According to Equations (3.35) and (3.36), for 200 MW VFT, when the rated frequencies of the systems on both sides are identical to each other, namely

50 or 60 Hz, the frequency difference of the systems on both sides of VFT is considered to be 1 Hz with a margin of 10%. Meanwhile, using the configuration capacity of the DC motor of the first VFT manufactured by GE as reference, in this book the rated power of DC motor is considered to be 5 MW.

9. *Rectifying circuit.* The rated power of the rectifying equipment is considered based on the rated power of DC motor with a 10% margin, namely 5.5 MW. The transformation ratio of the rated voltage of the rectifying transformer is 20:4; the DC output voltage is 5 kV; the voltage on the low-voltage side of the step-down transformer is 4 kV.

5.2.2 Simplified Asynchronous Interconnection System

In order to verify the basic functions of a VFT such as startup, frequency regulation, grid connection, and power regulation, in this book we use a VFT to connect two simplified systems with different frequencies in simulation research to reduce the impact of the regulation of other elements on the VFT simulation process so as to more directly reflect the operation and control characteristics of VFTs.

The wiring diagram of this simplified asynchronous interconnection system is shown in Figure 5.1. Two single-machine single-load systems of different frequencies are interconnected through a set of VFTs. In Figure 5.1 the frequency of the systems on both sides of VFT are 50 and 60 Hz, respectively (sometimes the rated frequency of the systems on both sides are regulated to be 50 Hz based on need). The rated voltage of the equivalent power source on both sides is 230 kV; the rated power is 240 MW; the internal impedance includes 0.05 H reactance and 1Ω resistance; and the load on both sides is 100 MW constant impedance load. In addition, the load bus is connected to the VFT stator winding and rotor winding through the transformer and circuit breaker with a transformation ratio of 230:20. The rated voltage of the VFT is 20 kV and its rated power is generally considered to be 200 MW.

5.2.3 Typical Four-Generator System

The system application functions of VFT mainly include system power flow optimization control, fault isolation, suppression of system power oscillation and frequency regulation, and so on. In order to fully study and demonstrate these functions of VFT, in this book we select the four-generator system with typical characteristics of

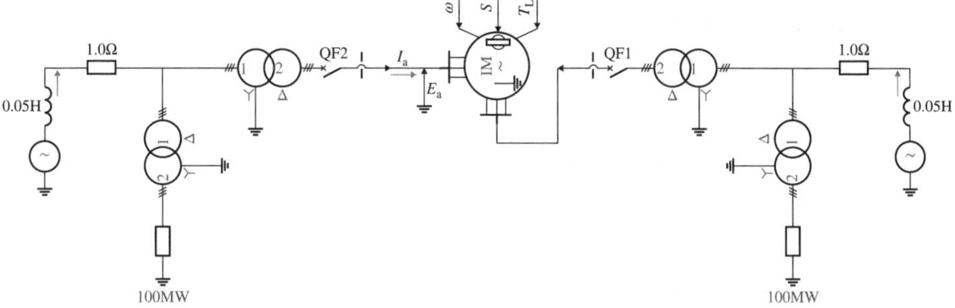

Figure 5.1 Wiring diagram of a simplified asynchronous interconnection system.

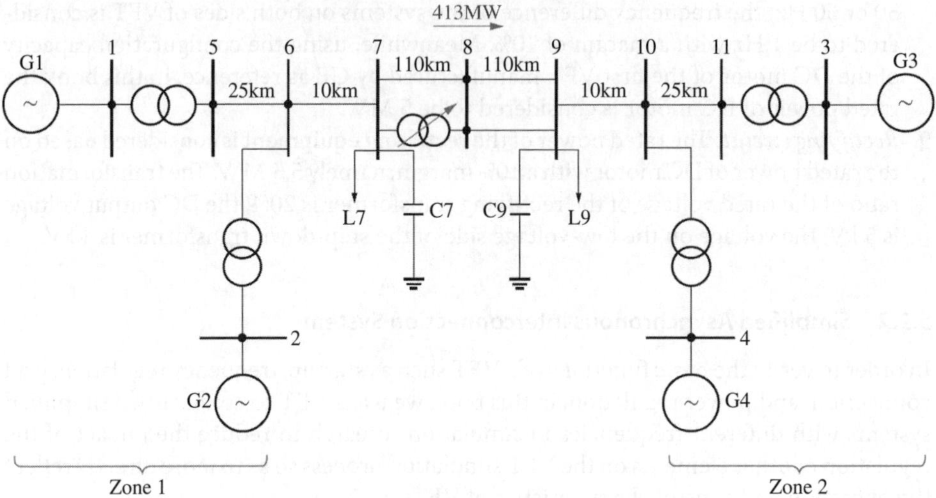

Figure 5.2 Wiring diagram of a four-generator electrical power system [5].

interconnected power grids including low-frequency oscillation as the research object, which is shown in Figure 5.2.

In Figure 5.2, two perfectly symmetrical systems are connected by two 110 km 230 kV double-circuit lines. Although this system is not large in scale, it can effectively reflect the basic characteristics of the interconnected system. This interconnected system is divided into two regions and there exist three main modes of oscillation. In each region there are two synchronous generator sets with a rated capacity of 900 MVA and a rated voltage of 20 kV. These synchronous generators have the same electrical parameters, but different inertias. In Region 1 $H = 6.5$ s while in Region 2, $H = 6.175$ s. At the rated capacity (MVA) and rated voltage (kV), the per unit parameters of the generators are [5]:

$$X_d = 1.8, \quad X_q = 1.7, \quad X_1 = 0.2, \quad X'_d = 0.3, \quad X'_q = 0.55$$

$$X''_d = 0.25, \quad X''_q = 0.25, \quad R_a = 0.002\,5, \quad T'_{d0} = 8.0s, \quad T'_{q0} = 0.4s$$

$$T''_{d0} = 0.03s, \quad T''_{q0} = 0.05s, \quad A_{Sat} = 0.015, \quad B_{Sat} = 9.6, \quad \psi_{T1} = 0.9$$

$H = 6.5$ (corresponding to G1 and G2); $H = 6.175$ (corresponding to G3 and G4); $K_D = 0$

The parameters of the excitation system are

$$T_A/T_B = 0.1 \; T_B = 10 \; K = 200 \; T_E = 0.05 \; E_{MIN} = 0 \; E_{MAX} = 5$$

$$C_{SWITCH} = 0 \; r_c/r_{fd} = 0$$

The parameters of the speed regulator are

$$R = 0.05 \; r = 0.75 \; T_r = 8 \; T_f = 0.05$$

$$T_g = 0.5 \; \text{VELM} = 0.2$$

$$G_{MAX} = 1 \; G_{MIN} = 0 \; T_W = 1.3$$

$$A_t = 1.1 \; D_{turb} = 0.5 \; Q_{NL} = 0.8$$

The per unit value of impedance of each step-up transformer at the reference capacity of 900 MVA and the reference voltage of 230 kV is 0.15 pu. The rated voltage of the transmission system is 230 kV. The line length is indicated in Figure 5.2. The per-unit values of the line parameters at the benchmark capacity of 100 MVA and the benchmark voltage of 230 kV are:

$$r = 0.0001 \ pu/km \quad x_L = 0.001 \ pu/km \quad b_C = 0.00175 \ pu/km$$

In system operation: Region 1 transmits 413 MW power to Region 2 and the running state of the generator is

$$G1: P = 700MW, \quad Q = 185M \ \text{var}, \quad E_t = 1.03\angle 20.2°$$
$$G2: P = 700MW, \quad Q = 235M \ \text{var}, \quad E_t = 1.01\angle 10.5°$$
$$G3: P = 719MW, \quad Q = 176M \ \text{var}, \quad E_t = 1.03\angle -6.8°$$
$$G4: P = 700MW, \quad Q = 202M \ \text{var}, \quad E_t = 1.01\angle -17.0°$$

The load and the reactive power supplied by the shunt capacitors at Node 7 and Node 9 are as follows

$$\text{Node 7} : P_L = 967 \ \text{MW}, Q_L = 100 \ \text{Mvar}, Q_C = 200 \ \text{Mvar}$$
$$\text{Node 9} : P_L = 2000 \ \text{MW}, Q_L = 100 \ \text{Mvar}, Q_C = 350 \ \text{Mvar}$$

The VFT is connected to the line between Node 7 and Node 8, and according to the research needs this system is adjusted appropriately. For example, sometimes the parallel line between Node 7 and Node 8 is disconnected and asynchronous interconnection is achieved only through VFT branches.

5.2.4 Large-Scale Complex Power System

In order to study the operating characteristics of a VFT in a large system, verify the adaptability of VFT power flow model and electromechanical transient model developed by PSASP in the large system simulation, and improve the simulation analysis ability of a VFT, in this book we use the Central China-North China-Northeast China synchronous interconnected system shown in Figure 5.3. Using the operation mode of the 2007 level year system before the Northeast China-North China (Gaoling) back-to-back DC transmission project was put into operation, this system has 5077 buses, 526 generators, 4268 AC lines, 3 DC lines, and 3641 transformers. One of the purposes of using this system is to study the function of VFT of suppressing

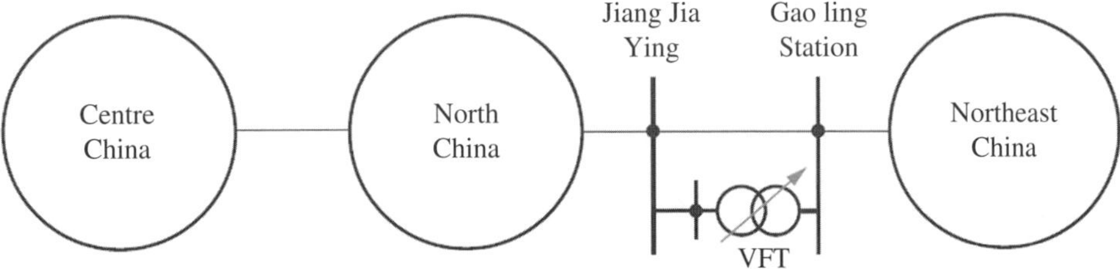

Figure 5.3 Schematic diagram of the wiring of the Central China-North China-Northeast China synchronous interconnected system (2007 level year).

large system low-frequency oscillation because this AC synchronous interconnection system will change to back-to-back DC asynchronous interconnection, low-frequency oscillation needs to be prevented.

In this plan, the Jiangjiaying Substation in North China Grid is synchronously interconnected with Gaoling Substation in the Northeast China Grid through the two-circuit 500 kV line while the Central China Grid is also connected with the North China Grid through the 500 kV line (since 2009, the North China Grid and Central China Grid have been synchronously interconnected through 1000 kV UHV). In this book, two 200 MW VFTs connected in parallel with each other are connected in series into the Gaoling side of the Jiangjiaying-Gaoling one-circuit line to study the control and regulation characteristics of VFT in the complex system.

5.3 Startup and Power Regulation of VFTs

In order to facilitate the following introduction, the single-line wiring diagram of the simplified asynchronous interconnected system is given in Figure 5.4.

5.3.1 VFT Power-On Process

Switching a no-load VFT is equivalent to the connecting a reactor with saturation characteristic. As shown in Figure 5.4, when the circuit breaker QF1 is closed and circuit breaker QF3 is opened, the circuit breaker QF2 is closed to electrify VFT stator winding. At this point, the voltage at Point C of stator winding is the same as the frequency of System I, namely 50 Hz. Due to the magnetic field coupling of VFT stator winding and rotor winding, the induced voltage with the same frequency can also be induced at Point D of VFT rotor winding.

Figure 5.5 and Figure 5.6 show the relevant voltage and current waveforms when the circuit breaker is closed in different voltage phase angle positions. E_{a1} is the A-phase

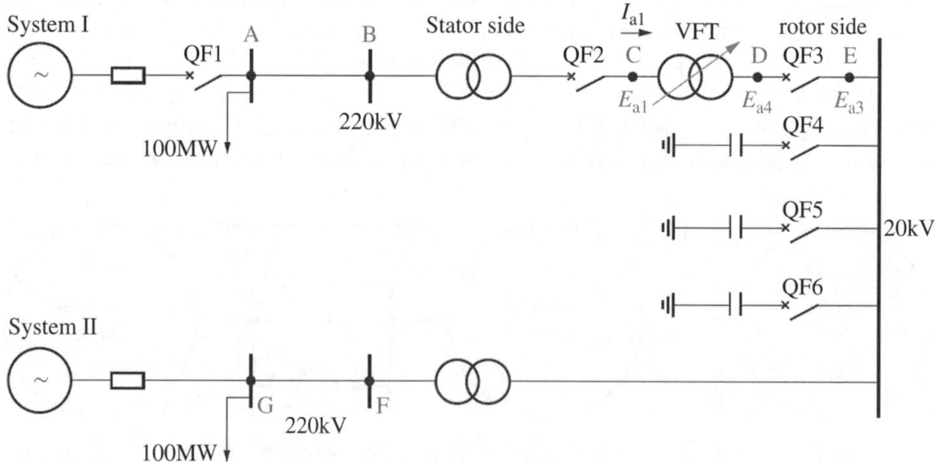

Figure 5.4 Single-line wiring diagram of the simplified asynchronous interconnected system.

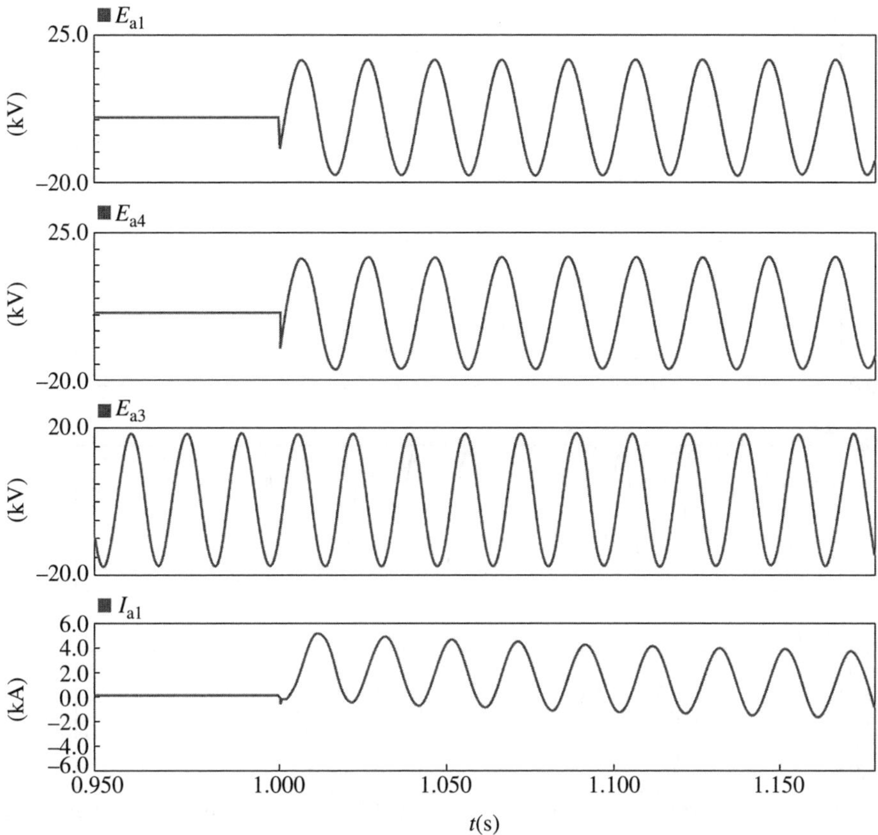

Figure 5.5 VFT power-on process simulation (without controlling circuit breaker closing phase angle).

voltage at Point C of stator winding; E_{a4} is the A-phase voltage at Point D of rotor winding; E_{a3} is the A-phase voltage at Point E of rotor winding and its frequency is 60 Hz; and I_{a1} is the stator winding current. After 1 s, close the circuit breaker on the stator side to make the excitation current pass through the VFT stator winding. At this point, the voltage frequency of both stator winding and rotor winding is 50 Hz. It can also be seen from Figure 5.5 that as the voltage is close to zero when closing, there is a large DC component in I_{a1} current. If the magnetic saturation characteristic of VFT is taken into consideration, there might appear to be a high inrush current under specific conditions [6]. At this point, the circuit breaker closing phase angle can be controlled to effectively reduce inrush current, as shown in Figure 5.6. In Figure 5.6, when the voltage sine wave is 90°, it is equivalent to the result of closing at the peak of voltage. As a result, controlling the circuit breaker closing phase angle can promote system and equipment security.

5.3.2 VFT Grid Connection Process

It can be seen from Figure 5.5 and Figure 5.6 that when the VFT is charged, E_{a3} and E_{a4}, the port voltages on both sides of circuit breaker QF3, have different frequencies and

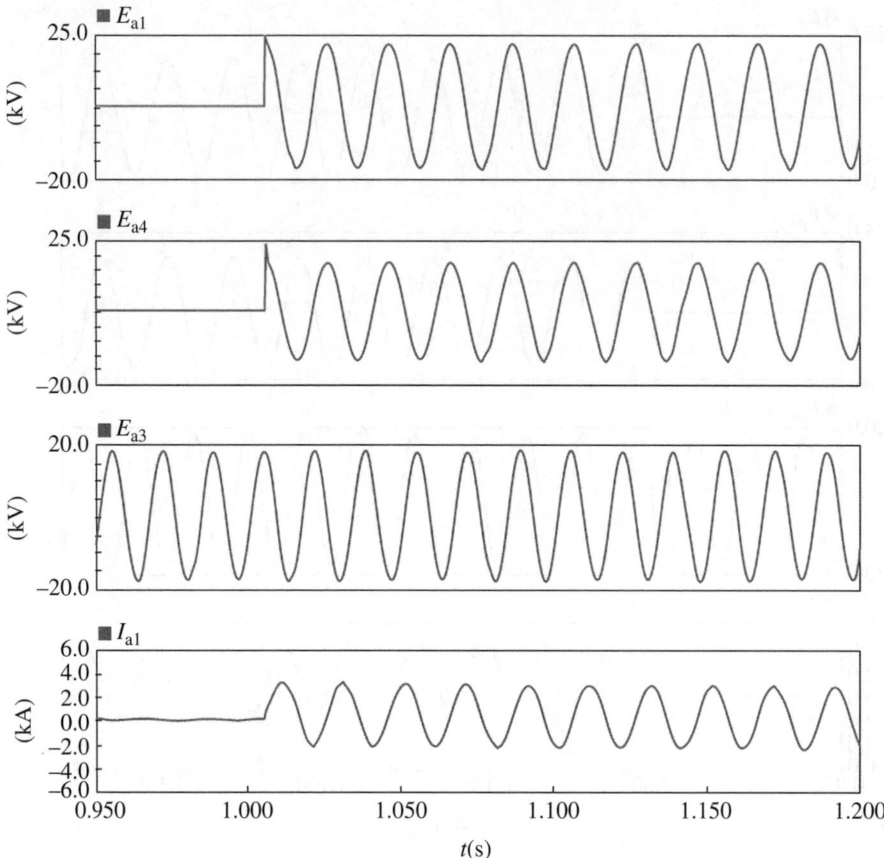

Figure 5.6 VFT power-on process simulation (controlling circuit breaker closing phase angle).

phases. If at this point the circuit breaker QF3 is closed, it will make a serious impact on the VFT and the electrical power system on both sides. As a result, before the interconnection of the two systems, the VFT rotor should be rotated to make the frequency of E_{a3} and E_{a4} identical to each other and the phase consistent with each other. Based on the synchronizing close control logic diagram shown in Figure 4.14, as shown in Figure 5.7 and Figure 5.8, at the time of 2 s, the control system starts the rotor speed control module and the closing phase angle control module, and after a while the control target of synchronous speed and the phase angle difference being lower than the limit is realized.

It can be seen from Figure 5.7 that the charging of VFT is over before 2 s; then the frequency control is carried out; driven by the DC drive system, the rotor begins to rotate and Frequency f_3 also rises simultaneously with the increase of the rotor speed. At 10.5 s, the electrical speed of the rotor is close to 10 Hz, equal to the frequency difference between the two systems on both sides. f_3 rises to 60 Hz, consistent with the frequency of the system on the rotor side, which indicates under the frequency control of VFT, the rotor speed compensates the frequency difference between the systems on both sides and that two power grids with different frequencies can achieve asynchronous

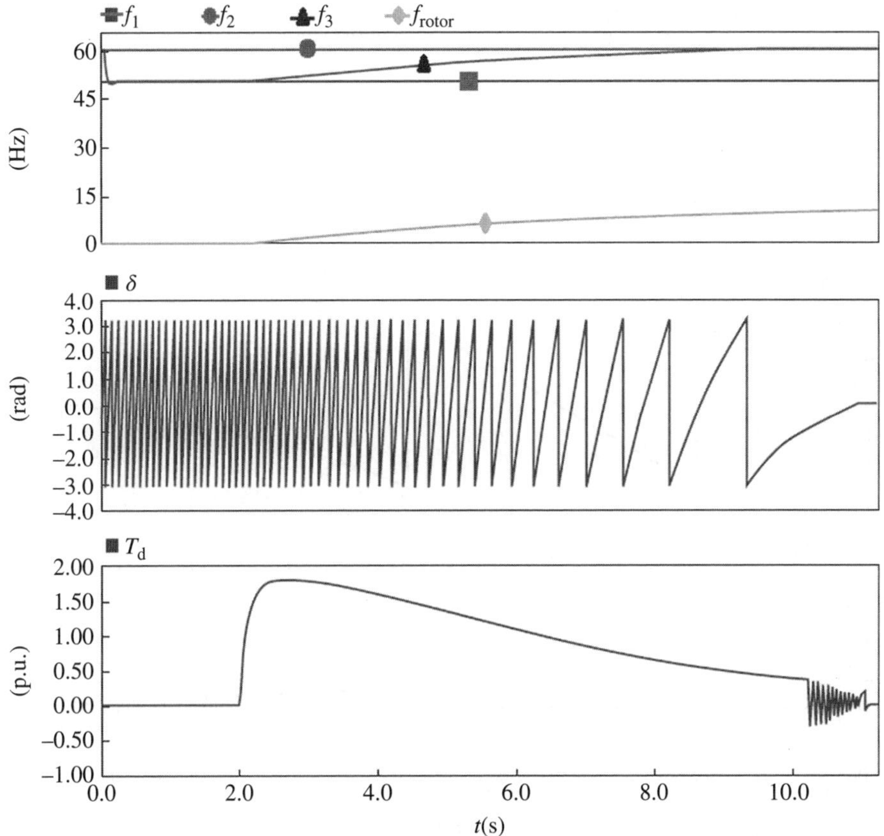

Figure 5.7 Relevant waveforms of a VFT starting grid connection with a frequency difference between systems on both sides of 10 Hz. f_1 = stator side system frequency; f_2 = rotor side system frequency; f_3 = frequency of the voltage at the circuit breaker terminal on the rotor side E_{a4}; f_{rotor} = electrical frequency rotor speed corresponds to rotor mechanical speed; δ = phase difference between the voltages on both sides of circuit breaker QF3; and T_d = rotor torque of the VFT (the same for Figure 5.8).

interconnection through VFT rotor. With the completion of the rotor frequency regulation, the VFT control system automatically starts the phase angle control module; the voltage phase angle control simulation waveforms are shown in Figure 5.7; at about 11 s, the phase angle meets the requirements. After that, the closing of circuit breaker QF3 is commanded to achieve grid-connected operation. The whole process takes about 9 s. Mainly because of the large inertia of the VFT rotor, it takes some time for the speed rising from zero to which 10 Hz corresponds.

For interconnected power grids with the frequency of the systems on both sides close to each other, the synchronizing time will be greatly reduced. Figure 5.8 shows the waveforms when the system frequency difference is 0.6 Hz. Synchronizing process including speed regulation and phase angle regulation takes 2.5 s. Of course, reasonably regulating the VFT control system can effectively reduce the time needed for synchronizing and phase angel regulation.

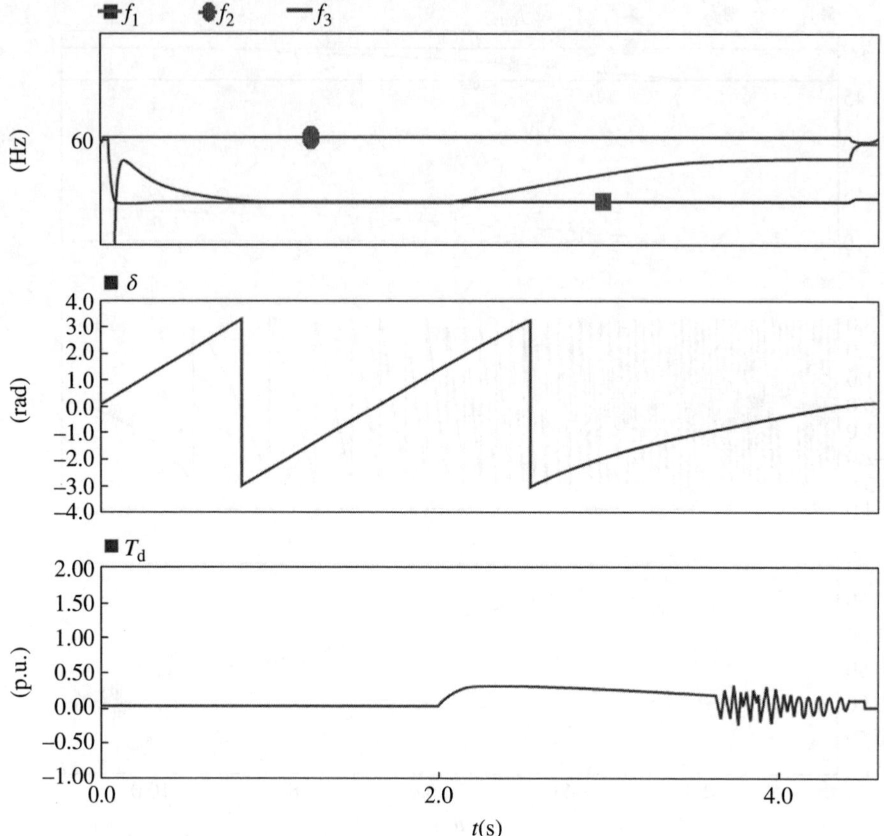

Figure 5.8 Relevant waveforms when VFT starts grid connection with frequency difference between systems on both sides of 0.6 Hz.

5.3.3 VFT Power Regulation

After the VFT is connected to the grid, its transmission power can be adjusted in a real-time manner according to dispatching commands. The climbing response and step response of power control are important indexes that characterize the system control performance and the most commonly used control method, which is the basis of realizing different control functions.

Shown in Figure 5.9 are the change of transmission power, current, voltage, and speed of VFT in the climbing control. The specific control process is as follows:

1. The power climbing control starts at the 12 s. The set value of the transmission power rises from 0 to 100 MW within 8 s and lasts for 10 s;
2. The set value of the transmission power reduces to 0 MW within 10 s and lasts for 10 s;
3. The power transmission direction is the opposite. Within 10s the transmission power becomes −100 MW and lasts for 10 s;
4. Within 10 s the transmission power is regulated to 100 MW.

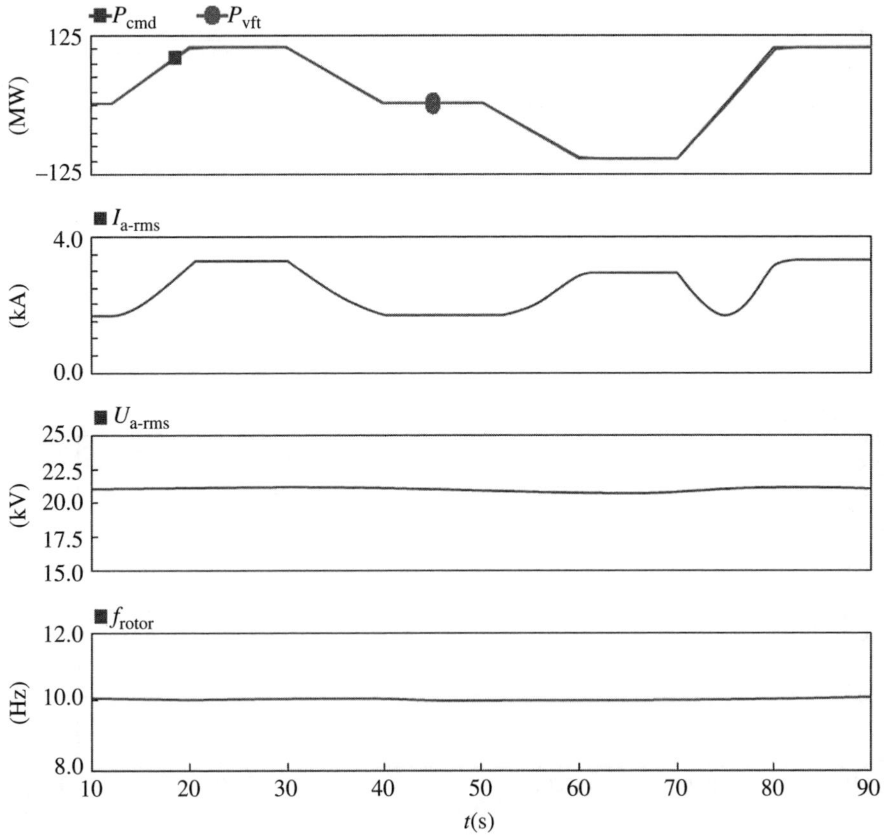

Figure 5.9 Waveforms of VFT transmission power climbing control. Where: P_{cmd} = command power; P_{vft} = VFT transmission power; $I_{a\text{-rms}}$ = VFT current; $U_{a\text{-rms}}$ = VFT stator side voltage; and f_{rotor} = electrical frequency corresponds to rotor mechanical speed.

It can be seen from Figure 5.9 that the transmission power of VFT can change with the command power with short delay. The winding current of VFT also changes with the power. However, the speed of VFT basically remains unchanged.

Figure 5.10 shows the change of transmission power, current, voltage, and speed of VFT in step control. The specific control process is as follows:

1. At 15 s the command power rises from 0 to 100 MW and lasts for 5 s;
2. At 20 s it reduces from 100 to 0 MW and lasts for 5 s;
3. At 25 s it reduces to −100 MW and lasts for 5 s; in 30 s it rises to 100 MW.

It can be seen from Figure 5.10 that the power step response time of a VFT is about 400 ms. The time delay is close to the measured step response result of the world's first VFT and the difference is within a reasonable range. Compared with HVDC, the VFT has no minimum transmission power limit. So its transmission power can be smoothly regulated within the designed range, but meanwhile we should realize that the step response speed of the VFT is slower than that of the HVDC, which is roughly tens of milliseconds.

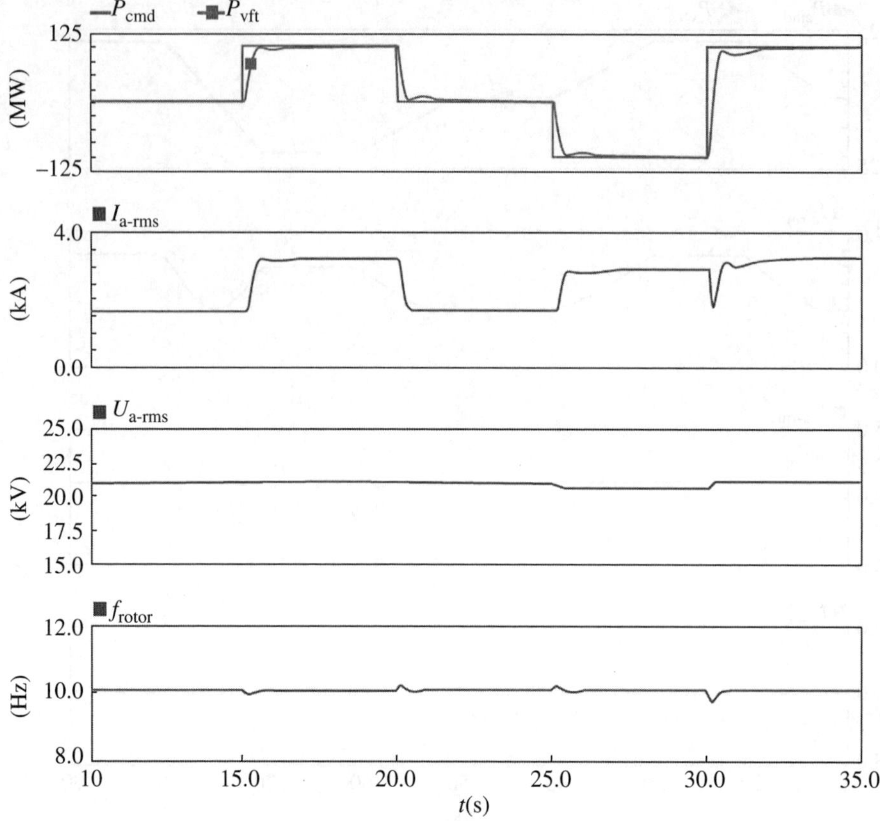

Figure 5.10 The waveforms of VFT transmission power step control.

Power control is the basis of the system application control of VFTs. This simulation calculation verifies the effectiveness of the VFT models as well as basic control methods and lays a foundation for research into the system application of VFTs.

5.4 Using VFTs to Regulate System Power Flow

5.4.1 Optimizing the Power Flow Distribution of Interconnected Systems

According to the basic control characteristics of a VFT, when the two systems are interconnected only through it, the transmission power of the cross section is completely controllable. If there are not only VFT, but also other AC interconnected branches between the two systems, the distribution of the cross-sectional power flow in different branches can be adjusted through the VFT within its rated power range.

In this section, we are going to take the typical four-generator system as an example and use the cross section power flow between Node 7 and Node 8 in the VFT control diagram shown in Figure 5.2. The distribution of its waveforms is shown in Figure 5.11. Judging from the waveforms, in the synchronous interconnection mode, the switching power between the systems P_{b8} is determined by the load demand of the systems on

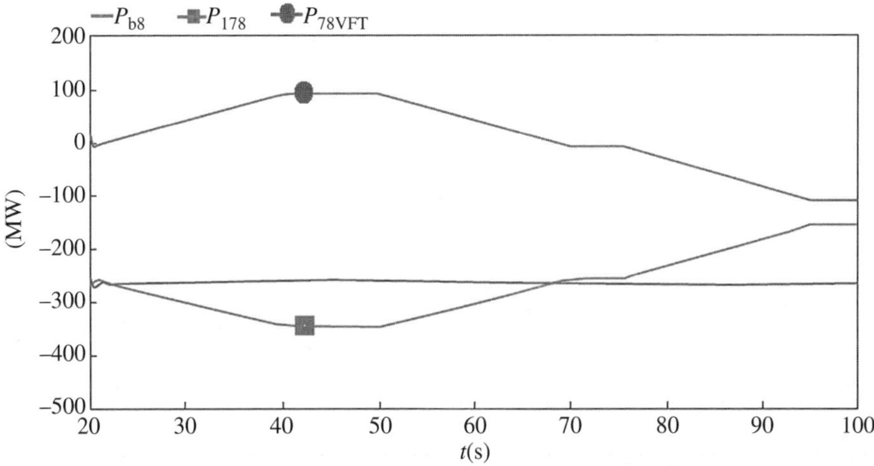

Figure 5.11 Waveform of using a VFT to control the distribution of cross-sectional power flow of interconnected systems. Where: P_{b8} = switching power of the section system at Node 8; P_{l78} = transmission power of the line between Node 7 and Node 8; and P_{78VFT} = transmission power of the VFT.

both sides. Before the frequency regulation machine is put into use, P_{b8} changes slightly. However, when the transmission power of the VFT branch P_{78VFT} is regulated, the power of the other parallel line P_{l78} will be automatically regulated.

5.4.2   Reducing System Power Transmission Loss

When the system power flow distribution changes, the system loss will also be adjusted accordingly. So changing the system power flow distribution through a VFT can also reduce the system loss. Figure 5.12 shows the total grid loss of the typical four-generator system $P_{\text{loss-net}}$ and the total power transmission loss between Node 7 and Node 8 when the VFT power flow changes. In Figure 5.12, the system loss is minimized when the VFT transmits 61 MW power from Node 8 to Node 7.

5.4.3   System Reactive Voltage Control

In the operational process, the VFT and the transformer leakage reactance on both sides will absorb some reactive power. For a weak system, with an increase of the absorbed reactive power, the system voltage will drop. In order to maintain the reactive power balance of the system, parallel capacitor banks of a certain capacity need to be put into operation during VFT power regulation. Meanwhile the low-voltage limit power module is specially designed for VFTs so that when the voltage is lower than a certain degree, the VFT power level will be limited.

Figure 5.13 shows the automatic switching of capacitor banks in the VFT power regulation process and the waveforms of the transmission power being automatically limited in the case of serious low voltage. In Figure 5.13, the transmission capacity of the VFT is 200 MW. At 30 s, an inductive load is connected to Bus 8 and the VFT bus voltage is reduced to 17.2 kV. As a result, three sets of 30 Mvar capacitors are sequentially put into

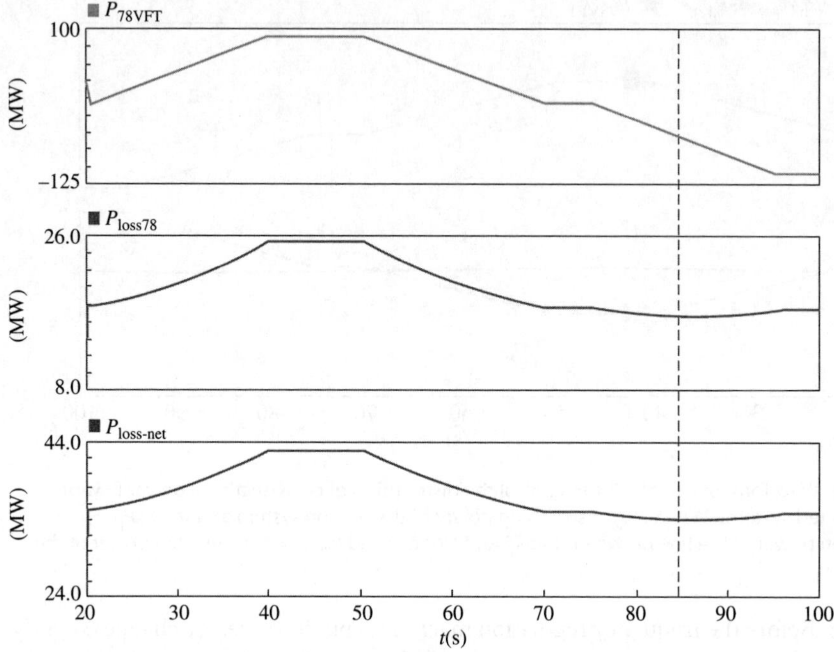

Figure 5.12 Using a VFT to optimize power transmission loss. P_{78VFT} = transmission power of VFT; P_{loss78} = power loss of the line between Node 7and Node 8; and $P_{loss\text{-}net}$ = transmission loss of the whole grid.

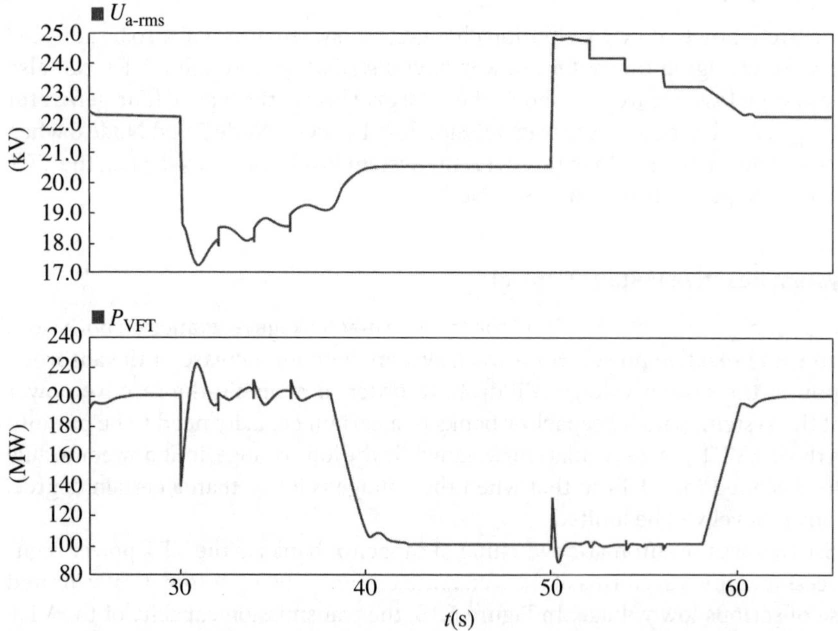

Figure 5.13 Waveforms of reactive power and voltage regulation simulation of a VFT and capacitor banks. $U_{a\text{-}rms}$ = voltage of the VFT on the stator side; and P_{VFT} = transmission power of the VFT.

use and the voltage gradually rises to around 19 kV. Finally, the VFT power flow is also reduced to 100 MW and the system voltage is restored to 20.5 kV. At 50 s, the inductive load is disconnected from Bus 8 and the system voltage rises to 24.7 kV. So, the three sets of 30 Mvar capacitors are sequentially withdrawn and the voltage is gradually reduced to around 23 kV. Finally, the VFT power flow is also restored to 200 MW and the system voltage is restored to 22 kV.

5.5 Characteristics of VFTs During a Fault Period

Because of the large rotor inertia of VFTs, a short duration fault or disturbance has a small impact on the running state of the VFT. As a result, the VFT has a strong ability to resist disturbance and faults, which is also another important physical characteristic. Now, we are going to use the typical four-generator system shown in Figure 5.2 as the research object to study the operational characteristics of a VFT and the system when single-, two-, or three-phase grounding faults occur in the system.

5.5.1 Single-Phase Short-Circuit Fault

Figure 5.14 shows single-phase grounding short-circuit fault simulation waveforms. At 20 s, the single-phase grounding short-circuit fault occurs at Node 7 of the load bus of the four-generator system and lasts for 0.1 s. When the fault is removed, the transmission power of the VFT is restored to normal levels within 30 ms. Meanwhile, during the fault period the current passing through the VFT changes slightly. The VFT bus voltage is about two-thirds the level before the occurrence of the fault and the transmission power is reduced to about 60% of previous conditions accordingly.

5.5.2 Two-Phase Short-Circuit Fault

Figure 5.15 shows two-phase grounding short-circuit fault waveforms. The two-phase short-circuit fault occurs and lasts for 0.1 s. When the fault is removed, the transmission power of VFT is also restored to normal levels within 30 ms. During the fault period, the transmission power is reduced to about a quarter of previous conditions accordingly.

5.5.3 Three-Phase Short-Circuit Fault

Figure 5.16 shows three-phase grounding short-circuit fault waveforms. The three-phase short-circuit fault occurs and lasts for 0.1 s. When the fault is removed, the transmission power of VFT is also restored to normal levels within 30 ms. During the fault period, the transmission power is reduced to around 0 accordingly.

This simulation shows that when a general fault occurs to the system, because of the large inertia of VFT and the large VFT loop impedance, it has a slight impact on the interconnected systems of the VFT during the fault period. When the fault is removed, the VFT will soon return to normal operation. However, in case of the same fault, serious commutation failure and even unipolar or bipolar blocking might occur to HVDC transmission. As a result, compared with HVDC, the VFT has a strong ability to resist faults.

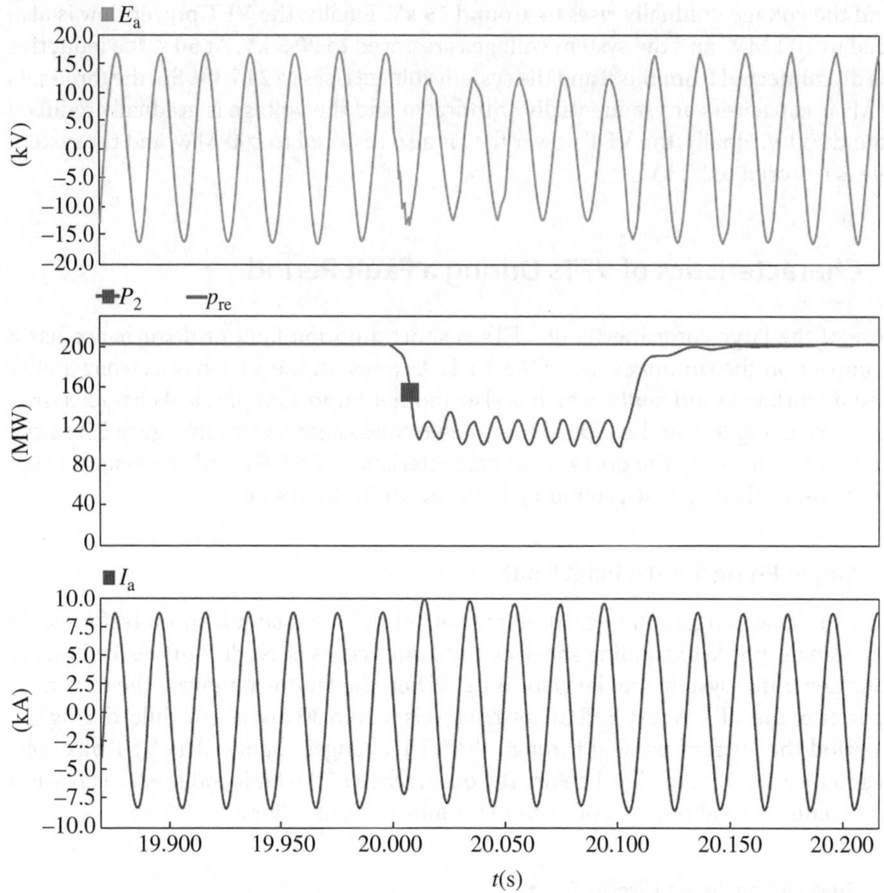

Figure 5.14 Single-phase grounding fault waveforms. Where: E_a = A-phase voltage of the VFT on the stator side; P_2 = transmission power of the VFT; P_{re} = set transmission power of VFT; and I_a = A-phase current of the VFT stator winding (the same in Figures 5.15 and 5.16).

5.6 Using VFTs to Regulate System Frequency

In using VFTs to realize the asynchronous interconnection of two power grids, the two power grids can serve as a standby for each other and provide power support during a fault and disturbance period, which is an effective frequency regulation means. Based on the four-generator system in Figure 5.2, asynchronous interconnection of Node 7 and Node 8 is realized only through the VFT branch. So, by appropriately regulating system load and generator output, the function of the VFT as a system frequency regulation means can be reflected. The specific simulation process is as follows:

1. At 10 s, the system is in the steady state; the outputs of the generators G1, G2, G3, and G4 are 737, 375, 183, and 43 MW, respectively, and G1 is the main frequency

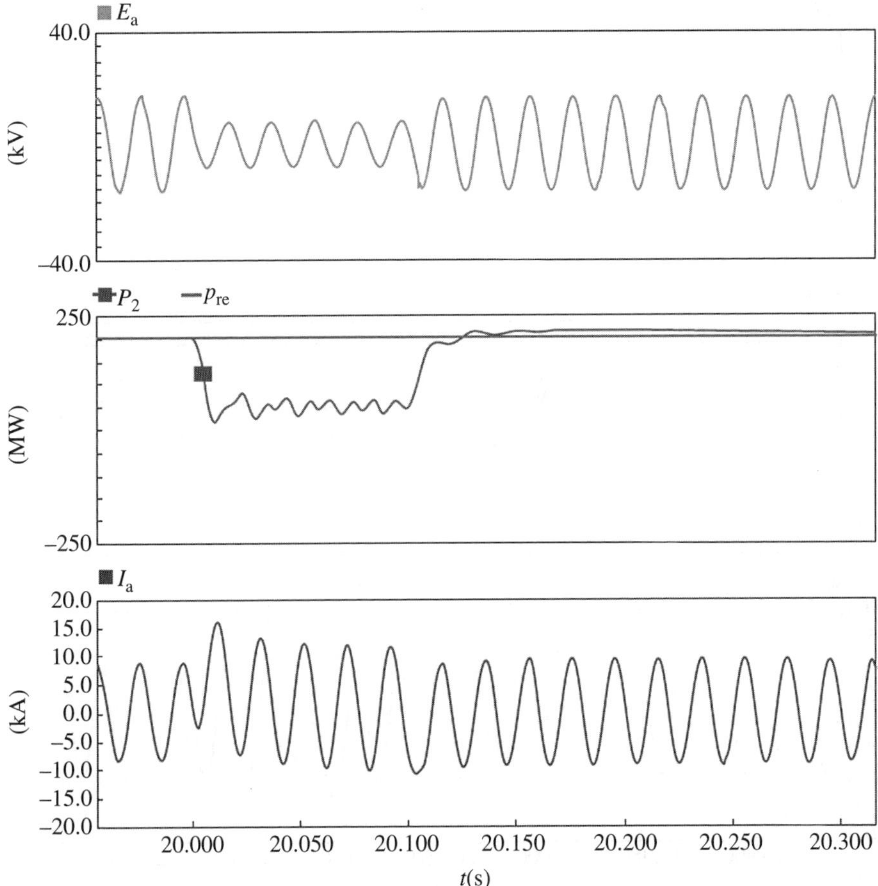

Figure 5.15 Two-phase grounding short-circuit fault waveforms. Where E_a, P_2, P_{re}, and I_a are same as those of Figure 5.14.

regulation generator. The loads of Node 7 and Node 9 are 1089 and 219 MW, respectively, and the transmission power of VFT is 0 MW.

2. At 30 s, the load of Node 9 increases by 150 MW while the VFT still transmits power of a given capacity. At this point the frequency of System II is reduced to 49.3 Hz.

3. At 40 s, the frequency regulation function of the VFT is put into use; the VFT transmits 74 MW power from System I to System II; the frequency of System II rises to 49.8 Hz while the frequency of System I is reduced to 49.9 Hz. As both frequencies are within the allowable range, the system frequency regulation target is realized. Figure 5.17 shows the waveforms of system frequency, equipment power, generator output, and VFT current in this process.

In this simulation, when the switching load or switching generator unit in the system leads to a rise or fall of the system frequency on this side, the VFT can be used to rapidly regulate the magnitude and direction of the transmission power to make the system fre-

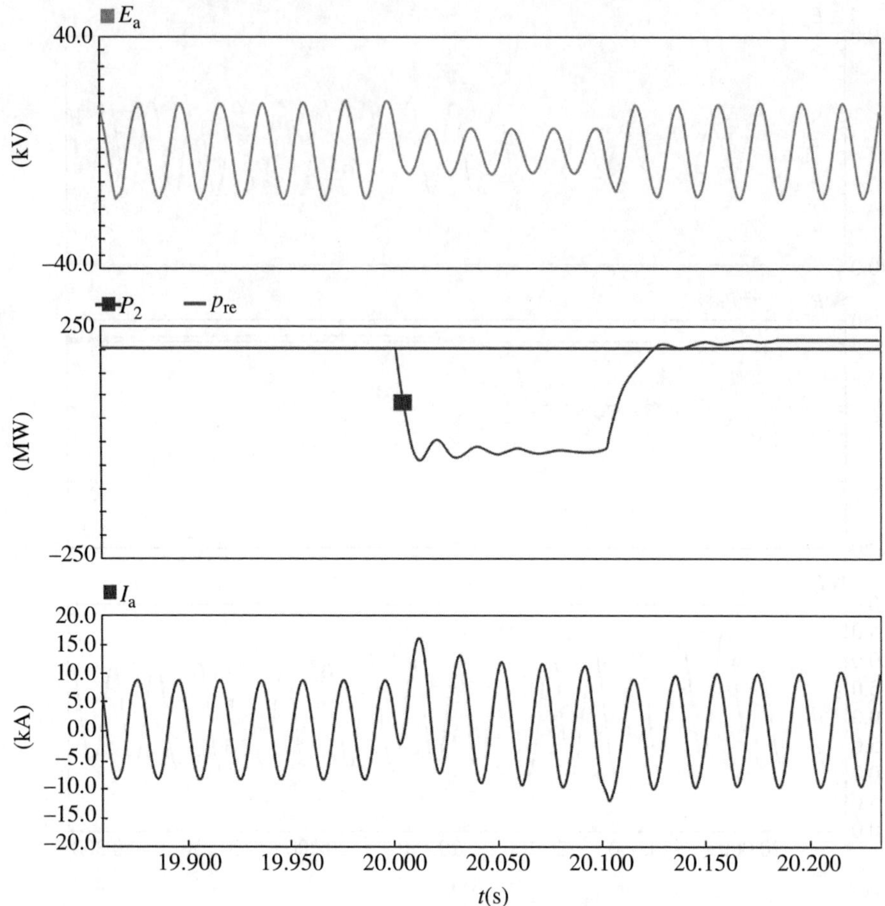

Figure 5.16 Three-phase grounding short-circuit fault waveforms. Where E_a, P_2, P_{re}, and I_a are same as those of Figure 5.14.

quency on this side fall within a reasonable range. In this process, the system frequency on this side might also be regulated appropriately. This is the basic principle of using a VFT to use the system on the other side as frequency regulation machine.

5.7 Using VFTs to Supply Power to Weak Power Grids and Passive Systems

High-voltage direct current transmission (HVDC) generally uses the current commutation principle and requires that the AC systems on both sides have strong voltage support capability. As a result, conventional DC is generally not suitable for supplying power to weak power grids. However, a VFT has no specific requirements for system voltage and has the ability to supply power to weak power grids and even passive systems. Now we are going to use the four-generator system shown in Figure 5.2 as the research object to connect Node 7 and Node 8 only through the VFT branch to study the power supply of VFTs to weak power grids and passive systems.

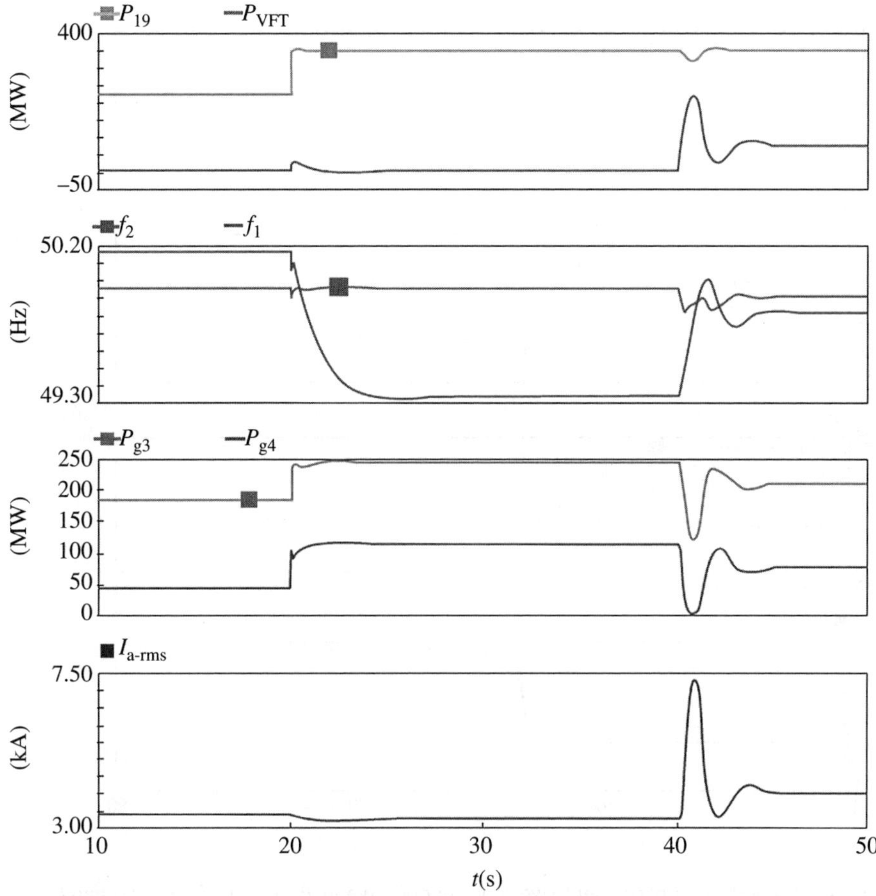

Figure 5.17 Using a VFT to regulate system frequency. Where: P_{l9} = load of Node 9; P_{VFT} = power transmitted by VFT from System I to System II; f_1, f_2 = frequency of System I and System II, respectively; P_{g3}, P_{g4} = output of G3 and G4, respectively; and $I_{a\text{-}rms}$ = VFT current.

5.7.1 Supplying Power to Weak Power Grids Losing Some Power

For the system shown in Figure 5.2, at 60 s, Unit G3 trips and is disconnected, losing 209 MW of output. The frequency regulation function of VFT and the conventional frequency regulation function of G4 are put into use to automatically increase the power transmitted to System II to 150 MW; the frequency and load of System II are controlled at 49.7 Hz and 227 MW, respectively. Figure 5.18 shows the waveforms of the system frequency, equipment power, generator output, and VFT current. In this process, the reactive power absorbed by VFT is increased while the bus voltage is reduced and the capacitor banks are put into use in batches.

5.7.2 Supplying Power to Passive Systems

For the system shown in Figure 5.2, at 80 s, Unit G4 trips and is disconnected losing 91 MW of output. The frequency regulation function of a VFT is put into use to automatically increase the power transmitted to System II to 200 MW; the frequency of System

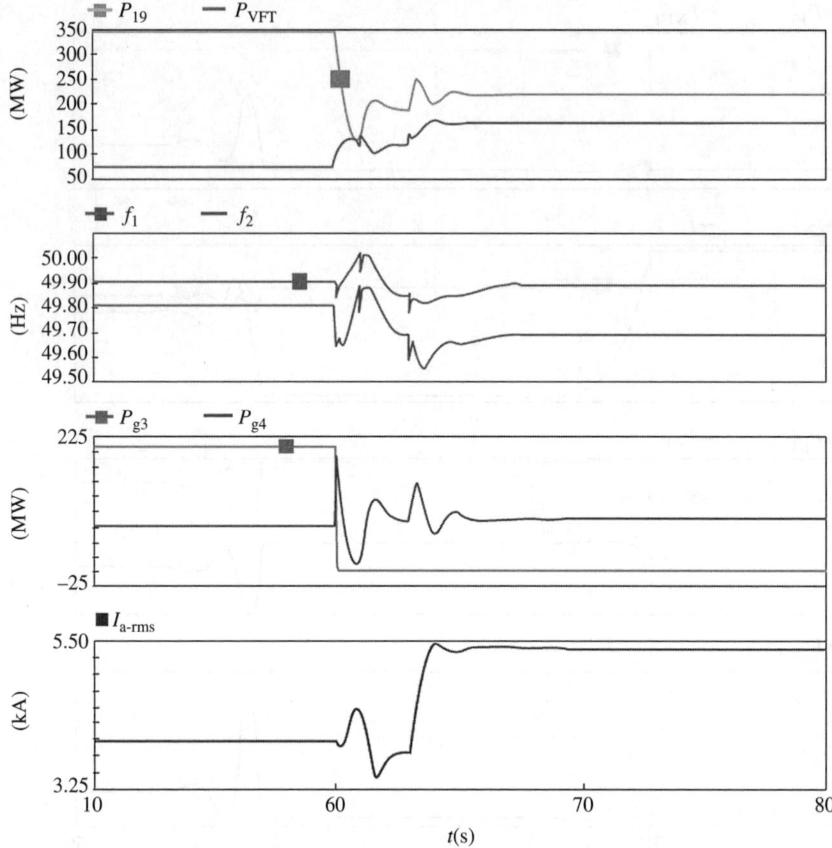

Figure 5.18 Using VFT to supply power to weak power grids. Where P_{l9}, P_{VFT}, f_1, f_2, P_{g3}, P_{g4}, and $I_{a\text{-rms}}$ are same as those of Figure 5.17.

II is controlled at 49.64 Hz while due to the impact of voltage drop of bus, the load is reduced to 178 MW. It can be seen from it that the VFT has the ability to supply power to passive systems. As a result, the VFT can be used for black-start power to realize the rapid recovery of faulty system. Figure 5.19 shows the waveforms of the system frequency, equipment power, generator output, and VFT current.

It can be seen from the simulation of power supply to weak systems and passive systems that the VFT has strong adaptability and can provide reliable power supply to weak systems and passive systems, which is of important practical significance to marginal interconnection between systems and providing black-start power for them.

5.8 Application of VFTs in a Large Complex Electrical Power System

Previously, we analyzed the operating characteristics of a VFT in a four-generator system. Now we are going to use PSASP to analyze the control and regulation characteristics of a VFT in the large complex power system. The studied system is

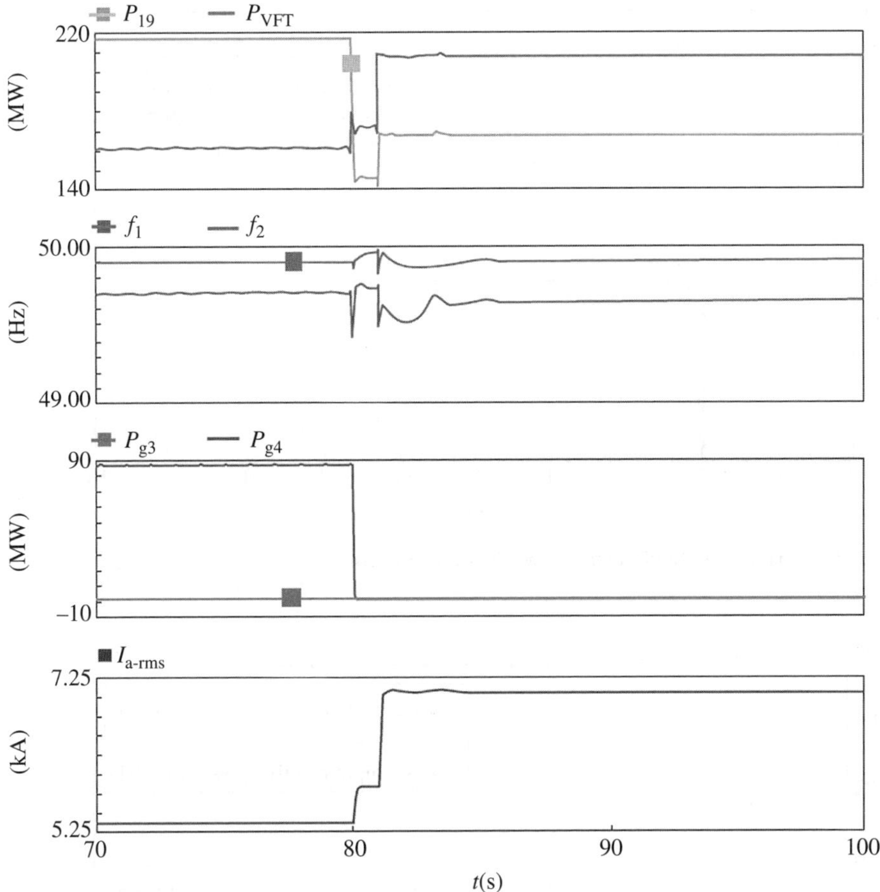

Figure 5.19 Using a VFT to supply power to passive systems. Where P_{l9}, P_{VFT}, f_1, f_2, P_{g3}, P_{g4}, and $I_{a\text{-}rms}$ are same as those of Figure 5.17.

the 2007 level year Central China-North China-Northeast China AC interconnected system, while the used model is the VFT power flow and steady-state model. In this model, a VFT is described as a dynamic element connecting two buses with the bus voltage as input variables and the bus current as the output variables, respectively.

5.8.1 Power Flow Control of VFTs in the Complex Electrical Power System

Regulating the tie line switching power of North China Grid and Northeast China Grid is shown in Figure 5.3. The relevant basic parameters are as follows:

$$S_B = 1000 \text{ MVA}, X_{VFT} = 1.0 \text{ pu}, U_{Gaoling} = 1.0551 \text{ pu}, U_{Jiangjiaying} = 0.994 \text{ pu},$$

$$K_1 = -1.0, \text{ a} = 0.003, \text{ b} = 0.10, T_2 = 0.08, K_2 = 1.20.$$

At 2 s, the set value of VFT active power increases from 0.1 to 0.3 pu. The active/reactive power response of VFT and the active/reactive power response of the tie line cross section are shown in Figure 5.20 and Figure 5.21, respectively. It can be seen from it that the VFT can rapidly regulate the power flow of interconnected systems.

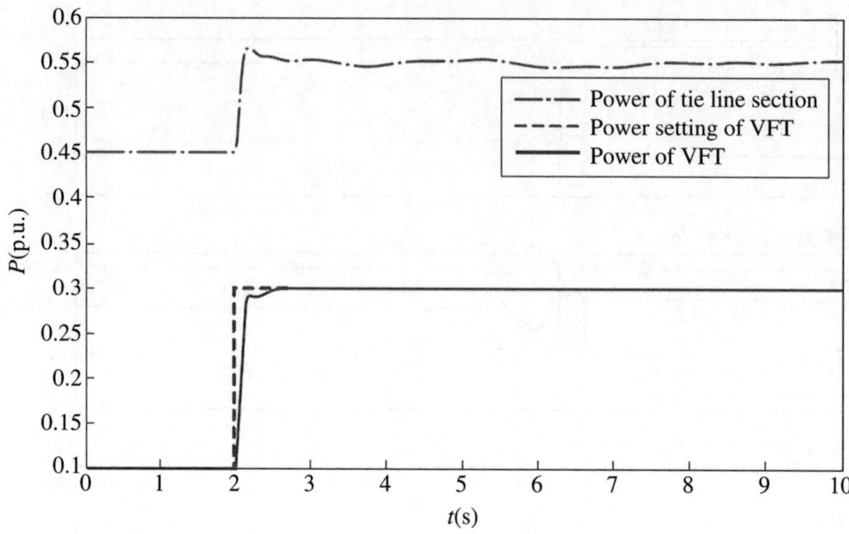

Figure 5.20 Step response of VFT active power in a large system.

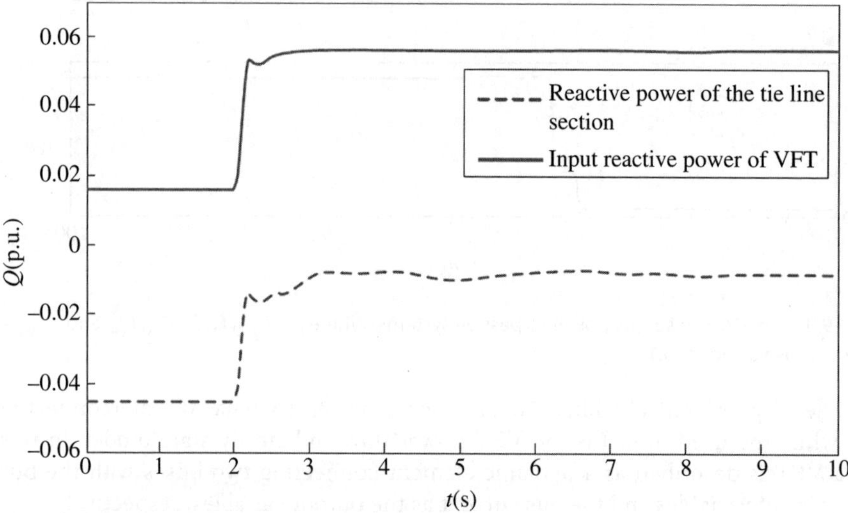

Figure 5.21 Step response of VFT reactive power in a large system.

5.8.2 Transient Stability of VFTs in a Complex Power System

Here, we use the Central China-North China-Northeast China interconnected system again as an example. In the simulation, a single-phase grounding fault that lasts for 0.1 s occurs to the Gaoling-Jiangjiaying AC tie line, which is connected in parallel with a VFT. The close-in generator power angle and VFT transient simulation waveforms are shown in Figures 5.22–5.24.

It can be seen from these waveforms that, when the fault occurs, the generator power angle, tie line power, and VFT power will also be changed. After that the waveforms of

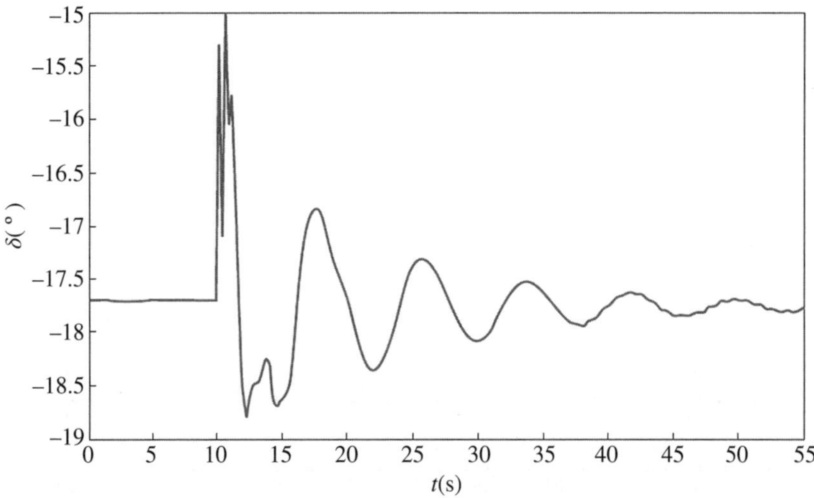

Figure 5.22 Suizhong power plant generator G1 power angle response.

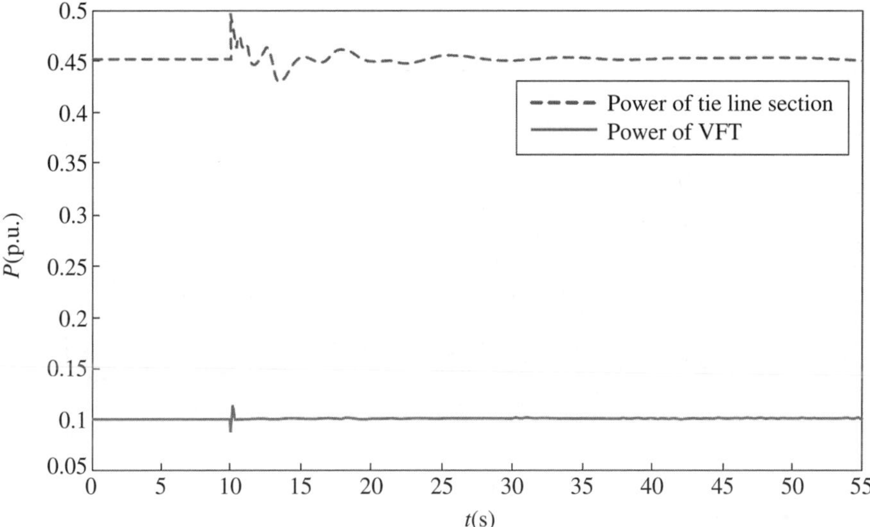

Figure 5.23 Tie line cross section and VFT active power response.

all variables gradually attenuate and tend to be stable. It can be seen from it that the model using a VFT can work stably in case of faults.

At the same time, we also study the system change when the no-fault trip occurs to parallel AC lines. In Figure 5.3 the no-fault trip occurs to the Gaoling-Jiangjiaying AC tie line that is connected in parallel with a VFT. The close-in generator power angle and VFT transient simulation waveforms are shown in Figures 5.25–5.27. It can be seen from Figure 5.25 that the VFT power changes slightly, which indicates that in case of a parallel circuit trip, the VFT can be used to regulate the phase shift angle to control the power within a predetermined range and it will not lead to high power transfer.

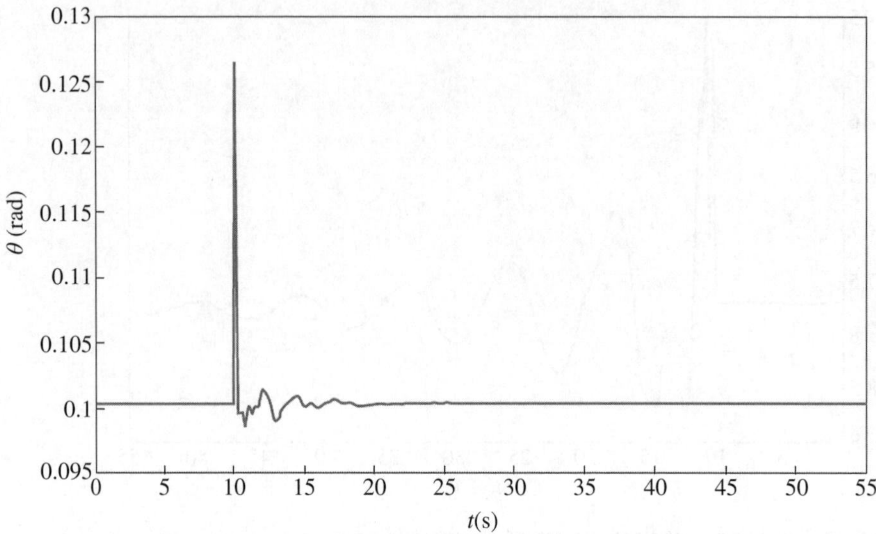

Figure 5.24 Angle difference between VFT stator winding and rotor winding.

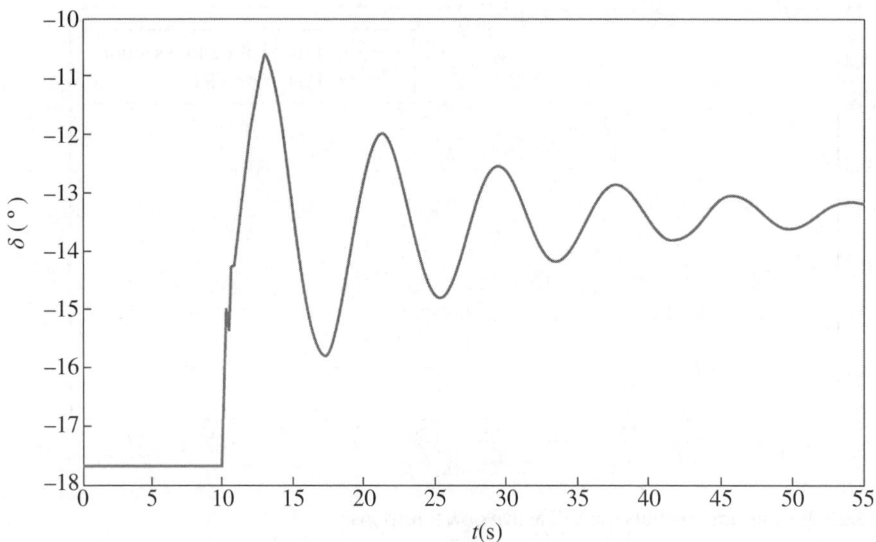

Figure 5.25 Suizhong power plant generator G1 power angle response.

5.9 Using VFTs to Suppress Low-Frequency Power Oscillation in the Electrical Power System

As a series flexible AC transmission system device, the VFT has the ability to flexibly regulate the cross-sectional power of the interconnected systems, which provides conditions for suppression of the system low-frequency oscillation using VFTs. In this section,

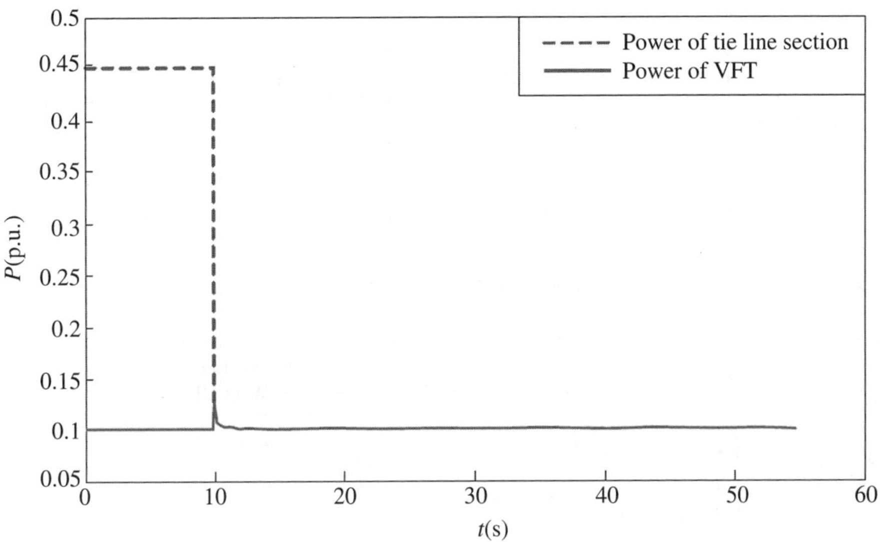

Figure 5.26 Tie line cross section and VFT active power response.

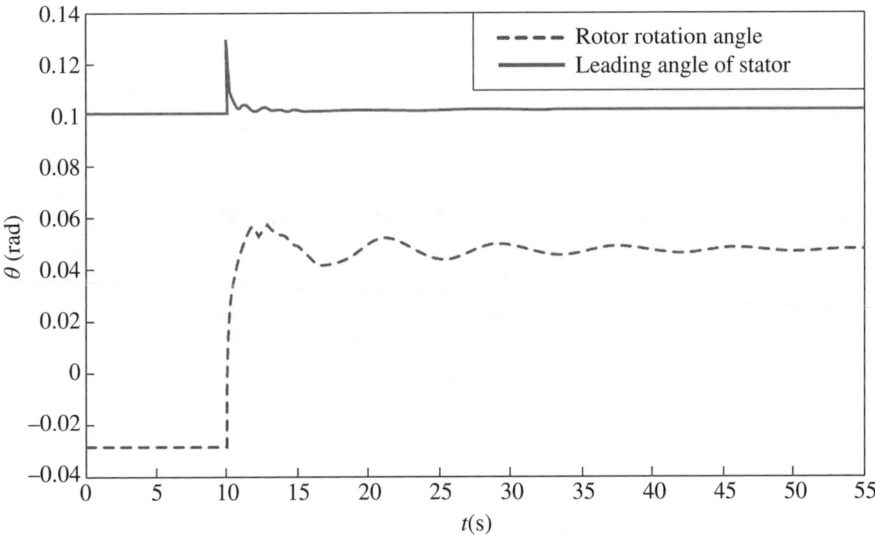

Figure 5.27 Response of angle difference between VFT stator winding and rotor winding.

we are going to use the control block diagram shown in Figure 4.17 to study the effect of using VFTs to suppress the system power low-frequency oscillation. In the simulation we use the simplified electrical power system shown in Figure 5.28 and the VFT is connected to a branch between Node B2 and Node B3. Figures 5.29 and 5.30 show the curves of generator power angle using VFTs and without using VFTs, and curves of generator power angle using a damping controller and without using a damping controller.

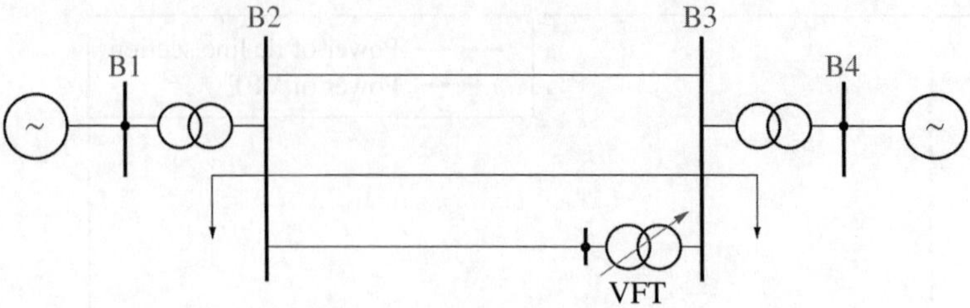

Figure 5.28 Schematic diagram of simplified system wiring.

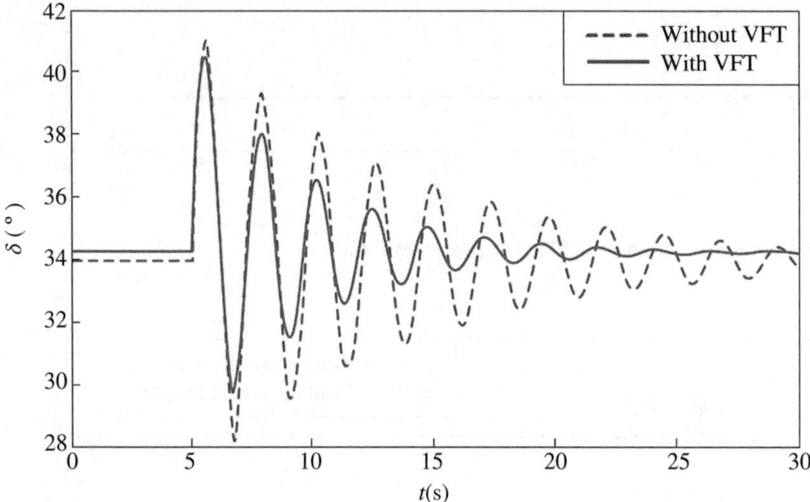

Figure 5.29 Curves of generator power angle using VFTs and without using VFTs (without an additional damping controller).

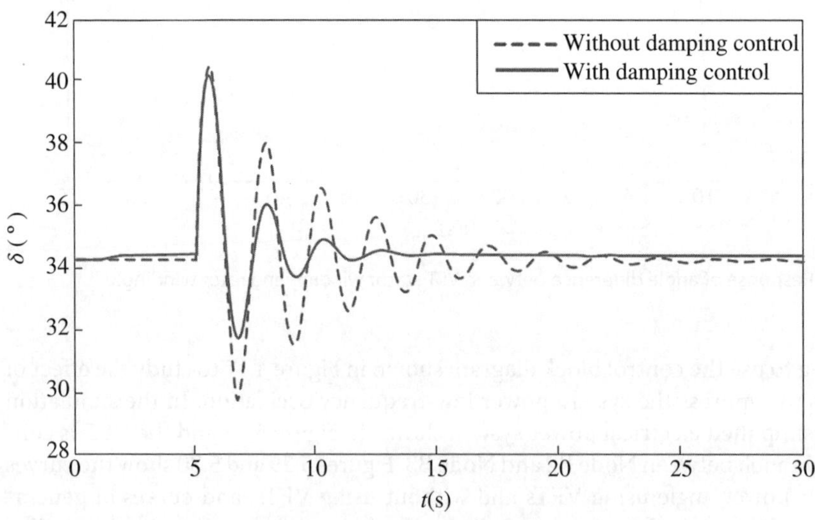

Figure 5.30 Curves of generator power angle with and without an additional damping control.

The general conclusion is that when VFTs are used, oscillation waveform attenuation accelerates while when the damping controller is used, and the power angle attenuation accelerates.

5.10 Summary

In this chapter, based on China's power grid situation and domestic as well as foreign experience in research into electrical power systems, we designed and proposed the basic parameters of a VFT and optimized and selected the simplified interconnected system, four-generator system, and large complex system for research into the operating characteristics and control functions of VFT. Based on this, we established the corresponding electrical power system digital simulation models using programs PSASP [7], EMTP [8], and PSCAD/EMTDC [9] to match with the previously established VFT ontology model, DC drive model, and control system model. The operating characteristics of VFT in different working conditions as well as faulty conditions were studied. From these, we make the following conclusions:

1. A VFT has very good power regulation characteristics; it is capable of smooth regulation and control within the design capacity range; with no minimum transmission power limit, it can realize bidirectional transmission of power.
2. In transmitting power, a VFT will absorb some reactive power. The amplitude of reactive power is equivalent to 30–40% of the active power, which has a certain impact on voltage stability in a weak system, but the reactive power balance of the system can be ensured through the use of capacitor banks and low-voltage limit power control.
3. A VFT is a physical system with a lot of inertia and a strong ability to resist faults and disturbance. Through reasonable design of the control system and parameters, it has sound recovery characteristics following any faults.
4. A VFT can supply power to weak systems and passive systems, which reflects the fact that it has strong system adaptability. Compared with the back-to-back DC system, the VFT has obvious advantages in this regard.
5. A VFT can be used to make the two interconnected power grids become frequency regulation power supplies for each other, so as to realize overall utilization of power generation capacity in the interconnected system and improve the system security and stability.
6. The research into the power flow and transient simulation of a VFT in large complex system further illustrates the adaptability of VFTs in large power systems.
7. As a series flexible AC transmission system device, the VFT has unique advantages in suppressing system power oscillation.
8. Through simulation and modeling using PSASP, EMTP, and PSCAD/EMTDC, and with systems of different scales as the research objects, we realize various control functions of VFTs and prove the technical advantages of the VFT when it is used for power grid interconnection. Meanwhile, we verify the correctness and effectiveness of VFT models and control systems established using PSASP, EMTP, and PSCAD/EMTD.

References

1 E. Larsen, R. Piwko, D. McLaren, D. McNabb, M. Granger, M. Dusseault, et al. Variable-frequency transformer – A new alternative for asynchronous power transfer, presented at *Canada Power, Toronto, Ontario, Canada*, 2004.

2 C. Gesong, Z. Xiaoxin. Digital simulation of VFTs for asynchronous interconnection in power systems. *IEEE/PES Transmission and Distribution Conference & Exhibition: Asia and Pacific Proceedings, Dalian,* China, 2005, p. 8.

3 R. Piwko, E. Larsenm C. Wegner. VFT: The new technology for asynchronous networking transmission power. *Southern Power System Technology*, 2006, 2(1): 29–34.

4 C. Gesong. *Modeling and Control of VFT for in Power Systems.* D. Phil Thesis, Beijing: China Electric Power Research Institute, 2010.

5 P. Kundur. *Power System Stability and Control.* Z. Xiaoxin and S. Yonghua. (trans.). Beijing: China Electric Power Press, 2001.

6 C. Gesong, Y. Yonghua, T. Yong. Studies on resonance phenomena caused by energization of transformers, *PowerCon 2002, Kunmin,* China, 2002.

7 China Electric Power Research Institute. Manual of power system analysis soft program (PSASP), Beijing: China Electric Power Research Institute.

8 China Electric Power Research Institute. Manual of power system electromagnetic transient program (EMTP), Beijing: China Electric Power Research Institute.

9 Manitoba HVDC Research Center. PSCAD/EMTDC User Guide, Manitoba: Manitoba HVDC Research Center.

6

Design of an Adaptive Low-Frequency Oscillation Damping Controller Based on a VFT

6.1 Overview

The interconnection and intelligence of large grids is an inevitable tendency in modern grid development. Low-frequency oscillation is an important factor influencing the safety of large grids and curbing the conveying capacities of interconnecting ties, so it is urgent to solve the low-frequency oscillations of interconnected grids by means of innovation of interconnection technologies. This chapter first introduces the impacts of low-frequency oscillations for power systems and analyzes as well as compares the operating principles of low-frequency oscillation suppression technologies such as PSS, DC modulation, FACTS, and variable-frequency transformers (VFT). And then introduces the Prony method and its way to identify power system oscillation modes as well as main parameter design. The basic method for damping controller parameter calculation is derived from the VFT's step regulation and system response function. The design idea of an adaptive low-frequency oscillation damping controller based on VFTs and the Prony method is formed. The method is finally simulated and verified with the help of a typical four-generator system and interconnected power system covering Northeast China, North China, and Central China. The result shows that the damping controller can adapt to the changes of power systems' condition, regulate, and optimize damping controller parameters automatically, thus suppressing the variable-frequency oscillations of power systems better with help of VFTs.

6.2 Impacts of the Variable-Frequency Oscillations of Power Systems and Corresponding Control Actions

With the construction of trans-regional grid interconnection projects, the dynamic stability of interconnected power systems and the impact of the interconnection of large grids on it are attracting increasing attention. Objectively, interconnected power systems usually have a regional oscillation mode with a usual oscillation frequency of 0.5 Hz to 2.0 Hz and an inter-regional oscillation mode with a usual oscillation frequency of 0.1 Hz to 1.0 Hz [1–3].

In the 1960s, the power systems in North America had low-frequency oscillations. Since the 1980s, with the increase of proportion of the large-unit quick-response

Variable Frequency Transformers for Large Scale Power Systems Interconnection: Theory and Applications, First Edition. Gesong Chen, Xiaoxin Zhou, and Rui Chen.

excitation systems in the grids in China and the extensive interconnection of grids, some power systems in China such as the Hunan power system (in 1983), the interconnecting ties of the interconnected power system connecting Guangdong and Hong Kong (in 1984), the interconnected power system connecting Guangxi, Guangdong, and Hong Kong (1985), the China Southern Power Grid interconnected power system (1994), the Ertan power transmission system in Sichuan and the Chongqing Power Grid (1998 and 2000), and the AC/DC transmission system in the China Southern Power Grid and Hong Kong Grid (in 2003) also had low-frequency oscillations, to some extent [2–5].

Research results show that the low-frequency oscillations of an interconnected power system usually depend on stability of the corresponding inter-regional oscillation mode. In certain operating conditions of a power system, the damping generated on the basis of a heavy load, a long transmission distance and an automatic voltage regulator is equivalent to or larger than the positive damping of the power system. In this case, it is possible that there are low-frequency oscillations with slow damping or even an increase and that the interconnected grid is disconnected when the power system is disturbed. Thus, improvement of dynamic stability of the inter-regional oscillation mode is the top problem for solving in the research into interconnected power system projects [6, 7].

According to years of studies and practices, the key to suppressing the low-frequency oscillations of a power system is to increase damping through the following actions [8–12]:

1. Use PSSs in a generator excitation system: This is now an action extensively used in power systems. In the grid interconnection between North China and Northeast China and that between Central China and North China, PSSs have been used for numerous units of power systems and their parameters have been optimized. However, the application of a PSS is limited: on the one hand, inter-regional low-frequency oscillation is a common result of multiple units, making the mutual matching of PSS parameters and taking different wiring methods of power system into consideration is a basis for effective suppression of low-frequency oscillations. Numerous parameter data are necessary in the design of a large power system, making PSS design more difficult. On the other hand, the PSS is decentralized at different power points, so its effect on low-frequency oscillations depends a lot on the structure of the power system, making it necessary to consider comprehensively when designing related units of the power system. However, in practical operation, the PSSs of some units usually quit operation due to different factors, which increases the probability of low-frequency oscillations in power system.

2. Use the DC modulation function: For a DC asynchronous interconnected power system or a DC/AC parallel and synchronous interconnected power system, a low-frequency oscillation damping controller may be added to its DC control loop to increase the damping of the power system in the low-frequency oscillation mode to suppress low-frequency oscillations.

3. Use FACTS devices such as controllable series compensators and static reactive compensators: Take Tianguang AC/DC parallel transmission system as an example. Its low-frequency oscillations can be suppressed with help of controllable series compensators and corresponding controllers. Usually, when a FACTS is used for

low-frequency oscillation suppression, items such as line active power and regional inertia central angle are usually used as additional control input signals.

4. Make use of the power regulation function of VFTs: VFT is a new choice for low-frequency oscillation suppression, working on a principle similar to that of DC modulation: Control the power transmission in interconnecting ties or sections to increase the low-frequency damping of the power system to suppress low-frequency oscillations. The PSSs for low-frequency oscillation suppression are decentralized at different power points. VFTs mainly control a section, are free from the impact of units disconnection or structural changes of power system, and have specific control objects and objectives, becoming a better solution for the low-frequency oscillations among large power systems. In the design of a low-frequency oscillation controller based on VFTs, active power of the section where the VFT located is used as an additional input signal; with oscillations of the active power, input, and output of the low-frequency oscillation controller are the increment of oscillations of the active power (ΔP) and the given transmission power increment $\Delta Pref$ of the VFT, respectively. The given transmission power, P'_{ref}, necessary for low-frequency oscillation suppression is the sum of $\Delta Pref$ and the given power P_{ref} of the VFT during normal transmission, thus compensating the increment of active power oscillations (ΔP) and suppressing transmission power oscillations of the interconnecting ties. See Figure 4.18 for the control loop of the VFT's low-frequency oscillation controller.

The design of a damping controller depends a lot on the corresponding power system. In the conventional parameter design of a damping controller, it is usually necessary to use a power characteristic value analysis method and know structure of the power system and parameters of the main parts, resulting in numerous data and calculation. Furthermore, structure and operation mode of the power system often changes. How to design more reasonable damping controller parameters in compliance with practical dynamic changes of the power system is now the emphasis and difficulty in damping controller design. Section 6.3 carries out dynamic tracking and analysis for small disturbance-based power system response characteristics with help of the power regulation function of VFTs and the Prony method, and then proposes low-frequency oscillation damping controller parameters that are compliant with structures and operating conditions of power systems.

6.3 Prony Method-Based Transfer Function Identification

The Prony method is a signal processing method for fitting uniformly spaced sampling data by means of the linear combination of an exponential function. It can be used for analyzing such characteristics of a measured signal as frequency, attenuation factor, amplitude, and phase. Compared with the conventional characteristic value analysis methods, the Prony method is a time domain calculation method for modal parameter identification and needs no calculation of the characteristic values of large power systems. Its system model order number is depending on the object and need of identification, and has a high identification accuracy for the dominant oscillation frequencies of power systems. It can be used for transfer function identification, which is important for the design, parameter regulation, control signal selection, and system model verification of controllers. Thus, it can be practically used in power systems [13–15].

For a single-input single-output linear power system, suppose that its transfer function has the following form:

$$G(s) = \sum_{i=1}^{n} R_i/(s - \lambda_i)$$

where R_i is the transfer function's residue at pole λ_i.

Suppose that $I(s)$ and $Y(s)$ are the input signal and the output signal, respectively; there is the equation

$$Y(s) = G(s)I(s)$$

Suppose that the input signal is the superposition result of some delayed signals, $I(s)$ can be calculated through the following equation.

$$I(s) = \frac{c_0 + c_1 e^{-sD_1} + c_2 e^{-sD_2} + \cdots + c_k e^{-sD_k}}{s - \lambda_{n+1}} \tag{6.1}$$

In Equation (6.1):

$$D_i < D_{i+1} (i = 1,2, \cdots, k - 1)$$

can be used for calculating output response of the power system as here:

$$Y(s) = (c_0 + c_1 e^{-sD_1} + c_2 e^{-sD_2} + \cdots + c_k e^{-sD_k}) \left\{ \sum_{i=1}^{n} \frac{R_i}{(s - \lambda_{n+1})(s - \lambda_i)} \right\} \tag{6.2}$$

The following equation can be reached by means of factorization.

$$Y(s) = (c_0 + c_1 e^{-sD_1} + c_2 e^{-sD_2} + \cdots + c_k e^{-sD_k}) \left\{ \frac{Q_{n+1}}{s - \lambda_{n+1}} + \sum_{i=1}^{n} \frac{Q_i}{s - \lambda_i} \right\} \tag{6.3}$$

In Equation (6.3):

$$Q_i = \frac{R_i}{\lambda_i - \lambda_{n+1}} \quad i = 1,2, \cdots, n \tag{6.4}$$

$$Q_{n+1} = -\sum_{i=1}^{n} Q_i \tag{6.5}$$

The output time domain response can be calculated as (6.6) by means of Inverse Laplace Transform.

$$
\begin{aligned}
y(t) = c_0 &\left[Q_{n+1} e^{\lambda_{n+1} t} + \sum_{i=1}^{n} Q_i e^{\lambda_i t} \right] u(t) \\
+ c_1 &\left[Q_{n+1} e^{\lambda_{n+1}(t - D_1)} + \sum_{i=1}^{n} Q_i e^{\lambda_i(t - D_1)} \right] u(t - D_1) \\
+ \cdots + c_k &\left[Q_{n+1} e^{\lambda_{n+1}(t - D_k)} + \sum_{i=1}^{n} Q_i e^{\lambda_i(t - D_k)} \right] u(t - D_k)
\end{aligned} \tag{6.6}
$$

When t is equal to or larger than D_k, Equation (6.6) can be changed into:

$$y(t) = Q_1 \left(\sum_{i=0}^{k} c_i e^{-\lambda_1 D_i} \right) e^{\lambda_1 t}$$

$$+ Q_2 \left(\sum_{i=0}^{k} c_i e^{-\lambda_2 D_i} \right) e^{\lambda_2 t} + \cdots + Q_{n+1} \left(\sum_{i=0}^{k} c_i e^{-\lambda_{n+1} D_i} \right) e^{\lambda_{n+1} t} \qquad (6.7)$$

In Equation (6.7), D_0 is zero.
For easy analysis, suppose:

$$\tau = t - D_k \geq 0, \quad v(\tau) = y(t + D_k),$$

Then:

$$v(\tau) = Q_1 \left(\sum_{i=0}^{k} c_i e^{\lambda_1 (D_k - D_i)} \right) e^{\lambda_1 \tau}$$

$$+ Q_2 \left(\sum_{i=0}^{k} c_i e^{\lambda_2 (D_k - D_i)} \right) e^{\lambda_2 \tau} + \cdots + Q_{n+1} \left(\sum_{i=0}^{k} c_i e^{\lambda_{n+1} (D_k - D_i)} \right) e^{\lambda_{n+1} \tau}$$

$$= \sum_{j=1}^{n+1} B_j e^{\lambda_j \tau} \qquad (6.8)$$

In Equation (6.8):

$$B_j = Q_j \left(\sum_{i=0}^{k} c_i e^{\lambda_j (D_k - D_i)} \right) \quad j = 1, 2, \cdots, n+1 \qquad (6.9)$$

Based on Equations (6.4) and (6.9), the following equation can be reached.

$$R_j = \frac{B_j (\lambda_j - \lambda_{n+1})}{\displaystyle\sum_{i=0}^{k} c_i e^{\lambda_j (D_k - D_i)}} \quad j = 1, 2, \cdots, n \qquad (6.10)$$

Suppose that the input signal parameters are given, the transfer function of the power system can then be obtained with the Prony method. The Prony method is mainly used for identifying the low-oscillation modes of power systems, so the pole frequency of the transfer function with given input signal and output signal is a low-frequency and the input signal in Equation (6.1) can be properly simplified to make the calculation easier. For example, suppose that λ_{n+1} is zero, the input signal will be the superposition result of some delayed signals. Usually, a rectangular pulse signal is used as the input signal. For example, if the input parameter is a rectangular pulse signal lasting for 0.1 s and with a certain amplitude, the input signal parameters will be as follows:

$$c_0 = 1, c_1 = -1, D_1 = 0.1, k = 1, \lambda_{n+1} = 0$$

The power system response has both a forced component and free component between 0 and 0.1 s with a free component response after 0.1 s. The related parameters of the free component after 0.1 s can be identified using the Prony method.

6.4 Low-Frequency Oscillation Damping Controller Design with VFTs and a Prony Method

The low-frequency oscillations of a power system are long-time active power low-frequency oscillations between the power system and its generator units, or between different regions of power system because of insufficient damping torques of the generator units caused by quick excitation, long-distance AC power transmission, and so on. Thus, using the active power signal of the VFT's section as input signal for the damping controller can realize timely and effective capture of low-frequency oscillations and improve effectiveness and robustness of the low-frequency oscillation damping control.

Therefore, the most important thing in damping controller design is to find a proper input disturbance signal and its response characteristics. As previously mentioned, a transfer function can be easily derived from the response characteristics of a square-wave pulse. VFTs have good regulation performance including step control. Thus, add an approximate square-wave disturbance to the power system and measure the power response curve of the section concerned with help of step power control function of the VFT. A transfer function $G(s)$ including a system oscillation mode and the damping characteristic can be obtained with the Prony method introduced in Section 6.3.

Parameters of the damping controller can be determined through transfer function parameters, and with the root locus pole assignment method and the phase compensation method. The root locus pole assignment method moves the dominant pole to the designated position on plane s by means of feedback element control. For a power system as in Figure 6.1, its open-loop transfer function and feedback element are $G(s)$ and $H(s)$, respectively.

The closed-loop transfer function of a power system as in Figure 6.1 is

$$G_L(s) = G(s)/(1 - G(s)H(s))$$

and the characteristic equation of a closed-loop power system is:

$$G(s)H(s) = 1 \tag{6.11}$$

Suppose that λ_d is an expected closed-loop power system's dominant pole on plane s, then λ_d must meet characteristic equation of the closed-loop power system and the following equation can be reached.

$$G(\lambda_d)H(\lambda_d) = 1 \tag{6.12}$$

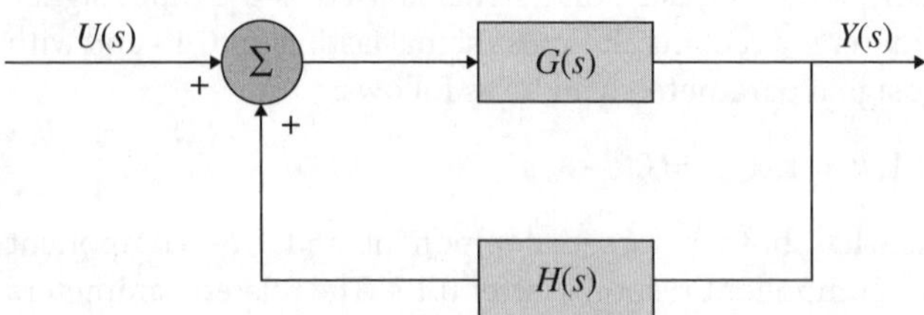

Figure 6.1 Structure of feedback control structure.

Namely, in the feedback element, λ_d meets both the amplitude condition and the phase angle condition as follows.

$$
\begin{cases}
|H(\lambda_d)| = \dfrac{1}{|G(\lambda_d)|} \\[2mm]
angle[H(\lambda_d)] = -angle[G(\lambda_d)]
\end{cases}
\tag{6.13}
$$

According to the given $G(s)$ and Equation (6.13), amplitude, and phase angle of the feedback element at dominant frequency λ_d can be determined. The corresponding parameters can be determined through a control structure similar to the feedback element in Figure 4.18.

What needs to be stressed is that the oscillation mode of a power system depends on its operating mode and structure and that, in particular, the disconnection of key generator units influences the oscillation frequency of a weak power system greatly. Thus, strictly speaking, any change of structure, wiring, or generator units of a power system needs an oscillation frequency change of the power system and corresponding parameter adjustment of the damping controller, otherwise the damping effect will be affected.

Step response is a basic function of VFTs. Thus, for a power system with VFTs, the VFT power step regulation function can help to generate an approximate square-wave disturbance signal for the power system actively and measure power response curve of the power system automatically. Moreover, for easier calculation, the low-frequency oscillation damping controller module can be locked for a short time. Compared with the curve in Figure 6.8, the curve in Figure 6.5, which is divergent, does not have a large amplitude, so it does not influence the power system in a short time and the damping controller module can be recovered quickly. On this basis, we can fit the forward transfer function $G(s)$ of the power system, calculate the dominant oscillation frequency and derive related parameters of the damping controller in compliance with the expected damping ratio and the given control loop of the damping controller (Figure 4.18). This makes automatic parameter adjustment of the damping controller of a VFT easy. See Figure 6.2.

6.5 Application of a VFT-Based Adaptive Damping Controllers in a Four-Generator Power System

6.5.1 System Overview

The four-generator power system in Figure 5.2 has weak or negative damping. Upon its disturbance, the low-frequency oscillations (about 0.5 Hz) between G1 and G2 or between G3 and G4 are increasing oscillations. For higher damping of the low-frequency oscillations of the power system, a circuit connecting VFTs in series is added between buses B6 and B7 as in Figure 6.3. It is planned to improve the power system's low-frequency oscillation damping and suppress its low-frequency oscillations by means of an additional damping controller of the VFTs.

6.5.2 Transfer Function Identification

The feedback element in Figure 6.1 is a damping controller. Its input signal is active power of the parallel AC circuit of the section where the VFTs are, and its output is

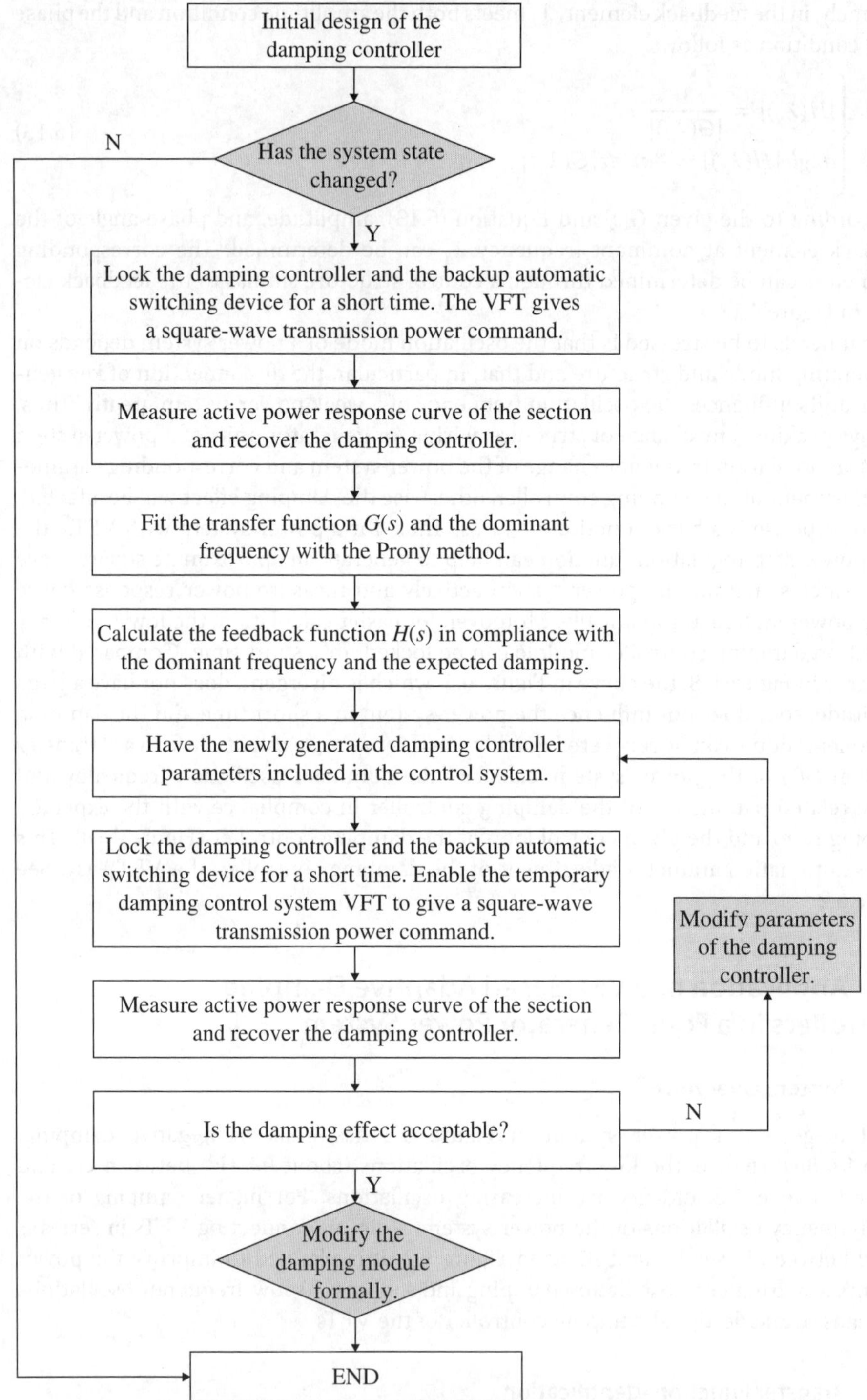

Figure 6.2 Design idea of the adaptive low-frequency oscillation damping controller based on a VFT and the Prony method.

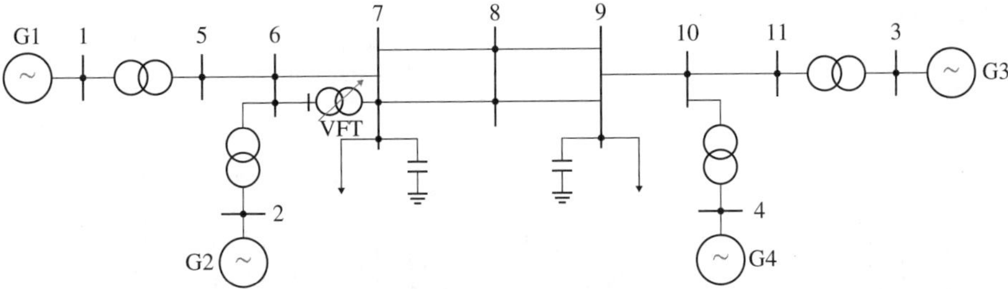

Figure 6.3 Access of typical four-generator system and a VFT.

Figure 6.4 Input signal.

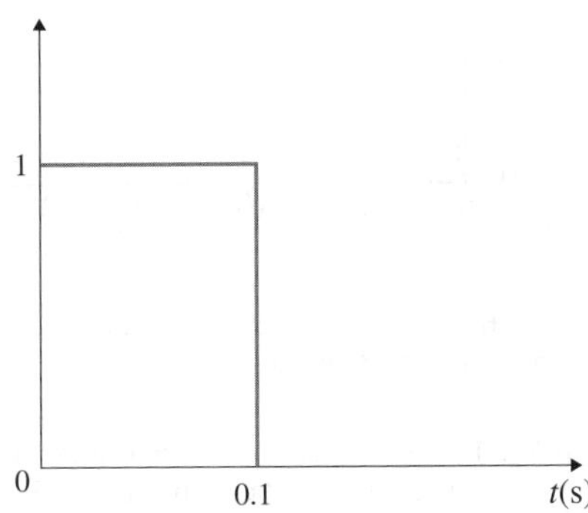

additional power reference value of the VFTs. So that the forward transfer function *G(s)* is the active power response of VFT's power reference signal to the parallel AC circuit. It can be identified with the Prony method following these steps.

1. Generate a rectangular pulse disturbance signal ΔP_{VFTRef} as in Figure 6.4 for the power reference value of the VFT. The amplitude is adjustable and the simulation lasts for 10–30 s. Measure increment ΔP_L of active power of the sections of buses 6 and 7.
2. Analyze increment ΔP_L with the Prony method, list all the oscillation modes and select the oscillation mode with weak or negative damping with a larger residue as the main oscillation mode of the power system.
3. Calculate the open-loop function including active power of the section and power change of the VFTs with the previous method and determine parameters of the damping controller through pole assignment and phase compensation.
   During disturbance by VFTs, 0.05~0.1 pu of rated power of the VFTs can be adopted to comply with the adjustable range of the VFT and avoid significant impact on the power system. The response power is possibly divergent, but the amplitude is too small to endanger the power system in a short time. See Figure 6.5. Furthermore, for the sake of safety, the low-frequency oscillation damping controller of the VFTs is locked for a short time while corresponding actions such as real-time monitoring are taken. When there are oscillations of a specified amplitude in the power system, the damping control circuit will be immediately put into operation.

In this research, a system simulation model as in Figure 6.3 is established with help of PSASP and a simulation module for the power flow and electromechanical transient of VFTs is established with help of the user-defining and self-programming functions of

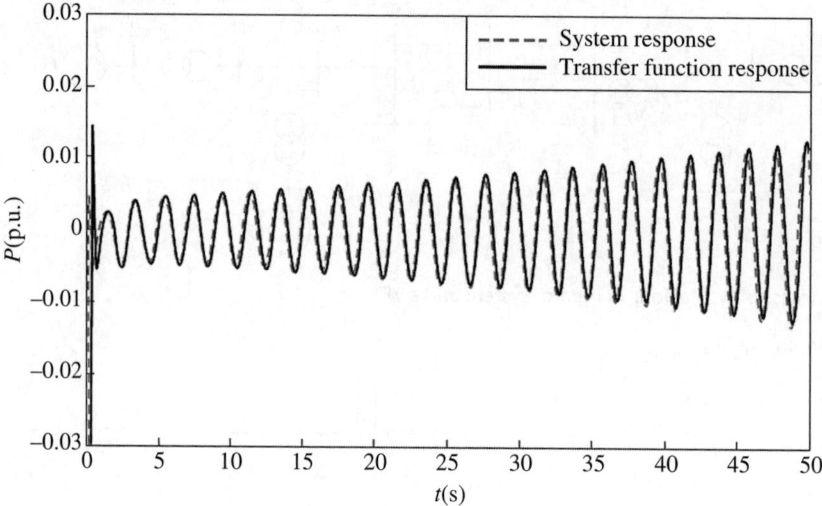

Figure 6.5 Comparison between system simulation data and data about Prony method-based identification and fitting.

Table 6.1 Characteristic root and residue of the transfer function obtained through identification with the Prony method.

Number	Characteristic Root	Residue
1	0.022 2 + 3.114 5i	−0.001 1 + 0.001 8i
2	0.022 2 − 3.114 5i	−0.001 1 − 0.001 8i
3	−0.249 9 + 6.249 3i	0.000 0 + 0.000 3i
4	−0.249 9 − 6.249 3i	0.000 0 − 0.000 3i
5	−0.989 9	−0.006 2
6	−7.882 5 + 8.430 8i	−0.219 4 + 0.184 1i
7	−7.882 5 − 8.430 8i	−0.219 4 − 0.184 1i

PSASP. After disturbance with an amplitude of 0.1 pu and a time width of 0.1 s is generated to the VFTs, a system response curve that is the dotted line in Figure 6.5 is obtained. Calculate the oscillation mode and the damping coefficient both in Table 6.1 and the open-loop transfer function in Equation (6.14) with the method before. Figure 6.5 shows the fitting effect.

According to Figure 6.5, upon disturbance of the power system as in Figure 6.3, active power of the AC interconnecting ties tends to have increasing oscillations, which endangers safety of the power system and needs necessary low-frequency oscillation suppression actions. According to Tables 6.1 and 6.2, the unstable oscillation mode is about 0.5 Hz.

$$G(s) = \frac{-4.214s^6 - 102.85s^5 - 359.41s^4 - 5008.5s^3 - 7146s^2 - 39316s - 39803}{s^7 + 17.21s^6 + 205.24s^5 + 1020.3s^4 + 7753.2s^3 + 13255s^2 + 56879s + 50037}$$

$$(6.14)$$

Table 6.2 Results of analysis using the Prony method.

Number	Amplitude	Damping Ratio	Frequency (Hz)
1	0.004	−0.007	0.496
2	0.004	−0.007	0.496
3	0.000 47	0.05	1.002
4	0.000 47	0.05	1.002
5	0.008 2	0.81	0.0
6	0.52	7.8	1.342
7	0.52	7.8	1.342

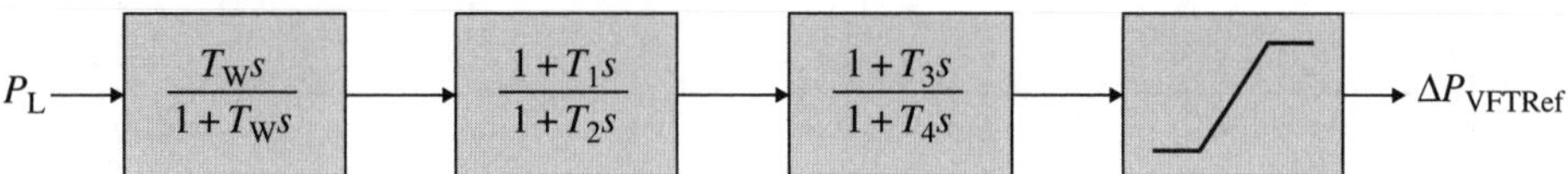

Figure 6.6 Structure of the damping controller.

6.5.3 Design of the Damping Controller

As in Figure 6.6, the damping controller has a structure similar to a PSS, which includes a DC blocking element and a lead/lag phase compensation element.

According to the dominant pole $\lambda_d = 0.022\ 2 \pm j3.114\ 5$, the pole configuration in Equation (6.13) and the phase compensation method, parameters of the damping controller can be determined as:

$$T_w = 5.0, T_1 = 0.137, T_2 = 0.1, T_3 = 0.137, T_4 = 0.1$$

6.5.4 Application Effect of the Damping Controller

For a power system such as in Figure 6.3, mount a VFT model between buses B6 and B7. Figure 6.7 shows structure of the VFTs. As in Figure 6.6, input signal of the damping controller is active power P_{67} of the parallel AC circuit.

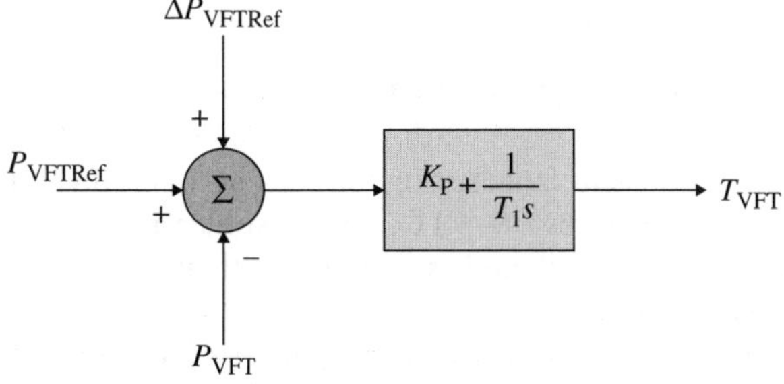

Figure 6.7 Controller structure of VFT.

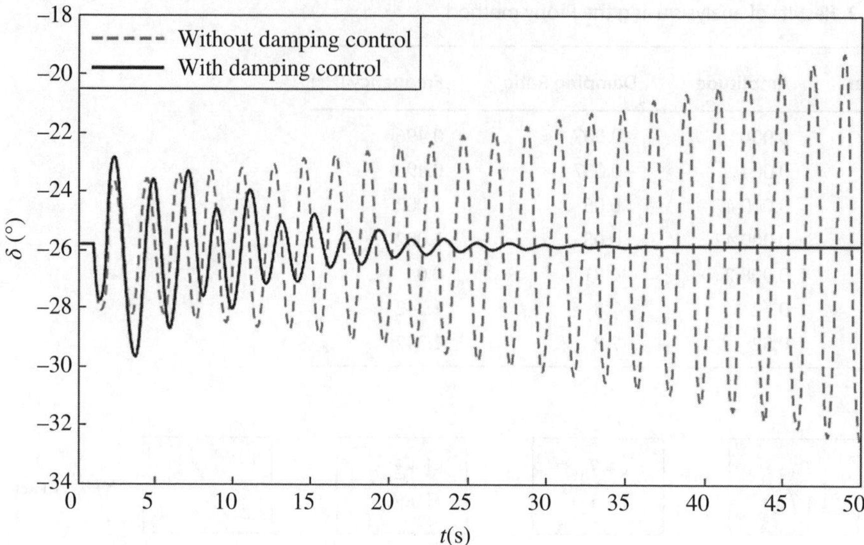

Figure 6.8 Comparison of simulation curves of power angles of generators G3 to G1.

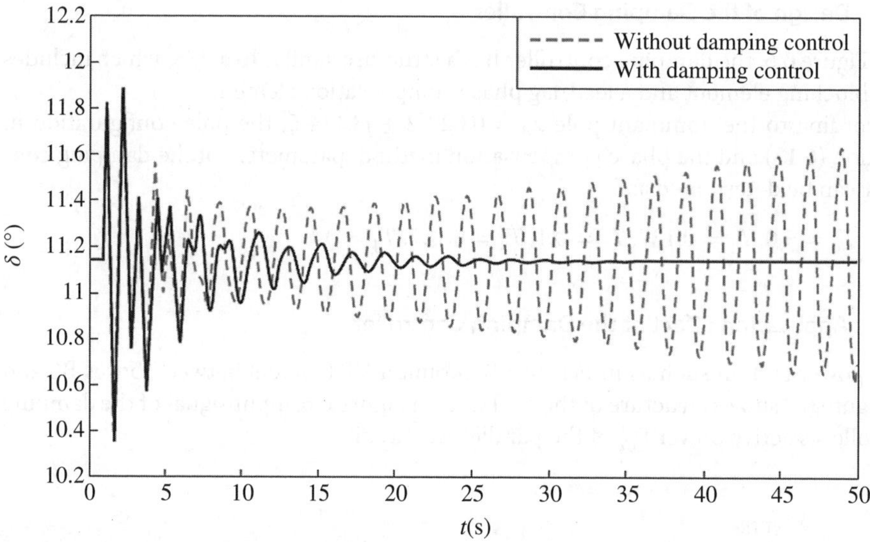

Figure 6.9 Comparison of simulation curves of power angles of generators G3 to G4.

Suppose that there is a three-phase transient fault lasting for 0.1 s between B8 and B9 of the AC circuit of the power system. Figures 6.8–6.11 compare simulation results when there is no damping controller and when there is a situation with an additional damping controller for the VFTs, respectively.

Table 6.3 indicates damping conditions in the low-frequency oscillation mode in the case of existence or absence of an additional damping controller, respectively. Low-frequency oscillation damping of the power system increases with help of the

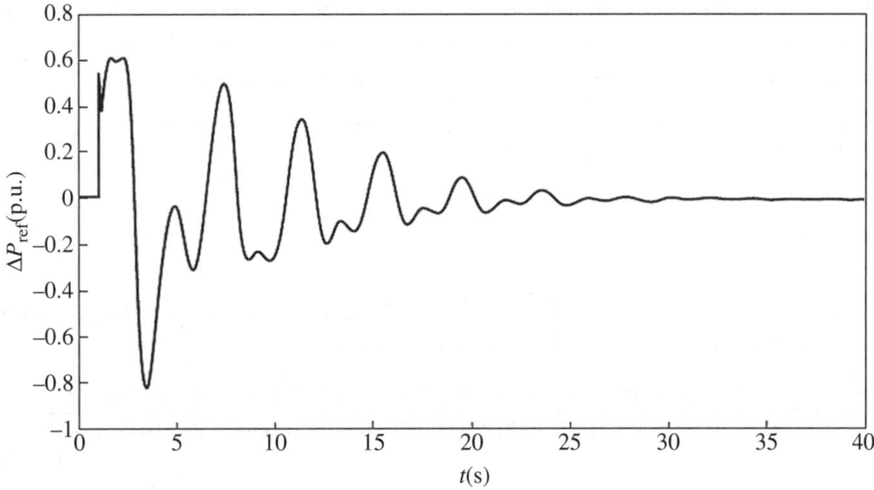

Figure 6.10 Output of the damping controller.

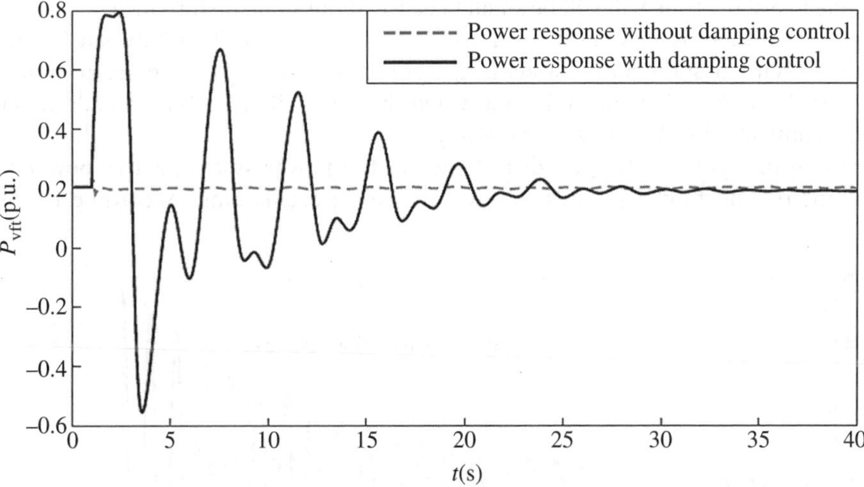

Figure 6.11 Comparison of active power response of the VFTs.

additional damping controller of the VFTs, ensuring significant suppression for both the inter-regional oscillation mode and the intra-regional oscillation mode. The waveforms in Figures 6.8–6.11 suggest a good damping effect of the damping controller too.

6.5.5 Parameter Design and Damping Effect of the Damping Controller After a Structural Change of the Power System

After the disconnection of generator G3 in a typical four-generator power system, oscillation mode of the power system will change. Thus, it is necessary to calculate the new

Table 6.3 Comparison of effects of existence or absence of a damping controller.

Without a Damping Controller		With a Damping Controller	
Damping ratio	Frequency (Hz)	Damping ratio	Frequency (Hz)
−0.007	0.495	0.05	0.486
0.05	1.001	0.10	1.07

parameters of the damping controller with the method in Section 6.5.4. The open-loop transfer function can be obtained through Equation (6.15).

$$G(s) = \frac{3.382s^6 + 10.69s^5 + 182.9s^4 + 502.2s^3 + 2119s^2 + 5837s - 262.5}{s^7 + 20.93s^6 + 121.1s^5 + 1120s^4 + 3810s^3 + 13060s^2 + 36210s - 54.34}$$

$$(6.15)$$

According to the comparison in Figure 6.12, the system simulation data and the data about Prony method-based identification and fitting are consistent. Obviously, the generator power angle of the power system has increasing oscillations. The oscillation amplitude in a short time will still be small because of the small disturbance.

Suppose that there is a three-phase transient fault lasting for 0.1 s between B8 and B9 of the AC circuit of the power system. Figures 6.12–6.15 compare simulation results when there is no damping controller and when there is a situation with an additional damping controller for the VFTs, respectively.

According to Figure 6.14, upon disturbance of the power system, active power of the AC interconnecting ties tends to have increasing oscillations. According to the

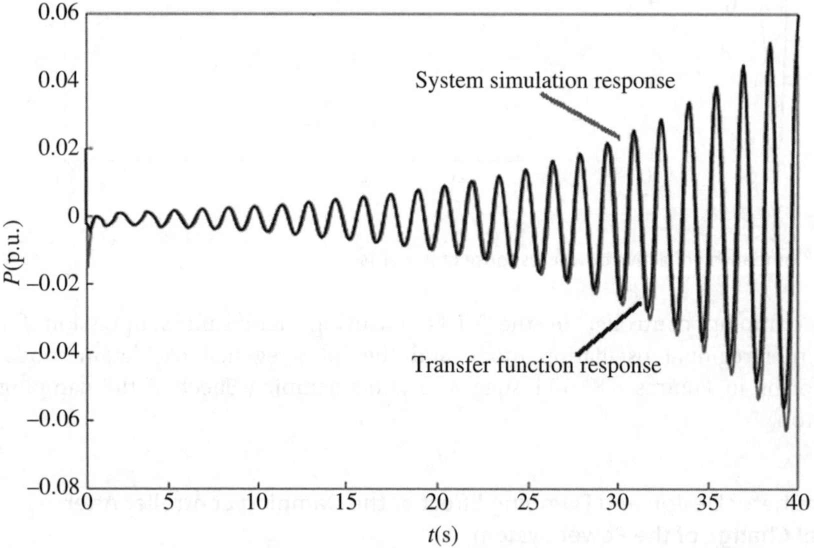

Figure 6.12 Comparison between system simulation data and data about Prony method-based identification and fitting.

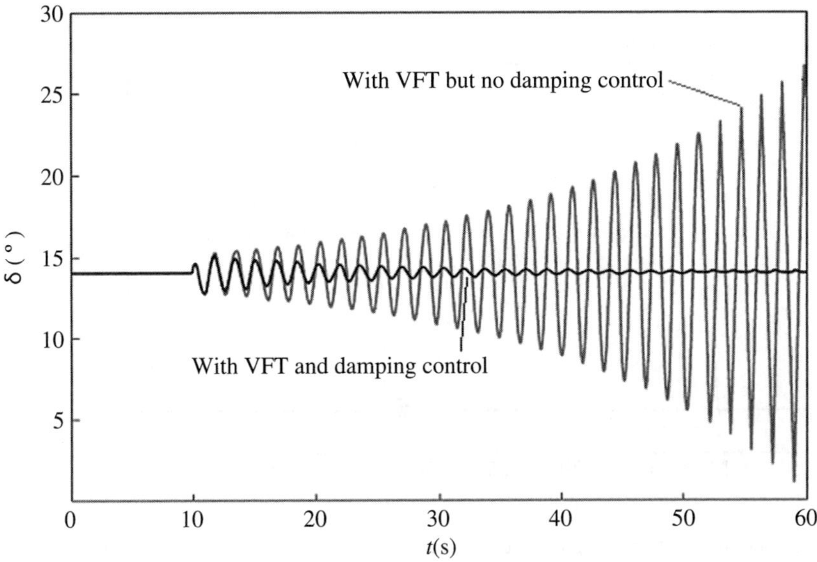

Figure 6.13 Comparison of simulation curves of power angles of generators G1–G4.

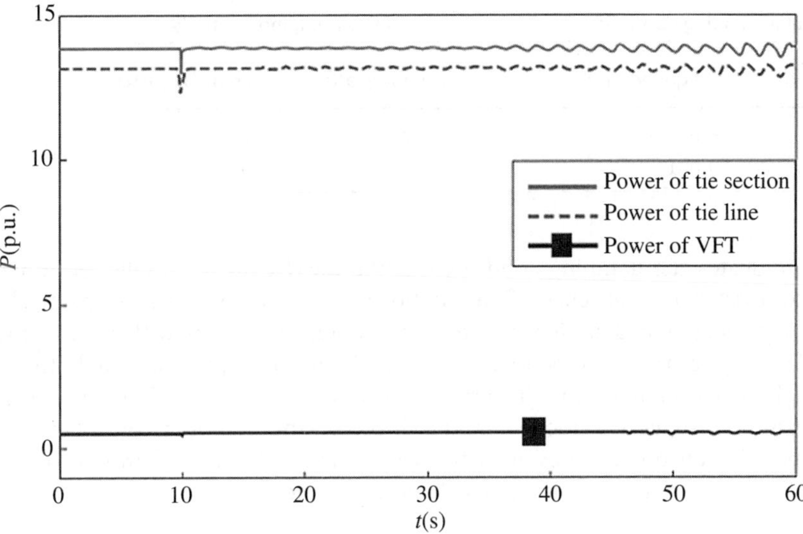

Figure 6.14 Power of the section, the parallel circuit, and the VFTs when these have no damping controller.

dominant pole $\lambda_d = 0.092 \pm j4.17$, and the pole configuration and the phase compensation method previously, parameters of the damping controller can be determined as:

$$T_w = 1.04, T_1 = 0.18, T_2 = 0.1, T_3 = 0.2, T_4 = 0.1$$

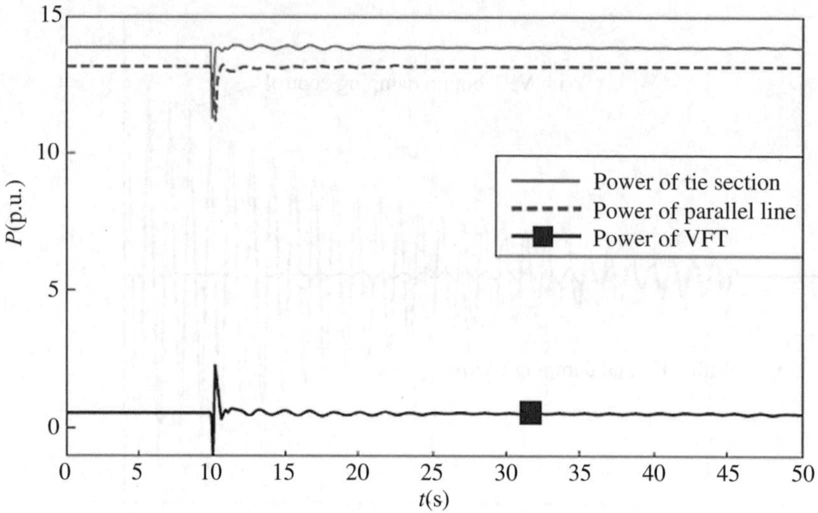

Figure 6.15 Power of the section, the parallel circuit, and the VFTs when these have a damping controller.

Table 6.4 Comparison of effects of existence or absence of a damping controller.

Without a Damping Controller		With a Damping Controller	
Damping ratio	Frequency (Hz)	Damping ratio	Frequency (Hz)
−0.049	0.58	0.072	0.59
0.27	1.001	0.39	0.94

Table 6.4 indicates the damping conditions in the low-frequency oscillation mode in the case of existence or absence of an additional damping controller, respectively. Low-frequency oscillation damping of the power system increases with help of the additional damping controller of the VFTs, ensuring significant suppression for both the inter-regional oscillation mode and the intra-regional oscillation mode. The waveforms in Figures 6.8–6.15 suggest a good damping effect by the damping controller too. Besides, although transmission power of the section is much larger than that of the VFTs, the VFTs suppress low-frequency oscillations significantly.

6.5.6 Adaptability of Power System Mode Identification with the Prony Method

Some people worry about effectiveness of identification using the Prony method because of such phenomena in the system mode being identified as disturbance (such as load application). Thus, simulation research has been done into the impact of disturbance. The simulation research results are as follows. Figure 6.16 shows the active

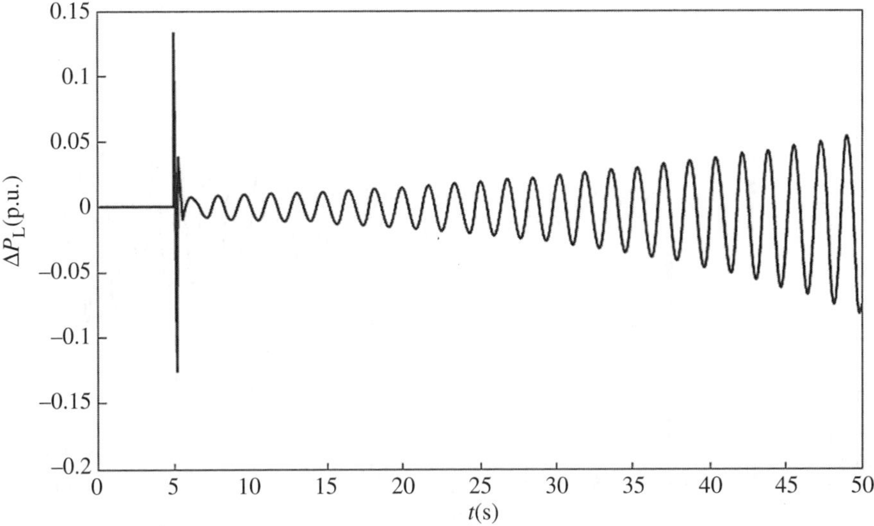

Figure 6.16 Active power increment response of the circuit when the power system has no disturbance.

Table 6.5 Characteristic root and residue of the transfer function identified with the Prony method when there is no sudden disturbance.

Number	Characteristic Root	Residue
1	0.049 5 + 3.668 4i	−0.003 5 + 0.000 6i
2	0.049 5 − 3.668 4i	−0.003 5 − 0.000 6i
3	−0.364 3 + 5.934 3i	−0.000 2 − 0.000 1i
4	−0.364 3 − 5.934 3i	−0.000 2 + 0.000 1i
5	0.081 6 + 7.337 3i	−0.000 1 + 0.000 2i
6	0.081 6 − 7.337 3i	−0.000 1 − 0.000 2i

power increment response of the circuit when the power system has no disturbance and Table 6.5 shows the characteristic root and residue of the transfer function identified with the Prony method. On this basis, it can be determined that dominant pole, damping ratio, and oscillation frequency of the low-frequency oscillations are 0.049 5±3.668 4i, −0.013 5, and 0.58 Hz, respectively.

Figure 6.17 shows the active power increment response of the circuit when the power system suffers disturbance. In an example of a four-generator power system, an impact load is imposed on Node 9 at the time of 10 s. Table 6.6 shows the characteristic root and residue of the transfer function identified with the Prony method. On this basis, it can be determined that dominant pole, damping ratio, and oscillation frequency of the low-frequency oscillation are 0.045 2 ± 3.657 0i, −0.012 4, and 0.58 Hz, respectively.

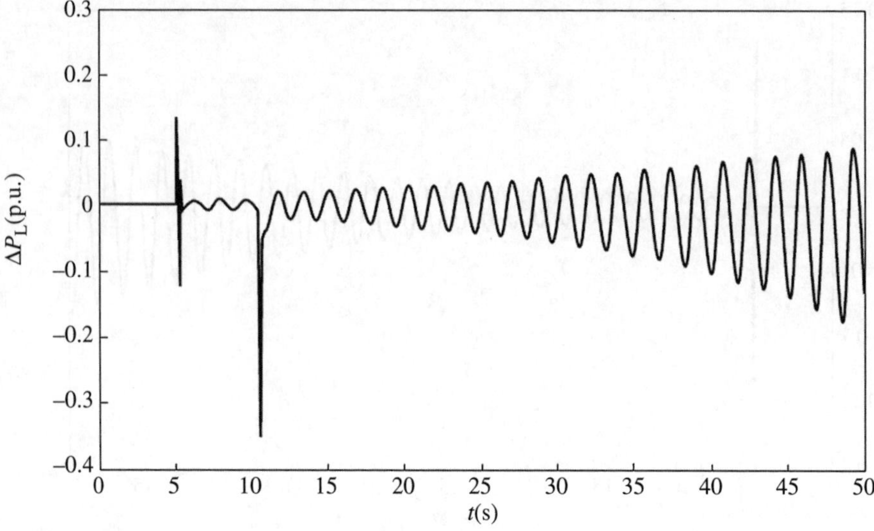

Figure 6.17 Active power increment response of the circuit when the power system suffers disturbance.

Table 6.6 Characteristic root and residue of the transfer function identified with the Prony method when there is sudden disturbance.

No.	Characteristic Root	Residue
1	−0.547 9 + 1.968 8i	−0.000 4 + 0.000 1i
2	−0.547 9 − 1.968 8i	−0.000 4 − 0.000 1i
3	0.045 2 + 3.657 0i	−0.003 5 − 0.001 9i
4	0.045 2 − 3.657 0i	−0.003 5 + 0.001 9i
5	−0.987 2 + 6.662 4i	−0.001 7 − 0.001 0i
6	−0.987 2 − 6.662 4i	−0.001 7 + 0.001 0i

It can be seen that the Prony method-based analysis results in the two circumstances that are very similar; it suggests that the Prony method can be immunized from a disturbance well.

6.6 Application of VFT-Based Adaptive Damping Controllers in Complicated Power Systems

6.6.1 Power System Overview

In this research, the interconnected power system covering Northeast China, North China, and Central China, shown in Figure 4.3, is adopted and a transmission line with serially connected VFTs is arranged between Gaoling Substation and Jiangjiaying Substation. The control objective is the active power change between these two substations. In the power system interconnection among Northeast China, North China, and Central China, main generation units of the original power systems are all provided with PSSs to suppress problems of the interconnected power system such as low-frequency oscillations. In this way, low-frequency oscillations of the interconnected power system can be solved well. For a good presentation of a VFT's function of low-frequency oscillation suppression, the PSSs of some units in the grids of Central China and Northeast China are temporarily disabled, thus forming a complicated interconnected power system with weak damping.

6.6.2 Transfer Function Identification

Input signal of the damping controller is an active power signal of the Gaoling-Jiangjiaying AC circuit that is in parallel with VFTs. Output of the damping controller is an additional power reference value of the VFTs. This means that the forward transfer function is the active power response of the power reference signal of the VFTs to the parallel AC circuit. The function can be identified with the Prony method.

According to Figure 6.18, upon disturbance of a power system as in Figure 4.3, low-frequency oscillations of the active power of the AC interconnecting ties have several modes such as a local oscillation mode and inter-regional oscillation mode. According to Table 6.7, the inter-regional oscillation mode is about 0.16–0.33 Hz. The component amplitude of 0.33 Hz is larger than the component amplitude of 0.16 Hz. The pole corresponding to this mode can be chosen as the dominant pole. The local oscillation mode is about 0.87–1.01 Hz.

An open-loop transfer function in Equation (6.16) can be obtained in this way. Figure 6.18 shows the result of comparison between system simulation data and data about Prony method-based identification and fitting.

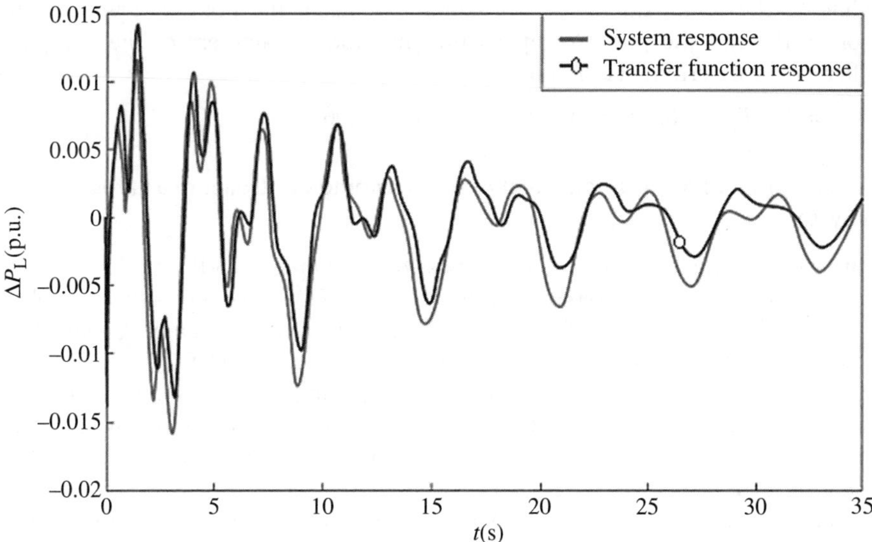

Figure 6.18 Comparison between system simulation data and data about Prony method-based identification and fitting.

Table 6.7 Characteristic root and residue of the transfer function identified with the Prony method in a practical complicated power system.

Number	Characteristic Root	Residue
1	$-0.026\,9 + 1.032\,9i$	$0.001\,7 + 0.001\,0i$
2	$-0.026\,9 - 1.032\,9i$	$0.001\,7 - 0.001\,0i$
3	$-0.088\,7 + 2.060\,0i$	$-0.005\,1 - 0.002\,9i$
4	$-0.088\,7 - 2.060\,0i$	$-0.005\,1 + 0.002\,9i$
5	$-0.164\,9 + 5.514\,2i$	$-0.003\,0 - 0.002\,9i$
6	$-0.164\,9 - 5.514\,2i$	$-0.003\,0 + 0.002\,9i$
7	$-0.292\,6 + 6.593\,9i$	$-0.000\,5 - 0.000\,6i$
8	$-0.292\,6 - 6.593\,9i$	$-0.000\,5 + 0.000\,6i$
9	$-0.328\,9 + 4.328\,5i$	$-0.000\,7 + 0.000\,9i$
10	$-0.328\,9 - 4.328\,5i$	$-0.000\,7 - 0.000\,9i$
11	$-0.676\,7 + 9.812\,5i$	$0.000\,7 - 0.002\,0i$
12	$-0.676\,7 - 9.812\,5i$	$0.000\,7 + 0.002\,0i$

$$G(s) = \frac{-0.178\,7s^{11} + 0.177\,5s^{10} - 29.66s^9 + 26.82s^8 - 1528s^7 + 1524s^6 - 2.965e^4s^5}{s^{12} + 3.157s^{11} + 198.5s^{10} + 437.6s^9 + 1.306e^4s^8 + 1.956e^4s^7 + 3.606e^5s^6 + 3.295e^5s^5}$$

$$\frac{+3.409e^4s^4 - 1.875e^5s^3 + 2.852e^5s^2 + 6.513e^4s + 1.272e^5}{+4.077e^6s^4 + 1.69e^6s^3 + 1.428e^7s^2 + 1.813e^6s + 1.097e^7} \tag{6.16}$$

6.6.3 Design of the Damping Controller

According to the dominant pole $\lambda_d = 0.088\,7 \pm j2.06$, the pole configuration in Equation (6.13) and the phase compensation method, parameters of the damping controller can be determined as follows:

$$T_w = 7.0, \; T_1 = 0.02, \; T_2 = 0.6, \; T_3 = 0.02, \; T_4 = 0.06$$

6.6.4 Simulation of Application Effect of the Damping Controller in a Large Power System

In a power system such as Figure 4.3, it is possible to connect a VFT model between the Gaoling and Jiangjiaying interconnecting tie's buses. Figure 6.6 shows the controller structure of the VFT and Figure 6.5 shows the input signal of the damping controller, which is the sum of active power of the parallel AC circuit and the VFT.

Suppose that there is a three-phase transient fault lasting for 0.1 s on the parallel AC circuit of the power system. Figures 6.19–6.21 compare simulation results when there is no damping controller and when there is a situation with an additional damping controller for the VFT, respectively.

According to the simulation results in Figure 6.19 and Figure 6.21, the application of the VFT has improved damping of the power system greatly, increasing the damping ratio from 0.015 to 0.023 (Table 6.8). The damping controller has increased the damping coefficient to 0.061 and reduced the amplitude greatly, suggesting great low-frequency

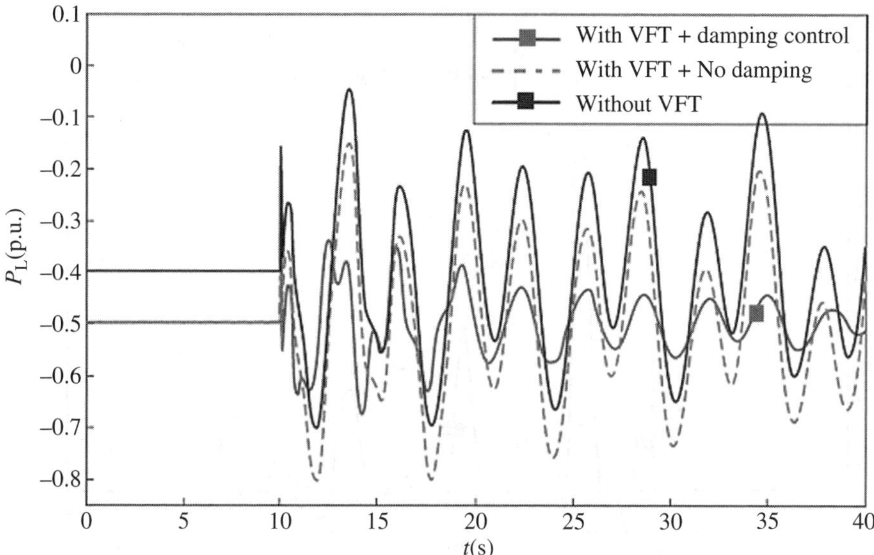

Figure 6.19 Active power waveform between Gaoling and Jiangjiaying.

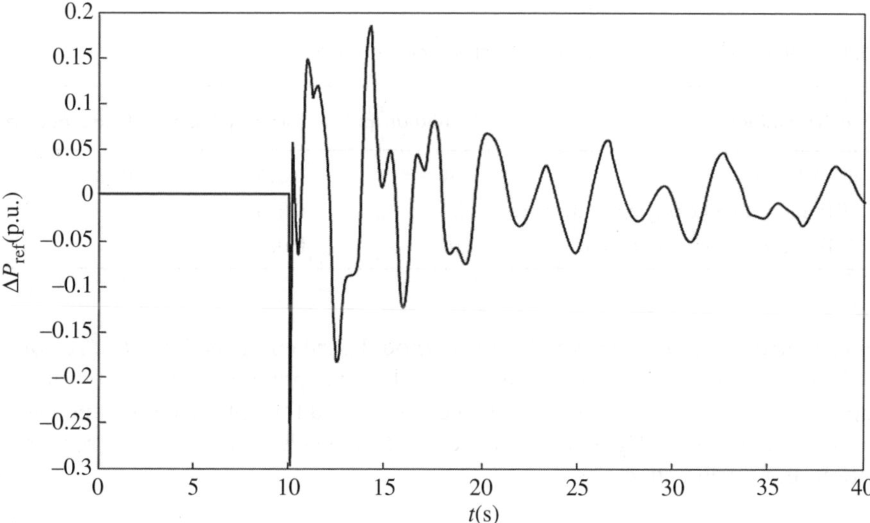

Figure 6.20 Output of the damping controller.

oscillation suppression of the VFTs. The low-frequency oscillations in the other modes of VFTs have a certain damping effect as well.

6.6.5 Parameter Design and Damping Effect of the Damping Controller After a Change of the Power System

In the case of connection of all necessary PSS, the interconnected power system covering Northeast China, North China, and Central China optimizes parameters of the

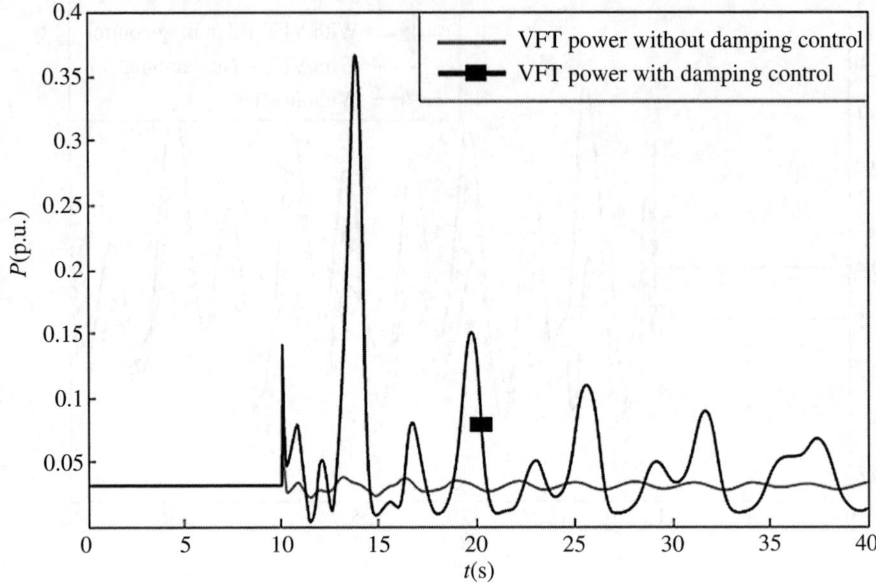

Figure 6.21 Comparison of active power response of the VFTs.

Table 6.8 Damping effect of the damping controller for low-frequency.

Calculation Condition	Amplitude (pu)	Damping Ratio	Frequency (Hz)
Without VFTs	0.24	0.015	0.33
With VFTs and without damping controller	0.25	0.023	0.33
With VFTs and a damping controller	0.16	0.061	0.31

damping controller automatically with the method described in Section 6.6.4. According to the comparison of the low-frequency oscillation suppression effect between the non-optimized parameters of the damping controller and the optimized parameters of the damping controller in Figure 6.22, the new parameters have a better low-frequency oscillation suppression effect.

6.7 Summary

Based on the control regulation function of a VFT, this chapter studied the design idea of a low-frequency oscillation damping controller based on a VFT and the Prony method: The Prony method is used to analyze the power system response signal generated by VFT power step control, and identify the transfer function and dominant oscillation mode under small disturbance. On this basis, the parameters of the corresponding damping controller were calculated for the dominant low-frequency oscillation mode. A simulation study was done for typical four-generator power systems and a practical

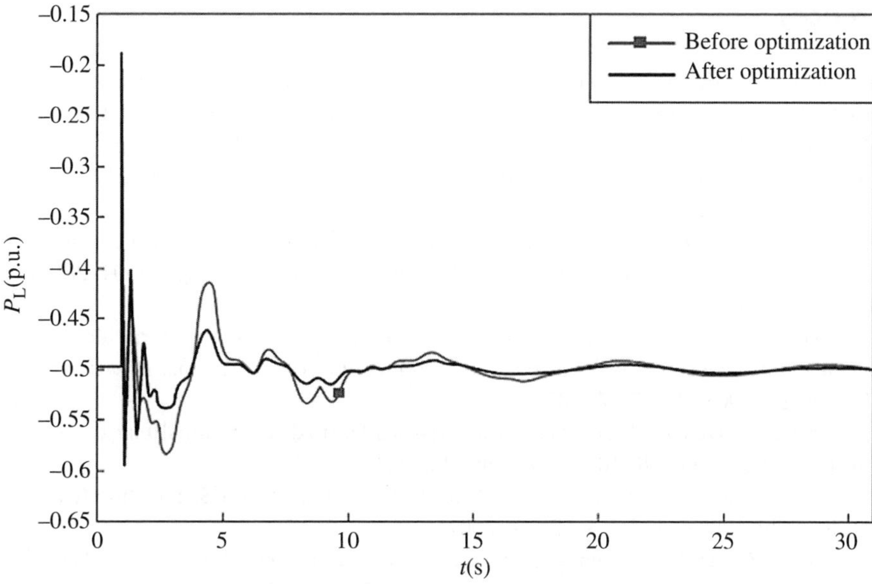

Figure 6.22 Comparison of the power angle simulation curves of generators G3 to G1 based on automatically adjusted parameters of the damping controller.

complicated power systems with help of these parameters. The conclusions are as follows:

1. The low-frequency oscillation frequencies and damping characteristic parameters can be analyzed easily with help of the power step regulation function of a VFT and the Prony method. This method applies to the disturbance of power systems well, providing a convenient design of adaptive damping controllers.
2. With the help of a damping controller, a VFT can suppress inter-regional oscillation mode of the interconnected power system, and generate a damping effect for the inter-regional oscillation mode. This suggests that a VFT can suppress the low-frequency oscillations of interconnected power systems well.
3. The Prony method is simple and quick because it does not need to solve complicated system equations or know specific structures and parameters of power systems, but needs to track the responses of power systems only to realize timely adjustment and improve adaptability to power system changes. A VFT can add disturbance to power systems because of its step regulation function, providing a way for active parameter adjustment for damping controllers and enabling us to establish power system damping controllers with better adaptability.
4. According to simulation results from four-generator power systems and practical complicated power systems with structural or parameter changes, the power system damping controllers designed based on a VFT and the Prony method can realize timely parameter adjustment for control systems and improve damping effects and safety as well as stability of grids. This further verifies the feasibility of adaptive damping controllers based on VFTs and the Prony method.

References

1 L. Rouco. Eigenvalue-based methods for analysis and control of power system oscillations. *IEE Colloquium on Power Dynamics Stabilization, Coventry*, UK, 1998.

2 Z. Fang, Z. Hongguang, L. Zenghuang, et al. The influence of large power grid interconnected on power system dynamic stability. *Proceedings of the CSEE*, 2007, 27(1): 1–7 [in Chinese].

3 Z. Fang, T. Yong, Z. Dongxia, et al. The influence of large power grid interconnected on power system dynamic stability. *Power System Technology*, 2004, 28(15): 1–5 [in Chinese].

4 D. Jixiang, H. Jianming, Y. Tianliang, et al. Analysis of low-frequency oscillation for Sichuan power grid in large scale interconnected power systems. *Power System Technology*, 2008, 32(17): 78–83.

5 B. Gunther, P. Dusan, R. Dietmar, et al. Lessons learned from global blackouts. *Electric Power*, 2007, 40(10): 75–81 [in Chinese].

6 P. Kundur. *Power System Stability and Control*. Z. Xiaoxin and S. Yonghua (trans.). Beijing: China Electric Power Press, 2001.

7 Z. Xiaoxin, G. Jianbo, S. Yuanzhang. *Basic Research on Operational Reliability of a Large Scale Interconnected Power Grid*. Beijing: Tsinghua University Press, 2008.

8 N. Mithularmnthan, C.A. Canizares, I. Reeve, et al. Comparison of PSS, SVC, and STATCOM controllers for damping power system oscillations. *IEEE Trans on Power Systems*, 2003, 18(2): 786–792.

9 X. Jingyu, X. Xiorong, T. Luyuan et al. Inter-area damping control of interconnected power systems using wide-area measurements. *Automation of Electric Power Systems*, 2004, 28(2): 37–40.

10 D. Mingqi, L. Wenying, Y. Juan, et al. Low-frequency damping method of interconnected power grids by increasing tie-lines. *Automation of Electric Power Systems*, 2007, 31(17): 94–98 [in Chinese].

11 C. Gesong. Discussion on the flexible AC/DC power transmission mode, *National Conference on Power System Technology*, 2001.

12 Y. Weijia, J. Ping, G. Wei. Discussion on additional signal selection of SVC damping control. Automation of Electric Power Systems, 2008, 20(2): 69–72 [in Chinese].

13 L. Guoping. *Analysis and Control of Power System Low-Frequency Oscillation Based on Prony Algorithm*. M. Phil Thesis, Hangzhou: Zhejiang University, 2004.

14 J.R. Smith, J.F. Hauer, D.J. Trudnowski. Transfer function identification in power system applications. *IEEE Trans. on Power Systems*, 1993, 8(3): 1282–1290.

15 C. Gesong. *Modeling and Control of VFT for in Power Systems*. D. Phil Thesis, Beijing: China Electric Power Research Institute, 2010.

7

Technical and Economic Characteristics of VFTs

7.1 Overview

Compared with the conventional phase shifts and back-to-back DC transmission technologies, VFT is a new device for the power transmission of interconnected power systems with both similar functions and different effects. Comparing the common points and differences of these devices or technologies is helpful for the designers of power projects to make appropriate decision in power system planning. Based on brief introduction to the phase shifts and back-to-back DC transmission technologies, this chapter analyzes and compares the operating principles, regulation function, impacts on power systems, and technical and economical characteristics of VFTs with these devices and technologies.

7.2 Comparison of the Technical and Economic Characteristics of VFTs and Phase-Shifting Transformers

7.2.1 Phase-Shifting Transformers

Phase-shifting transformers are devices for regulating voltage phase angles and controlling the transmission power of transmission lines or between power systems. In the power flow calculation of a power system, a phase shifting transformer can be simulated through a reactance leakage device Z_T (or X_T when the loss is not considered) connected in series with an ideal transformer with a transformation ratio of k. Figure 7.1 introduces the equivalent circuit of the connection of a phase shifting transformer with a single-generator infinite-bus power system.

If phase shift function of a phase-shifting transformer is not considered in a steady-state calculation, this power system will have the same power angle characteristic as a conventional power system (Equation (7.1) and Figure 7.2); if the phase shift function of the phase-shifting transformer is considered, power angle characteristics of this power system and a conventional power system are as shown in Equation (7.2) and Figure 7.3 (transformation ratio k: 1).

$$P = \frac{EU \sin \theta}{X_s + X_t} \tag{7.1}$$

$$P = \frac{EU \sin(\theta + \alpha)}{X_s + X_t} \tag{7.2}$$

Variable Frequency Transformers for Large Scale Power Systems Interconnection: Theory and Applications, First Edition. Gesong Chen, Xiaoxin Zhou, and Rui Chen.

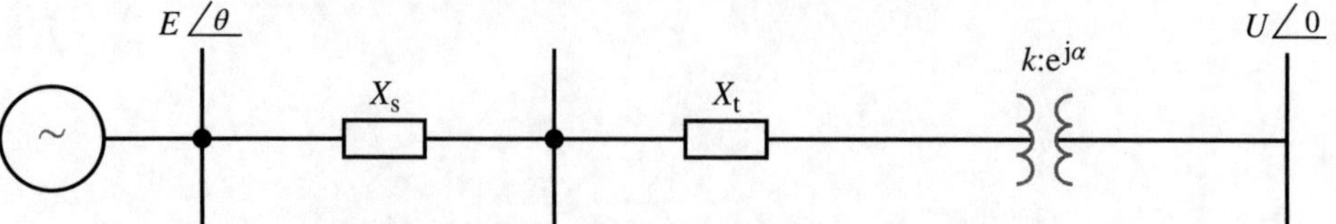

Figure 7.1 Equivalent circuit diagram of the connection of a phase shifting transformer with a single-generator infinite-bus power system. Where X_s = equivalent reactance of the power source; X_t = short-circuit reactance of the transformer; $E\angle\theta$ = electromotive force and angle of the equivalent power source; $U\angle 0.$ = voltage and angle of the infinite bus; α = phase-shifting angle; and k = transformation ratio.

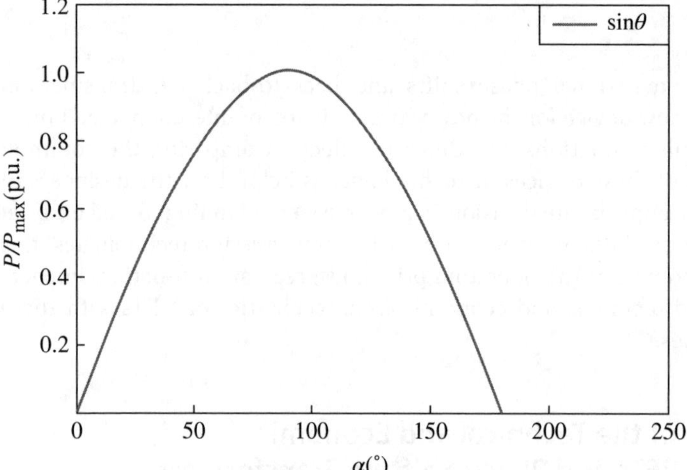

Figure 7.2 Power angle characteristics of the power system without a phase shift function.

According to the Equations (7.1) and (7.2), the power angle characteristics of a power system changes as follows [1] after a phase shifting transformer is installed:

1. If the reactance influence of a phase shifting transformer is not considered, amplitude of the maximum inter-system transmission power will remain unchanged, suggesting that the phase-shifting transformer cannot increase the inter-system steady-state stability limit.

2. The power angle corresponding to the maximum transmission power can change in a certain range, thus increasing the steady area of a power system. Suppose that the phase-shifting transformer can make the phase-shifting angle between $-\alpha$ and α, the θ angle range corresponding to the power limit will be as in Equation (7.3) and Figure 7.3.

$$\theta \in \left(\frac{\pi}{2} - \alpha, \frac{\pi}{2} + \alpha\right) \tag{7.3}$$

In the formula, α is the maximum phase-shifting angle can be realized by the phase-shifting transformer.

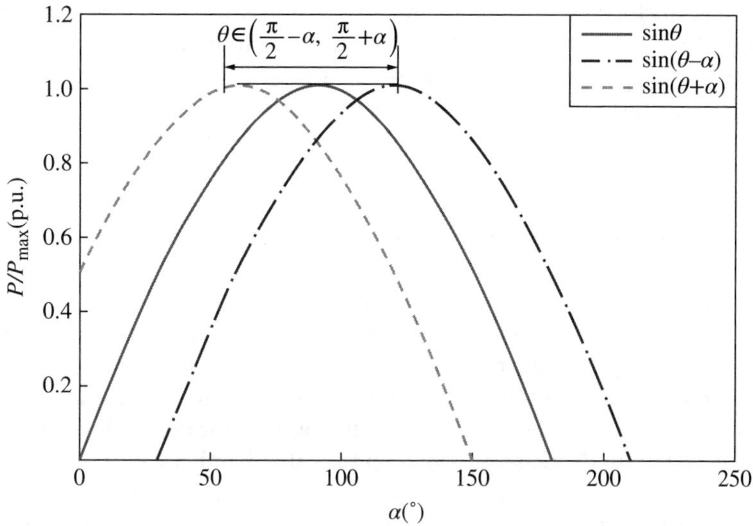

Figure 7.3 Power angle characteristics of the power system with a phase shift function. (Suppose that the phase-shifting transformer can make the phase-shifting angle between $-\alpha$ and α.)

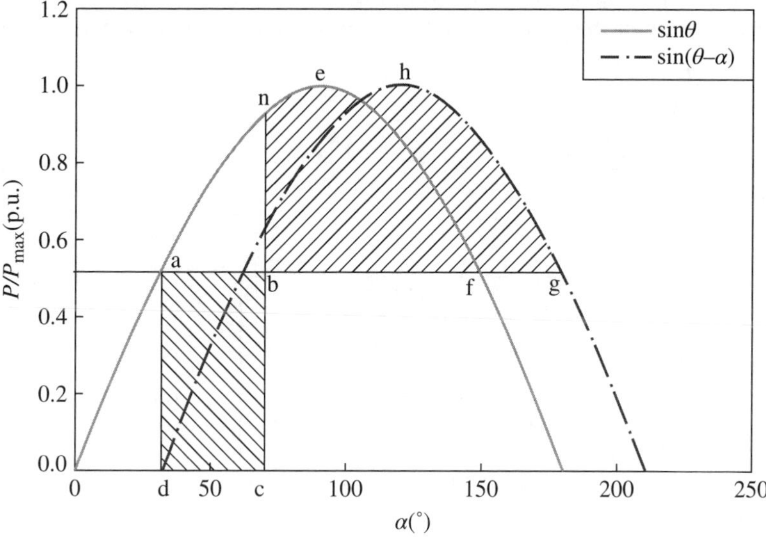

Figure 7.4 Schematic diagram of improvement of the transient characteristics of power system by a phase shifting transformer.

3. The phase-shifting transformer can improve transient characteristics of a power system. Take a single-generator infinite-bus power system for example. Upon a fault of the power system, a phase shifting transformer can increase deceleration area of the generator significantly to meet the stability requirement that maximum deceleration area of the generator be larger than its acceleration area (Figure 7.4). For a complicated power system, a phase shifting transformer has the same effect in transient characteristic improvement.

4. The phase-shifting transformer can increase inter-system transmission power. In the actual operation of a power system, a safety margin needs to be considered. Take a single-generator infinite-bus power system for example. If a phase shifting transformer is installed, the deceleration area is increased from S_{bnef} to $S_{\mathrm{bnef}} + S_{\mathrm{efgh}}$, the actual stability margin will be improved greatly, thus improving the inter-system transmission power properly and keeping it steady. This is the basic principle for a phase shifting transformer to increase the inter-system transmission power.

5. A phase shifting transformer can also improve the dynamic characteristics of a power system. Dynamic control over phase-shifting angle of the phase-shifting transformer can increase damping in power system oscillation and suppress oscillation of the power system quickly [2]. Figure 7.5 is an example of power system oscillation suppression by a phase shifting transformer.

6. A phase shifting transformer can also be used for controlling power distribution of paralleled transmitting channels: On one hand, it can optimize the power flow distribution of a power system to decrease grid loss; on the other hand, it can control the power flows of key branches to avoid overload, or, to some extent, prevent the transfer of a large power to another branch to cause a chain reaction. Figure 7.6 is a line diagram showing the connection of a phase shifting transformer to a loop power system. Equation (7.4) shows the power distribution of different branches of the loop power system. Obviously, adjusting the phase-shifting angle of a phase shifting transformer can control distribution of the power interchange between grids effectively [3].

$$P_1 = \frac{U_{\mathrm{m}} U_{\mathrm{n}} \sin \delta}{X_1}$$

$$P_2 = \frac{U_{\mathrm{m}} U_{\mathrm{n}} \sin(\delta + \alpha)}{X_2} \tag{7.4}$$

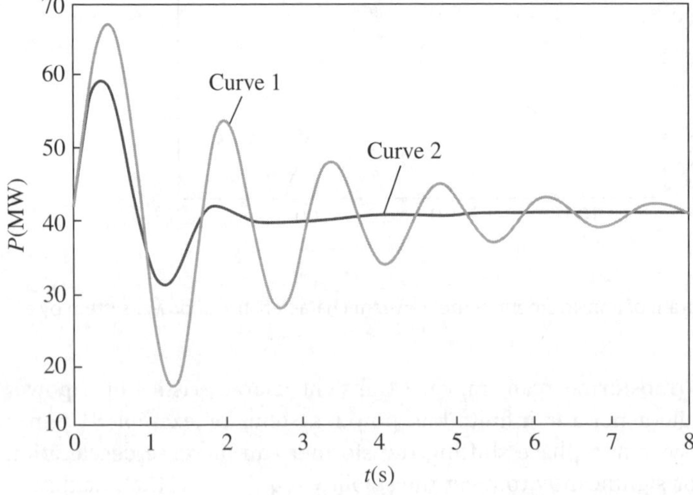

Figure 7.5 Example of improvement of the dynamic characteristics of a power system by a phase shifting transformer. Note: Curve 1 has a phase shifting transformer while Curve 2 has no phase-shifting transformer [2].

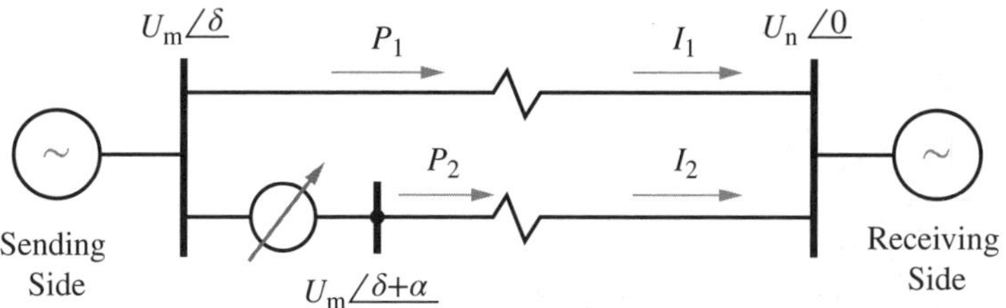

Figure 7.6 Schematic diagram of a phase shifting transformer to improve the power distribution of a looped power system. Where: $U_m \angle \delta$ = bus voltage and angle of the transmitting end; $U_n \angle 0$ = bus voltage and angle of the receiving end; $U_m \angle \delta + \alpha$ = voltage and angle on the phase-shifting transformer circuit side; P_1 = transmission power of line 1; P_2 = transmission power of line 2; I_1 = current of line 1; and I_2 = current of line 2.

7.2.2 Structures and Types of Phase-Shifting Transformers

Phase-shifting transformers are mainly used for the steady-state power flow control and dynamic stability control of power system. They present an early, mature power flow control technology. The first power phase-shifting transformer was put into operation in the US in the 1930s. The power phase-shifting transformers now in operation in the world are mainly based on the technology of transformer tap changing and, structurally, mainly include single- and dual-core ones. Single-core phase-shifting transformers are mainly used for small-capacity power systems of 110 kV or below; dual-core phase-shifting transformers can be used for the power systems with higher voltages and larger capacities. Figure 7.7 shows the wiring diagram of a dual-core phase-shifting transformer.

The earlier phase-shifting transformers were usually controlled by mechanical drives or mechanical switches. However, with the development of power electronic technologies, the static phase-shifting transformers based on thyristors and cut-off power electronic devices have now been developed and used. Viewed from their control methods and design principles, the phase-shifting transformers used in power systems mainly include worm wheel driven motor type phase-shifting transformers, mechanical-regulation tapped phase-shifting transformers, thyristor controlled phase-shifting transformers (TCPSTs), thyristor switched phase-shifting transformers, multi-bridge pulse width modulation (PWM)-based static phase-shifting transformers, and so on. These phase-shifting transformers are introduced next.

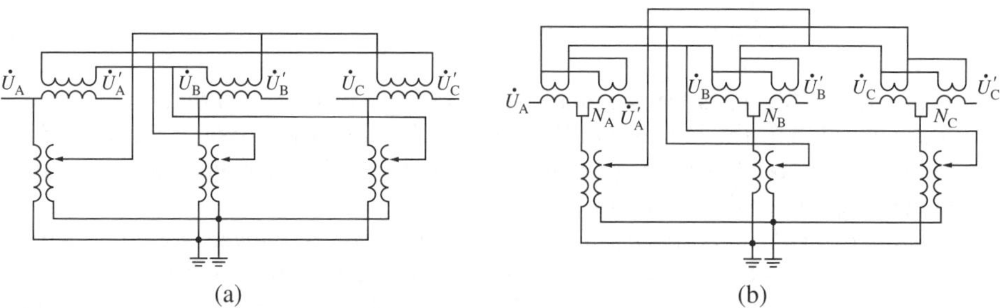

Figure 7.7 Two wiring diagrams of a dual-core phase-shifting transformer. (a) Wiring diagram A and (b) wiring diagram B.

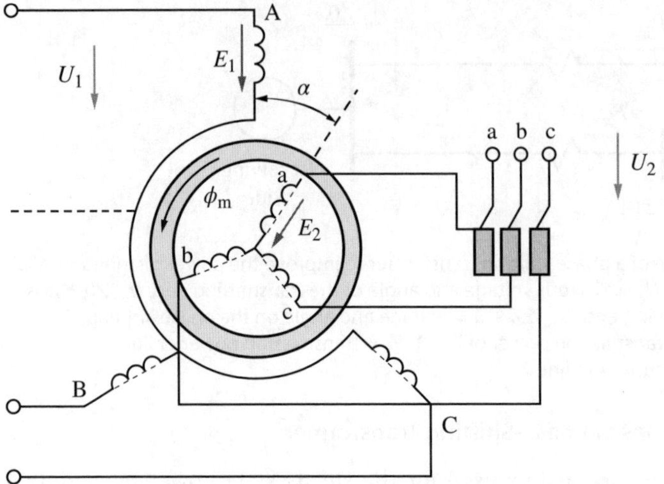

Figure 7.8 Equivalent circuit diagram of a worm wheel driven motor type phase-shifting transformer.

1. *Worm wheel driven motor type phase-shifting transformers*: Such a phase shifting transformer is similar to a clocked-rotor wound three-phase asynchronous motor. Usually, the stator winding serves as the primary winding and the rotor winding as the secondary winding. The rotor shaft of the phase-shifting transformer has a set of worm gears. The worm gears can be rotated to make rotor of the phase-shifting transformer rotate in a certain range relative to the stator. The product of the rotating angle and the winding pole pairs is the maximum phase-shifting angle of the phase-shifting transformer. After the primary winding on the stator is connected with a three-phase AC power source, the rotating magnetic field generated in the air gap will sense electromotive forces E_1 and E_2 from the primary winding and the secondary winding, respectively. E_1 and E_2 are in direct proportion to effective turns of the two windings, respectively, and the phase depends on the relative locations of axes of the two windings. For example, if the spatial positions of exes of the two windings have an electrical angle difference of α, and the leakage impedance voltage drops of the two windings are not considered, an equivalent circuit diagram as in Figure 7.8 will be obtained. The voltage relation of the two windings is introduced in Formula (7.5).

$$U_1 \approx E_1$$

$$U_2 \approx E_2 = \frac{E_1}{n_{sr}} e^{j\alpha} \tag{7.5}$$

In Equation (7.5), n_{sr} is transformation ratio of the primary winding and the secondary winding. If the location of the rotor is changed, the phase of the secondary winding voltage relative to the primary side voltage can be changed while the output voltage magnitude will remain unchanged.

2. *Mechanical-regulation tapped phase-shifting transformers*: Viewed from their wiring types and their control overvoltage phase angle regulation and amplitude regulation, these phase-shifting transformers include vertical-mechanical-regulation tapped phase-shifting transformers, horizontal-mechanical-regulation tapped phase-shifting transformers, and inclined-mechanical-regulation tapped phase-shifting

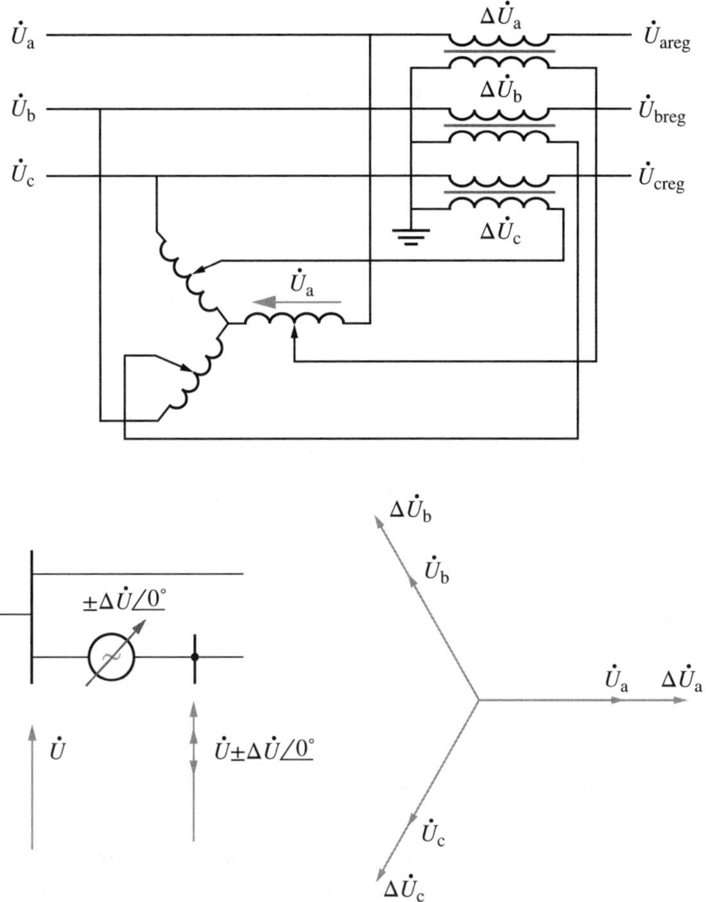

Figure 7.9 Wiring diagrams of a vertical-mechanical-regulation phase-shifting transformer [1].

transformers, with their wiring diagrams in Figures 7.9–7.11 [1]. Vertical-mechanical-regulation tapped phase-shifting transformers do not change voltage phases but voltage amplitudes only. The voltage increment of a horizontal-mechanical-regulation tapped phase-shifting transformer and the regulated voltage have a phase difference of 90°; the result changes both voltage amplitude and voltage phase angle; the regulated voltage, the voltage increment and the voltage after regulation form a right-angled triangle relation in the vector space. A vertical-mechanical-regulation tapped phase-shifting transformer can regulate the values and amplitudes of voltages, which controls voltages more flexibly. A conventional phase-shifting transformer is controlled by a mechanical regulator. Usually, a mechanical phase-shifting transformer is subject to step regulation and its control speed is at a second or minute level, making it fail to meet the quick regulation need of power systems in normal conditions. Furthermore, the contact of a mechanical phase-shifting transformer tends to result in equipment aging because of sparks generated by friction, thus increasing the effort needed for equipment maintenance.

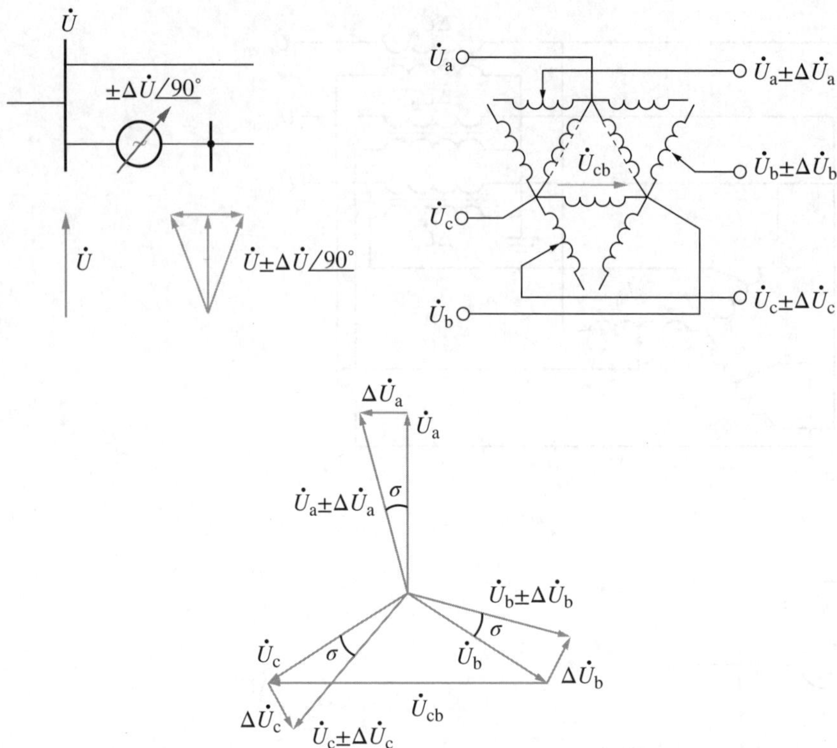

Figure 7.10 Wiring diagrams of a horizontal-mechanical-regulation phase-shifting transformer [1].

3. *TCPSTs*: Since the 1970s, some researches have been studying TCPSTs by combining thyristors and phase-shifting transformers. According to the studies, TCPSTs can increase the transmission power flows of interconnecting ties, suppress disturbance, improve the stability and oscillation damping of power systems, control bus voltages and interconnecting tie power flows, and so on. TCPSTs have the advantages of quick and stepless regulation on. However, TCPSTs need to consume reactive power; they are usually used with reactive compensators in practical operation; they have high harmonic contents, influencing power quality to some extent. Figure 7.12 is the circuit diagram of a TCPST. By means of their continuously controlled firing angles, the thyristors are conducted in turn to compensate the voltage fundamental component. This regulation is continuous and stepless but complicated, depending a great deal on the load characteristic of a power system. Furthermore, this regulation generates a harmonic component in a power system, making it necessary to install a filter with a certain capacity.

4. *Thyristor switched phase-shifting transformers*: In consideration of the control complexity and harmonic influence of TCPSTs, a simplified thyristor switched phase-shifting transformer with a circuit diagram as in Figure 7.13 has been proposed. Essentially, a thyristor is used as a mechanical-regulation switch to switch on or off the thyristor module. It solves problems in mechanical regulation such as quick response and contact wear and avoids the control complexity and harmonic influence of thyristors. However, this control is step regulation rather than stepless regulation. For better control performance, a multi-level series (such as 3^n) is usually

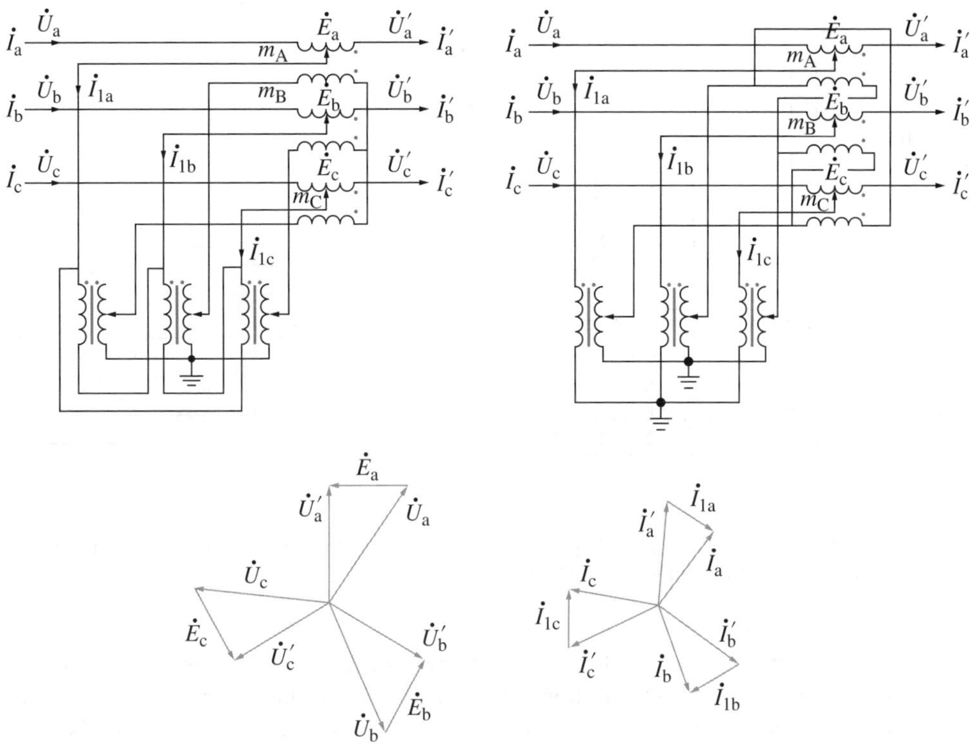

Figure 7.11 Wiring diagram of an inclined-mechanical-regulation phase-shifting transformer [1].

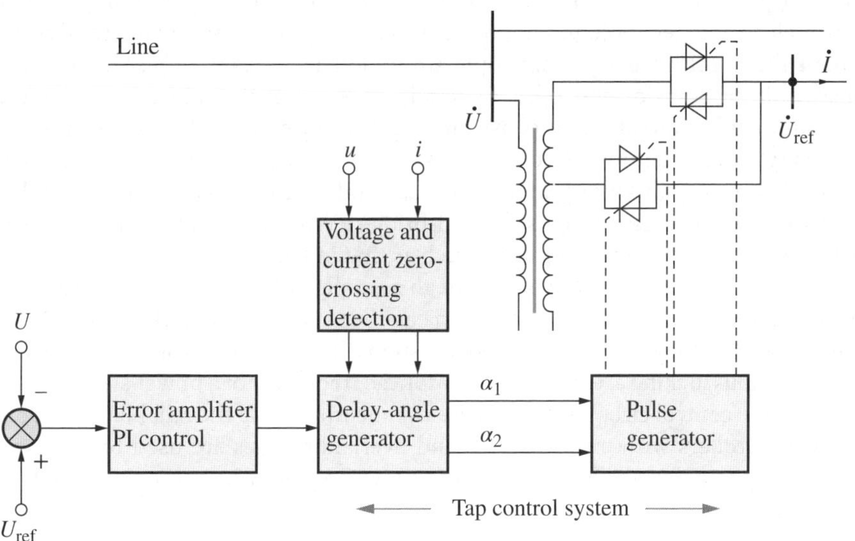

Figure 7.12 Schematic diagram of a TCPST [1].

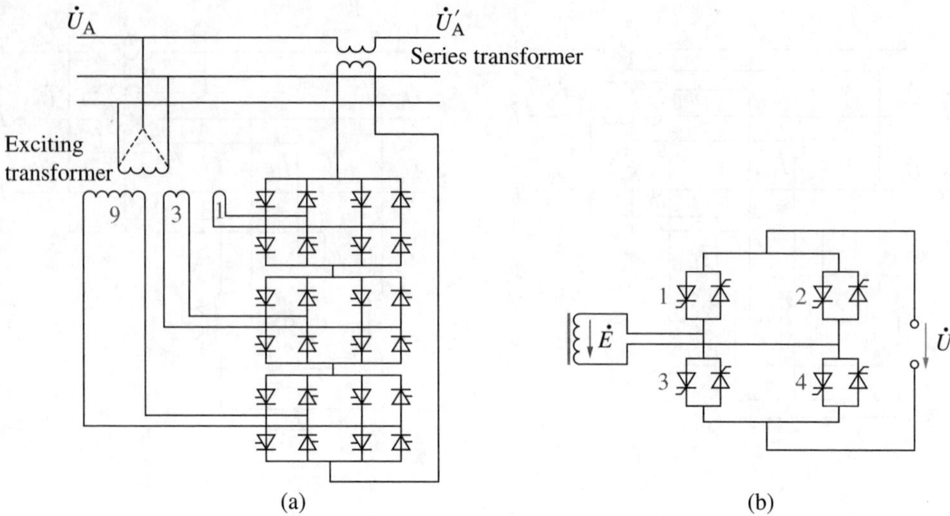

Figure 7.13 Schematic diagram of a thyristor switched phase-shifting transformer. (a) Wiring diagram and (b) different conduction methods of thyristors.

used for voltage compensation, in a bid to provide more levels by means of different combinations of limited windings. As in Figure 7.13, the power system provides three winding number choices, that is, 1, 3, and 9 only, but 27 regulation levels in all, including $-13, -12, -11, -10, -9, -8, -7, -6, -5, -4, -3, -2, -1, 0, 1, 2, 3, 4, 5, 6, 7, 8, 9, 10, 11, 12$, and 13, can be provided by means of different methods of conducting and locking of the thyristors.

5. *Multi-bridge PWM based static phase-shifting transformers* [4]: With the application of power electronic technologies in power systems and the increased depth of studies about FACTS in recent years, people are no longer satisfied in just the control of phase-shifting transformers over the steady-state power flows of power systems. As a result, studies about static phase-shifting transformers with quick responses are drawing more attention. Since the 1990s, power electronic technologies have been developing quickly among large-power turn-off devices and new circuit topologies, and some people have proposed multi-bridge PWM based static phase-shifting transformers with a schematic diagram such as that in Figure 7.14.

In such a phase-shifting transformer, through a simple duty ratio, the phase-shifting angle regulation is continuous, flexible, and easy because of the PWM direct AC/AC conversion technology, which has a quick response to PWM control. Phase locking and synchronous locking are not necessary for the frequency of a power system, thus simplifying the control circuit structure greatly and improving the reliability. In addition, AC controllers with multi-parallel and overlaid bridges are used to keep the phase-shifting transformer from generating a low-frequency harmonic, thus reducing the filter input.

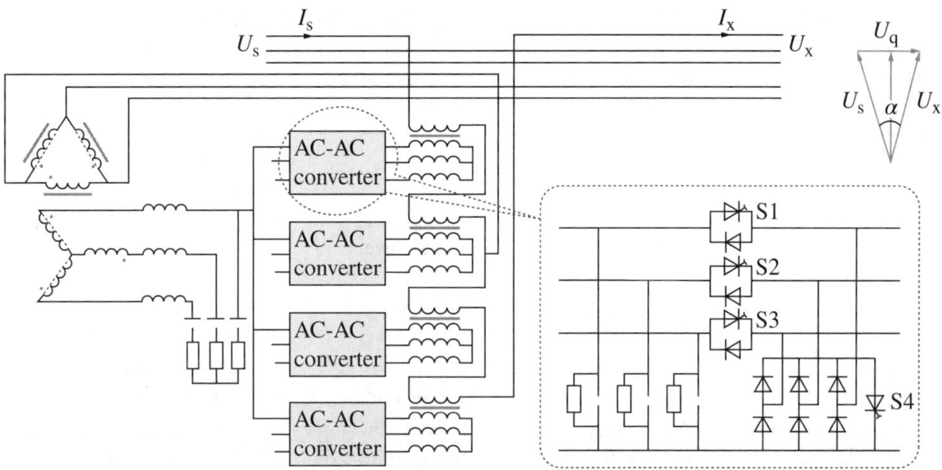

Figure 7.14 Schematic diagram of a four-bridge PWM-based static phase-shifting transformer [4].

7.2.3   A Comprehensive Comparison of VFTs and Phase-Shifting Transformers

Characteristics of the phase-shifting transformers of different types were introduced briefly in Section 7.2.2. This section introduces the main similarities and differences of VFTs and phase-shifting transformers on the basis of the studies about VFTs previously and the conclusions in related documents [5–14].

1. A VFT is a phase-shifting transformer with a function of stepless regulation (0–360°). It applies to the phase-shifting control in synchronous grids and the power transmission control between asynchronous grids as well.
2. Compared with a worm wheel driven motor type phase-shifting transformer or a mechanical phase-shifting transformer, a VFT has a quicker control speed and can realize extensive stepless regulation for a phase shifting angle and meet the power flow control needs of a power system. A mechanical phase-shifting transformer can realize step regulation at only a second or minute level.
3. Compared with a TCPST or a PWM based static phase-shifting transformer, a VFT has a slower control speed but better damping characteristics, no harmonic component, and it is unnecessary to install devices such as a filter.
4. Compared with a thyristor switched phase-shifting transformer, a VFT can realize stepless regulation and better damping characteristics but a slower control speed.

In general, as an internal power control device for synchronous grids, a VFT has functions similar to a phase shifting transformer; viewed from its principle, it is more similar to a worm wheel driven rotor type phase-shifting transformer, but it has better dynamic characteristics and, more importantly, applicability to the asynchronous interconnection. See Table 7.1.

Table 7.1 Comparison of the main performance of a VFT and a phase shifting transformer.

Type	Phase Regulation Method	Step Response Speed	Applicable Field
VFT	Stepless regulation (0–360°)	<400 ms	Synchronous or synchronous interconnection
worm wheel driven motor type phase-shifting transformer	Stepless regulation (in a certain range)	Second level ~ minute level	Synchronous power systems with low-speed control
Mechanical phase-shifting transformer	Step regulation (in a certain range)	Second level ~ minute level	Synchronous power systems with low-speed control
TCPST	Stepless regulation	Tens of milliseconds	Synchronous power systems with quick regulation
Thyristor switched phase-shifting transformer	Step regulation (in a certain range)	Tens of milliseconds	Synchronous power systems with quick regulation
PWM based static phase-shifting transformer	Stepless regulation	Millisecond level	Synchronous power systems with quick regulation

7.3 Comparison of the Technical and Economic Characteristics of VFTs and DC Transmission Systems

7.3.1 DC Transmission Systems

DC transmission systems are mainly used for long-distance and large-capacity power transmission and the interconnection of asynchronous grids as well as grids with different frequencies. At present, the high-voltage DC (HVDC) projects in service have a maximum rated voltage of ±800 kV and a maximum transmission capacity of 8000 MW. A DC transmission system mainly includes a converter station, DC line, DC filter, AC filter, reactive compensator, converter transformer, smoothing reactor, protection system, control system, and so on. DC transmission systems mainly include two-terminal DC transmission systems and multi-terminal DC transmission systems. Figure 7.15 shows the wiring diagram of a two-terminal DC transmission system. The core equipment of a DC transmission system is a converter station, which is the key for the DC transmission system to realize AC-DC-AC conversion. The first link is rectification while the last link is inversion. For a DC transmission system of ±660 kV or below, a 12- or six-pulse converter with 12 or six converter valves is usually used; for a DC transmission system of ±800 kV or above, a dual-12-pulse converter is usually used. A converter valve is usually composed of a half-controlled thyristor, a fully controlled IGBT, and so on.

7.3.2 Development and Types of DC Transmission Systems

Since 1954 when a DC transmission line (±100 kV, 20 MW) with a submarine cable length of 95 km was successfully put into operation in Sweden, HVDC transmission technologies have been developing quickly around the world and many projects using

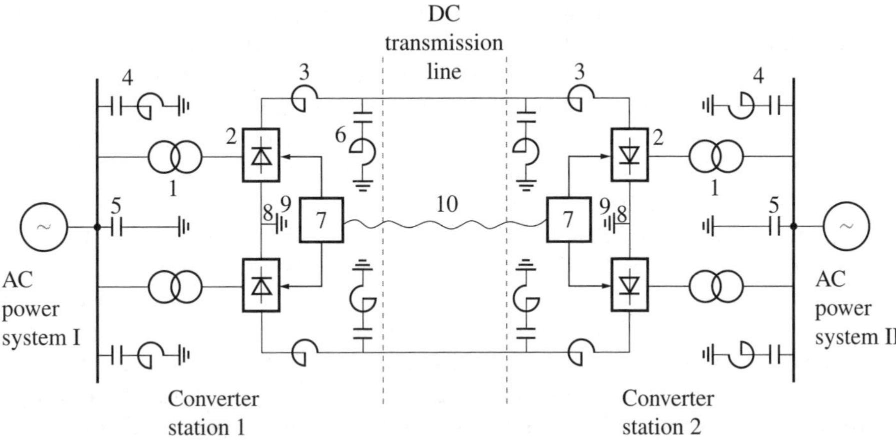

Figure 7.15 Wiring diagram of a two-terminal DC transmission system. 1 = Converter transformer; 2 = converter; 3 = smoothing reactor; 4 = AC filter; 5 = capacitor; 6 = DC filter; 7 = control and protection system; 8 = earth electrode line; 9 = earth electrode; and 10 = telecontrol communication system.

these technologies have been put into operation. The quick increase in the transmission capacities of HVDC transmission systems is a result of the continuous improvement of converter valve technologies and power system technologies, including the replacement of mercury-arc valve converters by thyristor valves, the replacement of forced air cooling by water cooling for valves, the modular design of valves and the application of technology to lift installation of valves, and the development and application of 6 inches high-voltage large-capacity thyristor valves. The main development in DC transmission systems includes the successful development of large-capacity converter valves, the computerized digital control and protection of valves and converter stations, the application of capacitative converter technology, the development and application of voltage control type DC valves, and the application of adjustable AC filters, active DC filters, and digital optical fiber transformers. Before the successful development of VFTs, DC transmission systems were the only technical means for the asynchronous interconnection of grids with different frequencies, playing an important role in long-distance power transmission and grid interconnection. At present, DC transmission systems mainly include conventional DC transmission systems, DC back-to-back transmission systems, capacitative converter DC back-to-back transmission systems, and flexible DC transmission systems [15–28].

1. *Conventional DC transmission systems*: To date, more than 100 HVDC transmission projects have been built around the world and typical ones include the Itaip DC transmission project ($\pm$600 kV, 6300 MW, 806 km) in Brazil and the Québec–New England multi-terminal DC transmission project (five converter stations) on. In China, DC transmission projects have been developing quickly along with multiple $\pm$500 kV DC transmission projects, mainly including the Gezhouba-Shanghai DC transmission project (1200 MW, 1064 km, in 1990), the Tianshengqiao-Guangzhou DC transmission project (1800 MW, 980 km, in 2001), the Three Gorges-Changzhou DC transmission project (3000 MW, 890 km, in

2003), the Three Gorges-Guangdong DC transmission project (3000 MW, 962 km, in 2004), lines 1 and 2 of the Guizhou-Guangdong DC transmission project (2 × 3000 MW, 880 km, in 2004 and 2007, respectively), and the Three Gorges-Shanghai DC transmission project (3000 MW, 1040 km, in 2006). Furthermore, the existing ±800 kV UHV DC transmission projects with three built lines include the Xiangjiaba-Shanghai DC transmission project (6400 MW, 1907 km, in 2010), the Yunnan-Guangdong DC transmission project (5000 MW, 1373 km, in 2010), and the Jinping-Sunan (7200 MW, 2058 km, in 2012). China is now a country with the maximum transmission scale, the maximum voltage level, the maximum single-project transmission capacity, and the most advanced technical level of DC transmission systems in the world. DC transmission technologies apply to long-distance and large-capacity power transmission. With progress in the West-East Power Transmission Project and the interconnection of large regional grids in China, there will be more DC transmission projects going into operation. Conventional DC converters have the following disadvantages because they use thyristor half-controlled power electronic devices based on the current conversion principle and controlling the conducting and cut-off of devices with help of the voltages of AC power systems.

a. In conversion, each converter consumes a lot of reactive power and the reactive power consumed by each converter station accounts for 40–60% of the DC transmission power, so the converter stations need to be provided with many reactive power compensators, which increases project costs; a converter commutation is realized by means of provision of a short-time phase-phase short-circuit current by grids that both the sending grid and the receiving grid should have sufficient capacities. When converter stations are connected with a weak AC power system, AC voltage disturbance tends to result in commutation failure. For higher stability and better commutation conditions for the dynamic voltage, synchronous phase modifiers, or static reactive compensators may be necessary.

b. For an AC power system, a converter station is a power-frequency load or power source and a harmonic current source as well, which transmitting high-frequency harmonic currents to the AC power system and resulting in distortion of the AC bus voltage. For the DC side, a converter station is a harmonic voltage source too. In order to limit the distortion of the AC bus voltage in the permissible range, the AC side needs a large-capacity filter and the DC side needs a smoothing reactor and a filter, which increases the project cost.

2. *DC back-to-back transmission systems*: Figure 7.16 shows the wiring diagram of such a transmission system. By now, multiple HVDC back-to-back transmission projects with a voltage range from ±100 kV to ±500 kV and a conversion capacity from 10 to 1500 MW have been built around the world. The typical DC back-to-back transmission projects include one that connects a 50 Hz grid and a 60 Hz grid in Japan and another one between China and Russia. As a special form of HVDC transmission, DC back-to-back transmission combines the rectifier stations and inverter stations for HVDC transmission in a converter station, and converting AC to DC, and then to AC again in a single position. Structures of its rectifier stations and the inverter stations and the facilities on the AC side are the same as those of a HVDC transmission system, so a DC back-to-back transmission system has the main advantages of a conventional HVDC transmission system and can realize asynchronous interconnection and interconnection of different AC grids. Compared with a conventional

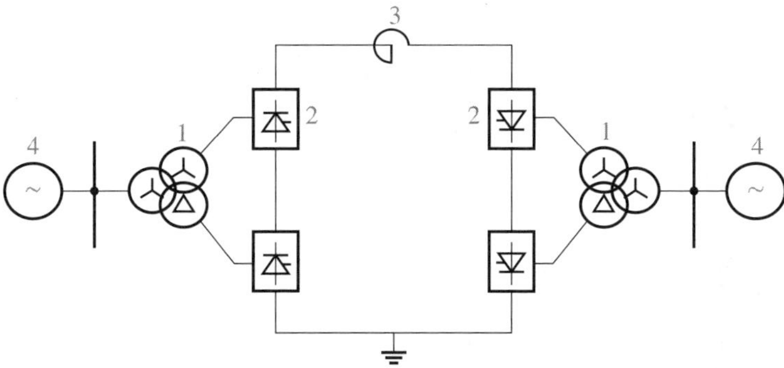

Figure 7.16 Wiring diagram of a DC back-to-back transmission system. 1 = converter transformer; 2 = converter; 3 = smoothing reactor; and 4 = two-terminal AC power system.

HVDC transmission system, a DC back-to-back transmission system has the following characteristics:

a. It has no DC line and its loss on the DC side is small;

b. The DC side can select low-voltage large-current operation to decrease insulation levels of related devices such as the converter transformer and the converter valve to cut down the cost;

c. The harmonic on the DC side can be fully controlled in the valve hall to avoid interference in the communication device;

d. It needs no DC devices such as earth electrodes, DC filters, DC lightning arresters, DC switch yards, or DC filters for its converter stations, so it needs a smaller investment.

3. *Capacitative converter DC back-to-back transmission systems*: A capacitative converter is connected with a power system through series capacitors rather than a conventional converter reactor. Figure 7.17 shows the wiring diagram of a capacitative converter. The capacitative converter method for DC transmission converter valves was proposed many years ago, it has been used in practical power systems in recent years after realizing effective protection for series capacitors. The Argentina-Brazil DC back-to-back transmission project (for connecting the Argentina grid and the Brazil grid using different AC voltage frequencies) put into operation in the year 2000 was the first practical industrial application project using capacitative converter technology. The capacitative converter technology makes HVDC transmission systems more applicable. It is obviously advantageous in connecting weak power systems and long cable lines, reduces the commutation failure of converter valves in a steady-state, and effectively suppresses the transient harmonics of lines. Furthermore, reactive compensation of the series capacitors is in direct proportion to transmission power of the power system, reducing the times of switching of the filter branch and the shunt capacitor reactive compensation branch greatly.

4. *Flexible DC transmission systems*: Figure 7.18 introduces a flexible DC transmission system. In recent years, large-power fully controlled power electronic devices such as gate turn-off thyristors (GTO), IGBT, integrated gate commutated thyristors (IGCT), emitter turn-off thyristors (ETO), and silicon carbide components have quickly developed. ABB constructed a new experimental DC transmission system in central Sweden in October 1997 with help of the voltage source converter (VSC)

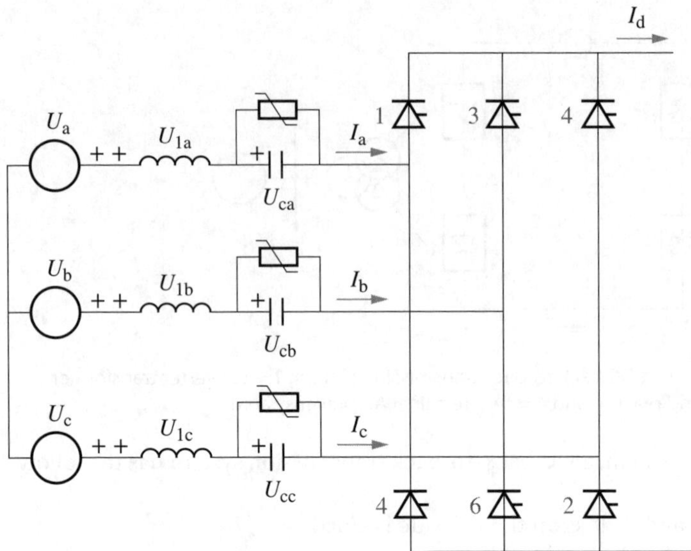

Figure 7.17 Wiring diagram of a capacitative converter.

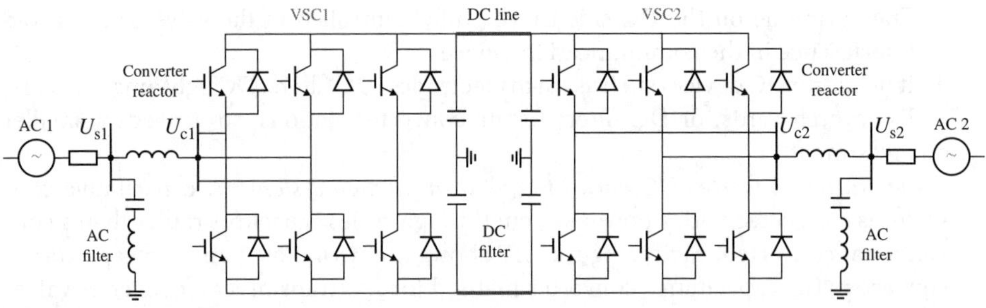

Figure 7.18 Wiring diagram of a flexible DC transmission system.

and PWM technologies of fully controlled devices. According to the experimental result, the new DC transmission system based on fully controlled devices has flexible control characteristics and can overcome the disadvantages of conventional HVDC transmission systems. In 2001, CIGRE established a working team (CIGRE B4–37) for the new DC transmission system. It has formally named a voltage source converter DC transmission system as VSC-HVDC.

Compared with a conventional HVDC transmission system, a VSC-HVDC transmission system has the following technical advantages [25–28]:

1. It uses fully controlled devices that operates in a passive inversion mode and can transmit power to isolated loads. So it overcomes a conventional HVDC transmission system's disadvantage of necessary access to an active network.
2. It employs PWM control technology, ensuring quick and continuous regulation for amplitude and frequency of the AC voltage generated by means of inversion. The frequencies with PWM waves are high, so the VSCs generate a few

harmonics. The characteristic harmonics generated can be reduced or eliminated by a small-capacity filter and there is no overvoltage on the AC side when the VSCs are not operating.

3. It employs quick power control, which realizes independent decoupling control over the active power and reactive power exchanged with an AC power system; its VSC converter stations can regulate their control objectives independently and without communication coordination.

4. In its power rating, VSC-HVDC can transmit inductive or capacitive reactive power to an interconnected AC power system quickly to serve as a static synchronous compensator (STATCOM) and a static synchronous series compensator (SSSC) to stabilize the AC bus voltage and the voltage stability of the power system.

5. In its power rating, VSC-HVDC can regulate quickly to realize active power exchange with an AC power system to provide an emergency power support for a failed region; its additional damping power control strategy can improve the low-frequency oscillation damping between it and an interconnected AC power system greatly, thus improving power angle stability of the AC power system.

6. In DC transmission power inversion, when the DC current has an opposite direction and the DC voltage has unchanged polarity, so multiple VSCs can form a multi-terminal DC network easily.

7. It employs the AC side current control strategy, which does not increase the short-circuit capacity of an AC power system.

8. It has no limit for the minimum DC current.

To date, multiple VSC-HVDC transmission systems have been constructed in Sweden, Denmark, USA, China, Australia and typical projects among them include the Cross-Sound VSC-HVDC transmission system in the US and the Murray Link VSC-HVDC transmission system in Australia.

VSC-HVDC transmission systems have overcome the disadvantages of conventional DC transmission systems, but they still have the following obvious disadvantages due to the existing limited power electronic technologies when compared with conventional HVDC transmission systems.

1. At present, a VSC-HVDC transmission system has an operating voltage and a transmission power of ±150 kV and 350 MW, respectively, much lower than the ±800 kV and 8000 MW of a conventional HVDC transmission system.

2. VSCs employ high-frequency PWM control technology and a bridge arm with many parallel and series valves, the power loss rate of a VSC converter station is now much higher than those of a conventional HVDC converter station.

3. If a DC line has no circuit breaker, when the DC line has a fault, a full-bridge rectifier made up of diodes in inverse-parallel connection with valves can still provide a fault current for the DC side even if the PWM control switch signal is locked by the VSCs. In this case, the circuit breaker on the AC side of the VSC-HVDC transmission system needs to trip off the DC fault.

As VSC-HVDC transmission systems are not applicable to long-distance and large-capacity transmission power, they have technically and economically advantageous for the transmission places with a minimum DC voltage of 150 kV and a maximum capacity of 200 MW because of their excellent control performance and operating characteristics. Therefore, VSC-HVDC is applicable to power transmission for passive remote loads, capacity increase of urban grids through AC cable networks, wind power interconnect to power systems, power transmission to islands or offshore platforms, interconnection with weak AC power systems, transmission to highly changeable loads, and so on.

With the development of power electronic technologies and increase of the voltages and currents of fully controlled large-power power electronic devices, the DC voltages, and transmission power of VSC-HVDC transmission systems will increase gradually. The valves in parallel and series connection with VSC bridge arms will reduce correspondingly, thus reducing loss rate of the switching power of converters to overcome the disadvantage of large losses of VSC-HVDC transmission systems. VSC-HVDC transmission systems can regulate the reactive power exchanged with AC power systems quickly and independently, which leads to technical advantages such as the dynamic characteristics of STATCOM and SSSC. VSC-HVDC transmission systems are expected to be used more in power systems for the purposes of regional power system interconnection, interconnection of asynchronous power systems, improvement of power angle stability and voltage stability of power systems, and so on.

7.3.3 A Comprehensive Comparison of VFTs and DC Transmission Systems

With similar functions, VFTs and DC transmission systems can both be used for the interconnection of asynchronous grids, but their technical routes are quite different. Before comprising the characteristics of their technical routes, conclusions of some studies about specific foreign projects are introduced next. Before installing a VFT at the Laredo Substation, AEP did numerous studies and analyses for different solutions about a VFT, a conventional DC back-to-back transmission system, and a VSC back-to-back transmission system for the power system (Figure 7.19). The studies and analysis mainly include steady-state analysis, stability analysis and operation reliability analysis, and a comparison of the characteristics of different types of devices [29, 30] (Figure 7.20).

The main conclusions of the study report finished in 2006 were:

1. According to analysis result about the stability in the fault state, a 150 MW DC back-to-back transmission system could meet the load level need of up to 2008; a 150 MW flexible DC back-to-back transmission system could meet the load level need of up to 2010; and a 100 MW VFT could meet the load level need of up to 2010. Obviously, even a small-capacity VFT can meet the need.
2. According to result of the black-start study, a DC back-to-back transmission system needs synchronous condensers while a VFT can support isolated loads.
3. According to analysis result about the influences on power systems, both a DC back-to-back transmission system and a flexible DC back-to-back transmission system possibly result in subsynchronous resonance for neighboring generator units, making it necessary to take the corresponding control strategy on the DC side and install corresponding protection actions on the generator side to prevent subsynchronous resonance; thanks to its larger rotor inertia, a VFT can avoid subsynchronous resonance.
4. According to the harmonic analysis, a conventional DC back-to-back transmission system needs many filters, a flexible DC back-to-back transmission system needs a few filters, and a VFT does not need filter because it generates no harmonics.

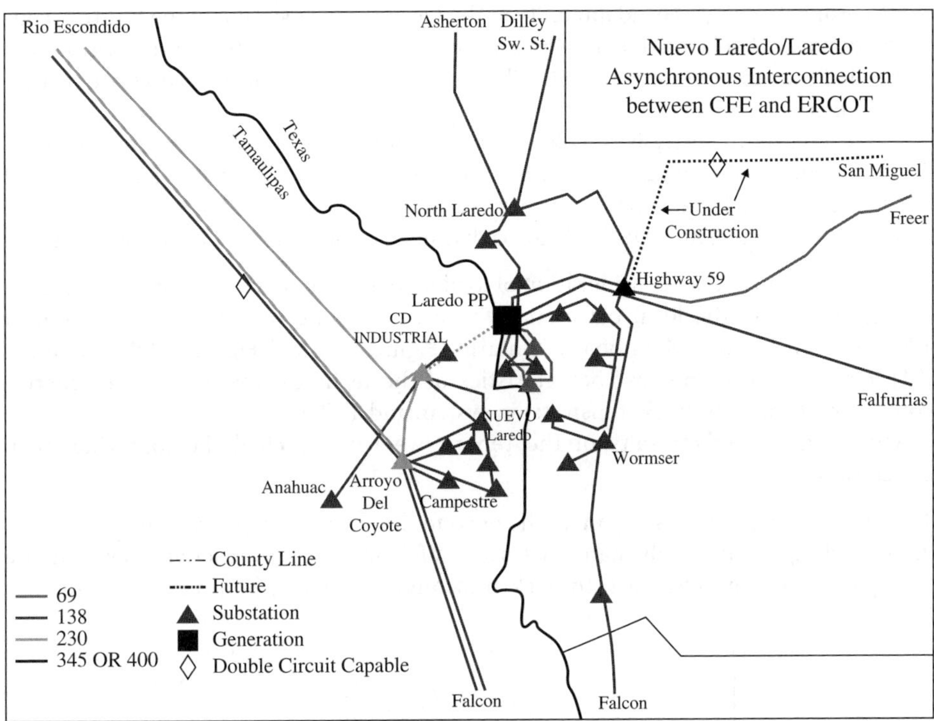

Figure 7.19 Wiring diagram of the power system of AEP of in Laredo [30].

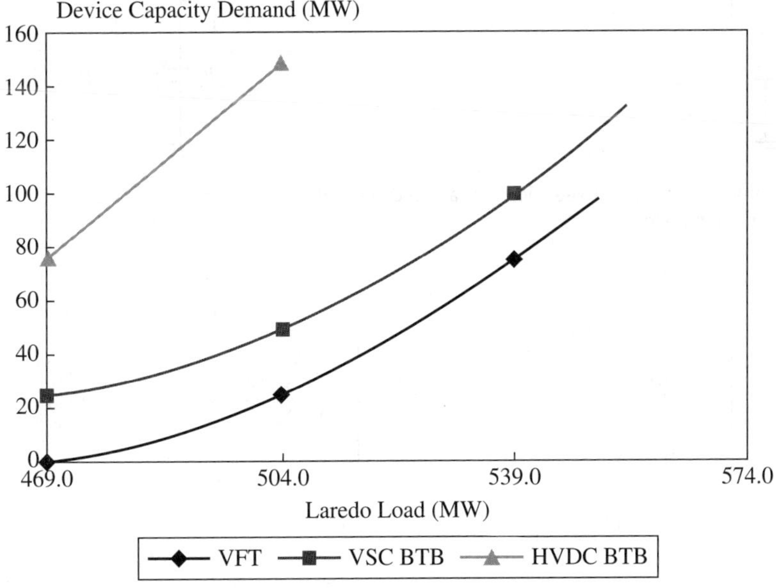

Figure 7.20 Transient stability limits in different transmission technologies [30].

5. According to analysis of adaptability in the Laredo power system in the future, a conventional DC back-to-back transmission system cannot meet the future development need of the power system, but a flexible DC back-to-back transmission system and a VFT can.

6. According to the analysis of voltage support of the Laredo power system , in level years 2007–2009, voltage support by STATCOMs was needed when a conventional DC back-to-back transmission system was used, but such a voltage support was not needed when a flexible DC back-to-back transmission system or a VFT was used.

Likewise, in an overall consideration of load increase and quick development of clean energy sources in the local place and with a background of power transmission of 1000–2000 MW from Manitoba to Ontario (Figure 7.21 and Figure 7.22), document [31] analyzes and compares characteristics of the technical routes of a capacitative converter DC back-to-back transmission system and a VFT.

According to study results of the report, we have reached the following main conclusions:

1. Both the two proposals can meet the need for large-capacity power exchange.

2. According to the steady-state analysis, a VFT absorbs less reactive power than a capacitative converter back-to-back transmission system; however, the difference

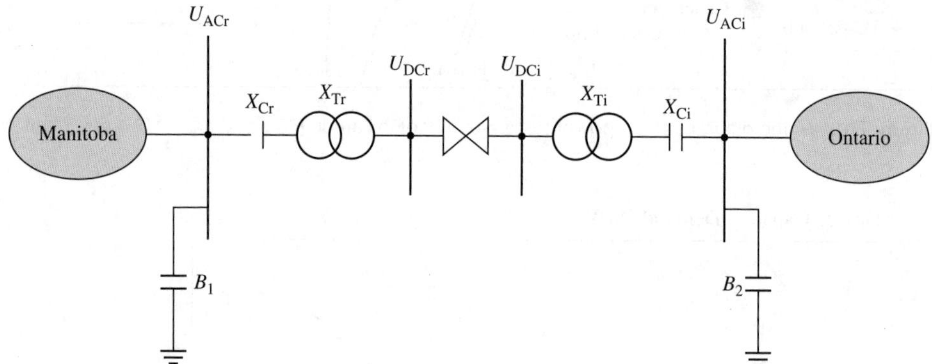

Figure 7.21 Grid interconnection between Manitoba and Ontario through a capacitative DC back-to-back transmission system.

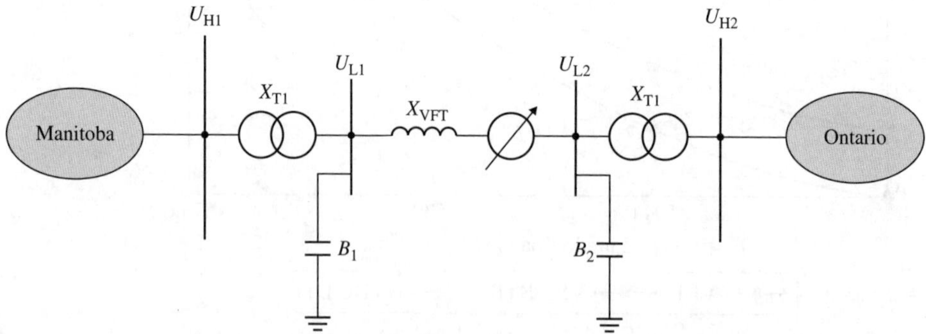

Figure 7.22 Grid interconnection between Manitoba and Ontario through a VFT.

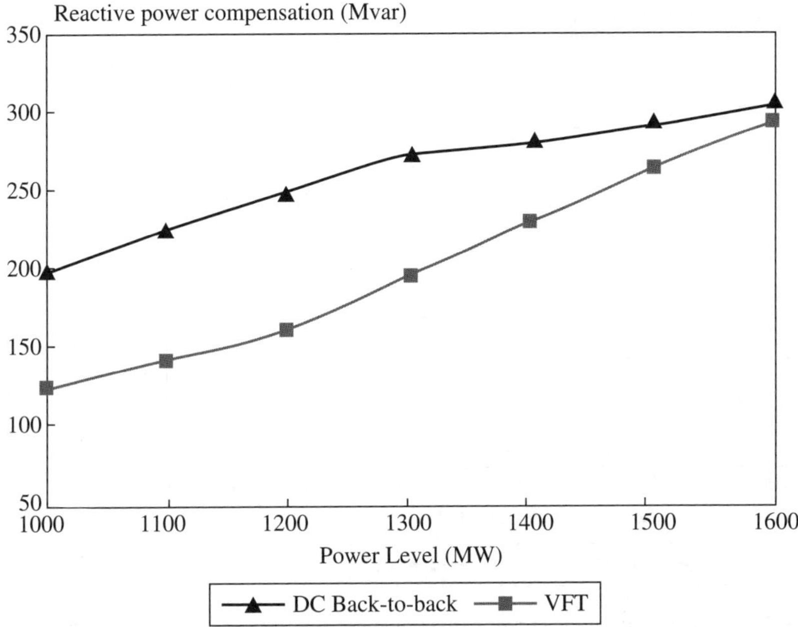

Figure 7.23 Reactive power consumption curves of the power system based on different transmission power.

decreases with increase of the transmission capacity. See Figure 7.23 and Figure 7.24 for the specific changes.

3. According to the dynamic analysis, in a fault and its recovery, a VFT has an initial response speed quicker than a capacitative converter back-to-back transmission system; a capacitative converter back-to-back transmission system recovers more quickly to the state before disturbance than a VFT. See Figure 7.25. In power step control, a capacitative converter back-to-back transmission system has a quicker regulation speed, but a VFT has a better natural damping effect. See Figure 7.26. Furthermore, a VFT control system can be optimized to shorten the duration of power step response to about 400 ms.

According to the analysis of the characteristics of VFTs and DC back-to-back transmission systems in specific projects and grid planning, VFTs are more advantageous. Generally speaking, both VFTs and DC transmission systems (including DC back-to-back transmission systems) are now important technical means for the asynchronous interconnection of grids: both of them can realize asynchronous interconnection of grids and stepless regulation as well as bilateral power transmission for active power flows, but they have their own characteristics in operating principles, technical realization, function indexes, and so on. Related contents are summarized as follows in accordance with the studies herein and related documents [1–32]:

1. Viewed from their operating principles, a VFT realizes spatial synchrony between the resultant magnetic field generated by its rotor winding current and the resultant magnetic field generated by its stator winding current, with help of the low-speed mechanical rotation of a rotary transformer. Stable and controllable inter-grid

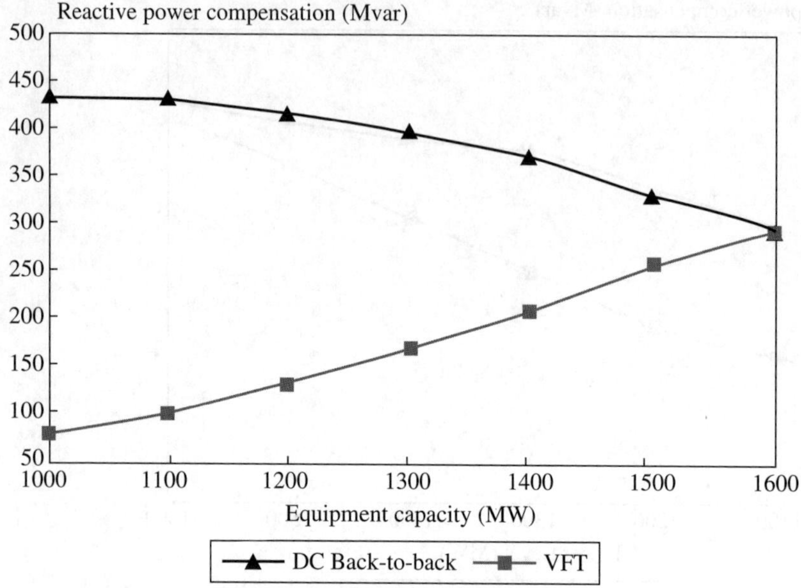

Figure 7.24 Reactive power consumption curves of the power system based on different equipment capacities.

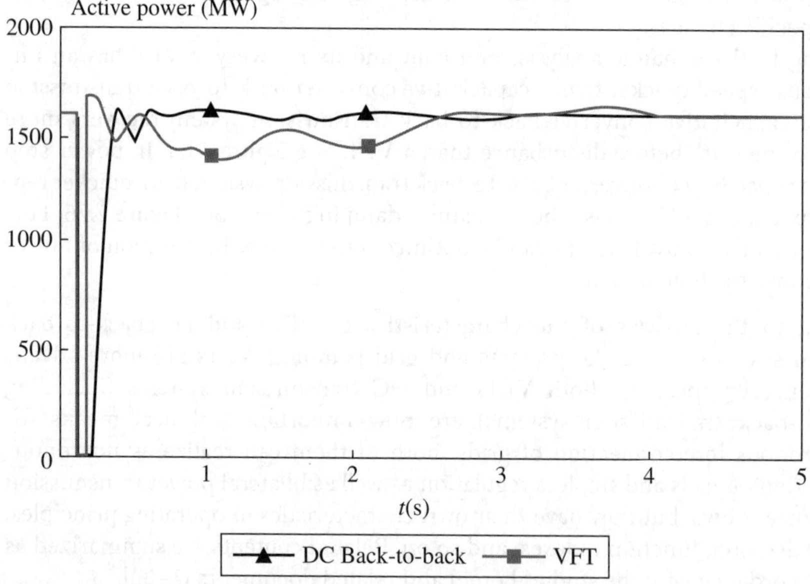

Figure 7.25 Characteristics of transmission power recovery of a VFT and a DC back-to-back transmission system in fault recovery.

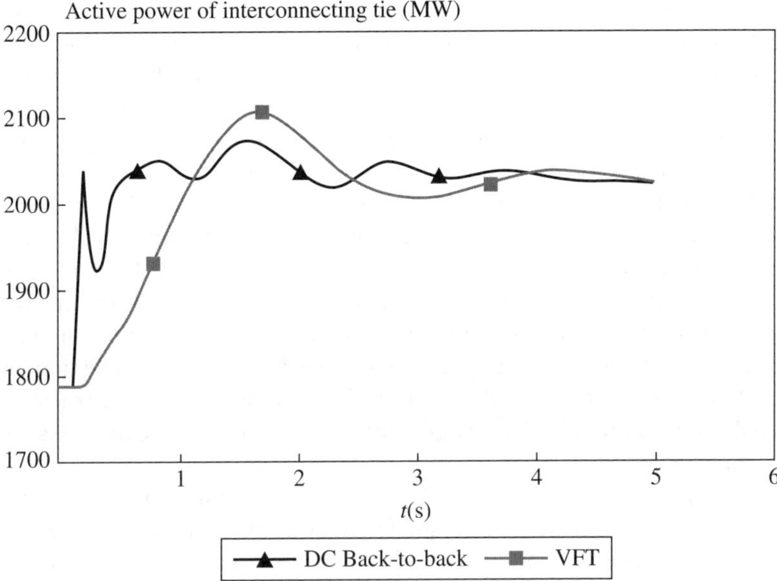

Figure 7.26 Power step response curve.

electromagnetic power exchange is realized by means of real-time control over phase shift of the two resultant magnetic fields. A DC back-to-back transmission system rectifies an AC into a DC through a power electronic device such as a rectifier, inverts the DC into an AC with the same or another frequency again on the other side, by means of control over DC current as well as voltage amplitude that realizes AC-DC-AC conversion and control for the power of two power systems. Obviously, a VFT has less intermediate operation.

2. Viewed from their technical structures, a VFT mainly uses conventional electrome-chanical technologies of turbine engines, transformers, DC motors, and so on and secondarily power electronic technologies such as rectification technologies as well as automatic control technologies; a DC back-to-back transmission system mainly uses power electronic technologies of thyristors, IGBTs, and secondarily AC/DC conversion control technologies. In spite of their later development than DC trans-mission systems, VFTs have conventional device and mature technologies.

3. Viewed from their response speed among their control performance, because of its large inertia, a VFT has a power step response time of hundreds of milliseconds, which is long enough to meet the control need of an ordinary power system but is much slower than a DC back-to-back transmission system where the power regu-lation speed is tens of milliseconds; viewed from their control ranges, a VFT has no dead zones in its design capacity range; a DC back-to-back transmission system has a limit for its minimum transmission power which is usually 10% of its rated power or, in other words, a DC back-to-back transmission system cannot operate in −10–10% of its rated current.

4. Viewed from their operation and maintenance, a VFT is a kind of conventional equipment with less components and a low maintenance and therefore has a relatively lower requirement for the quality of maintenance personnel. A VFT's

devices have a rich operational experience, high reliability, and longer lives. A DC back-to-back transmission system has more components including many devices for purposes such as valve cooling, overvoltage protection, and filtering leading to more serious aging of power electronic devices, so it has a high maintenance requirement and a comparatively higher requirement for the quality of technicians.

5. Viewed from their capacity expansion, a VFT has a unit design capacity of 100–300 MW; a DC back-to-back transmission system has a unit design capacity of several thousands of megawatts. A VFT is a modular equipment, which is an easier candidate for capacity expansion by means of a simple parallel connection of old and new equipment, so it has high adaptability for power systems already constructed and capacity can increase gradually; a DC back-to-back transmission system can expand its capacity by means of methods such as parallel connection or series connection, but it needs overall design in the beginning of its construction because its old and new modules have large mutual influence (e.g., the filter design needs to be adjusted greatly).

6. Viewed from reactive compensation, a VFT can not only transmit active power but also exchange reactive power between systems and, during operation, it (including the step-down transformers on its two sides) absorbs reactive power, which is about 40% of the transmission power; a DC back-to-back transmission system can transmit active power only and, during operation, its two sides need AC filter banks with the same capacity as the transmission power. Thus, the reactive compensation capacity necessary for a DC back-to-back transmission system is 2–2.5 times of that necessary for a VFT.

7. Viewed from their active power losses, the active power losses of a VFT mainly include a no-load loss, a winding resistance loss, a rectifier circuit resistance loss, and a forced air cooling loss of a small actual value around 1%; the active power losses of a DC back-to-back transmission system mainly include a no-load and a winding resistance loss both of its converter transformer, a thyristor conversion loss (the IGBT of a VSC-HVDC transmission system has a larger loss), and a loss of its valve and transformer cooling system. The loss of an ordinary back-to-back converter station is about 1.5%, while for a VSC-HVDC transmission system it is above 3%.

8. Viewed from their costs, GE said the two 100 MW VFTs developed for Québec have a cost similar to the cost of a 200 MW DC back-to-back transmission system of the same capacity. However, it is believed that VFTs will have lower costs with extensive application.

9. Viewed from its overload capacities, a VFT has an overload capacity mainly depending on its thermal stability and enabling it to operate for more than 30 minutes when the power is 105 MW; a DC back-to-back transmission system has an overload capacity depending on the overload capacities for its converter valve, converter transformer, and so on. Relatively speaking, a VFT has a larger short-time overload capacity while a DC back-to-back transmission system has a smaller overload capacity and a shorter overload operation time [32].

10. Viewed from land occupation, studies show that a group of 100 MW VFTs cover 2400 m^2 (30 × 80 m) while a DC back-to-back transmission system of the same capacity covers an area that is 1–2 times larger.

11. Viewed from their influences on power systems, a VFT generates no harmonics according to the principle; a DC back-to-back transmission system generates certain harmonic component in conversion and inversion and therefore needs many filters to avoid endangering the safe operation of the generators and the system equipment, and it may result in a risk of subsynchronous resonance for neighboring generators in specific cases due to its controller and many capacitors.
12. Viewed from fault elimination, a VFT is a big inertial body with great adaptablity to faults and a quick recovery speed after them; a DC back-to-back transmission system may suffer from converter commutation failure or even bipolar locking in faults of an AC power system. A VFT has a large short-circuit impedance and its short-circuit current in a fault is 1.5–2.5 times of the rated current, generating no large influence on the power system.
13. Viewed from their system applications, a VFT can realize the interconnection of strong power systems and power transmission for weak power systems or even passive power systems as well, and serve as a black-start power source; a DC back-to-back transmission system needs great voltage support for the power systems on two sides connected by it. In spite of its lower requirement for reactive power support, a VSC-HVDC transmission system has much higher operational loss and construction cost.

In general, VFTs and HVDC back-to-back transmission systems have their own characteristics and apply to different occasions. Generally speaking, a VFT can be used for a power system with a small transmission capacity less than hundreds of thousands of kilowatts, a lower requirement for the system regulation response speed and a system with weak power source; a DC back-to-back or a VSC-HVDC transmission system can be used for a power system with a large transmission capacity, a higher requirement for the quick response speed, while DC back-to-back needs strong power sources on both sides. With the application of multiple VFTs, some power users have begun to use VFTs as an important means for power system interconnection in future. Table 7.2 compares the main performance of a VFT and DC back-to-back transmission systems of different types.

7.4 Summary

1. In the power transmission and power flow control of synchronous grids, VFTs have functions similar to power phase-shifting transformers and obvious advantages including stepless regulation of a large phase-shifting angle in a large range (0–360°), no harmonics, great damping characteristics, and so on.
2. In the interconnection and power transmission of asynchronous grids, VFTs have functions similar to DC back-to-back transmission systems and advantages including no dead zones of regulation, no subsynchronous resonance risk, great damping characteristics, a simple system structure, a long life, and so on. However, VFTs have a smaller transmissions scale and a slower regulation speed than DC back-to-back transmission systems but a smaller operation loss than flexible DC transmission systems.

Table 7.2 Comparison of main performance between a VFT and a DC back-to-back transmission system.

Type	Technical Route	Regulation Characteristic	Influence on Power System	Loss	Applicable Field
VFT	The knowledge of kinematic mechanics and electromagnetism are combined to realize AC-AC conversion.	Without dead zones and with a step response speed slower than 400 ms	Without harmonics or a subsynchronous resonance risk	<1%	Interconnection of asynchronous grids, particularly weak grids or even passive grids
Conventional DC back-to-back transmission system	AC-DC-AC conversion is realized by means of the leakage reactances of converter transformers and the rectification and inversion of thyristor valves.	With dead zones and with a step response speed of tens of milliseconds	With harmonics, a subsynchronous resonance risk, and necessity to install filters and other related protective devices	About 1.5%	Interconnection of asynchronous grids with great voltage support and large-capacity power exchange
Capacitative converter DC back-to-back transmission system	AC-DC-AC conversion is realized by means of the capacitances of series capacitors and the rectification and inversion of thyristor valves.	With dead zones and with a step response speed of tens of milliseconds	With harmonics, a subsynchronous resonance risk, and necessity to install filters and other related protective devices	About 1.5%	Interconnection of asynchronous grids, particularly weak grids
Flexible DC back-to-back transmission system	AC-DC-AC conversion is realized by means of the leakage reactances of converters, the voltage source converters of fully controlled devices and PWM technology.	Without dead zones and with a step response speed of tens of milliseconds	With a few high-frequency harmonics, a subsynchronous resonance risk, and necessity to install filters and other related protective devices	About 3%	Interconnection of asynchronous grids, particularly weak grids

Thanks to their excellent regulation performance, VFTs have become another alternative technology for the interconnection of asynchronous grids following the DC transmission system technology. They are important in power systems and have been used in some regions in North America.

References

1 X. Xiaorong, J. Qirong. *Principle and Application of Flexible AC Transmission Systems*. Beijing: Tsinghua University Press, 2006.

2 P. Yonglong, W. Renzhou, L. Zhuo. Phase shifter control for transient stability of power system. *Journal of North China Electric Power University*, 1997, 24(2).

3 Z. Fang, Z. Hongguang, L. Zenghuang, et al. The influence of large power grid interconnected on power system dynamic stability. *Proceedings of the CSEE*, 2007, 27(1): 1–7 [in Chinese].

4 W. Peng, J. Yanchao, L. Zhuo. Static phase shifter based on multi bridge PWM AC control technique. *Proceedings of the CSEE*, 1999, 19(6).

5 Z. Liangsong, H. Huijun, P. Bo. Experimental research on the application and dynamic model of phase shifter in transient stability. *Journal of Central China University of Science and Technology*, 1997, 25(2).

6 X. Shizhang. *Electrical Machinery*. Beijing: Machinery Industry Press, 1988.

7 E. Larsen, R. Piwko, D. McLaren, et al. Variable-frequency transformer, e.g. a new alternative for asynchronous power transfer. Presented at *Canada Power, Toronto, Ontario, Canada*, 2004

8 H. Dayu. First "face to face" type of interconnecting device for asynchronous power grids and its remarkable significance. *Electric Power*, 2004, 37(10): 4–7 [in Chinese].

9 H. Dayu. A new type of "face to face" interconnecting device for asynchronous power grids based on VFT. *Electrical Equipment*, 2007, 8(7): 116–118 [in Chinese].

10 D. Nadeau. A 100-MW variable frequency transformer (VFT) on the Hydro-Québec TransÉnergie Network: The behavior during disturbance. *Power Engineering Society General Meeting*, 2007.

11 J.J. Marczewski. VFT applications between grid control areas. *Power Engineering Society General Meeting*, 2007.

12 P. Marken, J. Roedel, D. Nadeau, et al. VFT maintenance and operating performance. *Power and Energy Society General Meeting: Conversion and Delivery of Electrical Energy in the 21st Century*, July 2008.

13 M. Spurlock, R. Oel Spu, D. Kidd, et al. AEP's Selection of GE Energy's variable frequency transformer (VFT) for Their grid interconnection projection between the United States and Mexico, *The North American Transmission & Distribution Conference, Montreal, Canada*, 2006.

14 C. Gesong. *Modeling and Control of Variable Frequency Transformer for Power Systems*. D. Phil Thesis, Beijing: China Electric Power Research Institute, 2010.

15 Zhejiang University. *HVDC Transmission*. Beijing: Water Conservancy and Electric Power Press, 1991.

16 S. Boliang. Modern HVDC transmission technology and HVDC valve technology. *High Voltage Technology*, 2003, 39(6): 47–50.

17 P.B. Michael, K.J. Brian. The ABCs of HVDC transmission technologies. *IEEE Power & Energy*, 2007, 5(2): 32–51.

18 S. Richard. HVDC options today, an underused and undervalued solution. *IEEE Power & Energy*, 2007, 5(2): 93–96.

19 G. Asplund. Application of HVDC Light to power system enhancement. *Power Engineering Society Winter Meeting*, 2000.

20 C.V. Thio, J.B. Davies, K.L. Ken. Commutation failures in HVDC transmission systems. *IEEE Transactions on Power Delivery*, 1996, 11(2): 946–957.

21 Z. Yi, Y. Xianggen. Comparative research on HVDC and UHV power transmission. *High Voltage Technology*, 2001, 27(4): 44–46.

22 Y. Yong. High voltage DC transmission technique and its future application. *Electric Power Automation Equipment*, 2001, 21(9): 58–60.

23 Z. Nanchao. Role of HVDC transmission in the power system development in China. *High Voltage Technology*, 2004, 30(11): 11–12.

24 C. Qian. *The operation and control of a new type of multi terminal HVDC systems.* Dissertation. Southeast China University, 2004.

25 W. Jun, Z. Yigong, H. Minxiao, et al. HVDC based on voltage source converter: A new generation of HVDC technique. *Power System Technology*, 2003, 27(1): 45–51.

26 H. Zhaoqing, M. Chenxiong, L. Jiming, et al. A new generation of HVDC technique – HVDC light. *Transactions of China Electrotechnical Society*, 2005, 20(7): 12–16.

27 S. Ruihua. *Research on Modeling and simulation and control strategy of new HVDC transmission system.* Dissertation. Beijing: China Electric Power Research Institute, 2007.

28 W. Xijun, B. Hailong, Y. Jun. High-voltage DC Flexible technology and its demonstration. *Engineering, Distribution & Utilization*, 2011, 28(2).

29 J. Adams, H. Chao, C. Custer, et al. Planning for uncertainty: NYISO planning process and smart grid. *Power and Energy Society General Meeting: Conversion and Delivery of Electrical Energy in the 21st Century*, 2008.

30 P. Hassink, V. Beauregard, R. O'Keefe, et al. Second and future applications of stability enhancement in Ercot with asynchronous interconnections. *Power Engineering Society General Meeting*, 2007.

31 B. Bagen, D. Jacobson, G. Lane, et al. Evaluation of the performance of back-to-back HVDC converter and VFT for power flow control in a weak interconnection. *Power Engineering Society General Meeting*, 2007.

32 P. Doyon, D. McLaren, M. White, et al. Development of a 100 MW variable frequency transformers, presented at *Canada Power, Toronto, Ontario, Canada*, 2004.

8

Summary and Prospects

8.1 Overview

VFT is a flexible AC transmission system connected in series, as well as a kind of new smart grid equipment. With functions such as stepless regulation of transmission power, suppression of the low-frequency oscillation of large power systems, and frequency regulation for power systems, it can realize asynchronous interconnection between grids with different frequencies, power transmission to weak grids and passive power systems, optimize and regulate the transmission power and distribution proportions of parallel transmission channels, and use a power system on one side as the black-start power source for another power system on the other side. Promotion and application of this technology helps to solve existing problems that may effect the safety of interconnected large grids such as low-frequency oscillation and power flow transfer, improve the transmission capabilities of interconnected grids, make grids more flexible in operation control, and power systems safer and more stable, which plays an important role in constructing strong, smart, and efficient modern large grids.

In this book, the basic theories, mathematical model, simulation method, control system, operating characteristics, and technical and economical characteristics of VFTs have been studied systematically. On this basis, the basic equations, simulation calculation models, control strategy, and block diagrams of VFTs were proposed; a simulation module for VFTs was established in the universal power system analysis software; a digital simulation study done with help of simplified power systems, typical power systems, and practical large power systems; the good performance of VFTs was verified; and the technical and economical characteristics of VFTs, phase-shifting transformers, and DC transmission systems were analyzed and compared.

8.2 Main Conclusions

8.2.1 Structure of a VFT System

Based on both kinematic mechanics and electromagnetism, and integrating technologies such as transformers, phase-shifting transformers, water turbine generators, doubly fed motors, DC drive, and grid interconnection, VFTs are efficient equipment for power transmission and control. A VFT mainly includes a rotary transformer, a DC drive motor, and a collector ring. The rotary transformer is for generating a rotating

Variable Frequency Transformers for Large Scale Power Systems Interconnection: Theory and Applications,
First Edition. Gesong Chen, Xiaoxin Zhou, and Rui Chen.

magnetic field in synchronous operation; the DC drive motor is for generating a torque for driving the rotary transformer rotor to rotate; the collector ring is for connecting the rotor winding with an external power system. Viewed from its spatial structure, the rotary transformer, DC drive motor, and collector ring of a VFT are usually arranged vertically, longitudinally, and coaxially along the down-up direction, which guarantees a large mechanical inertia, high stability, and so on.

A VFT system includes a VFT, a DC rectification, and drive control device, step-down transformers, a reactive compensation capacitor bank, a circuit breaker, and so on. The VFT is for the power exchange between asynchronous grids; the DC rectification and drive control device provides a DC power source and a control signal for driving DC drive motor of the VFT to rotate; the step-down transformer is mainly for decreasing a high power system voltage to the related voltage of the VFT; the reactive compensation capacitor is mainly for compensating leakage reactance of the VFT and the reactive power absorbed by the step-down transformers; the circuit breaker is for switching on or off the VFT and the compensation capacitor and connecting or disconnecting interconnected grids.

8.2.2 Operating Principle of a VFT

A rotary transformer is the core part of a VFT. Control over its motion states (including speed and phase) is the key to realize functions of a VFT. Viewed from its operating principle of a VFT, rotation of the rotary transformer is used to dynamically compensate frequency difference of the power systems on two sides of the VFT, and realize stepless regulation for equivalent phase difference of the voltages on the two sides, which form a rotating magnetic field with an adjustable phase shift and the same frequency in the magnetic gap space of the VFT and realize dynamic stepless regulation for transmission power of the VFT.

Viewed from its electrical characteristics, a VFT can be roughly regarded as an equivalent electrical circuit connecting an ideal phase-shifting transformer with an adjustable range from 0° to 360° and an equivalent reactor in series. According to the circuit law, the transmission power of a VFT is in direct proportion to both product of the voltage amplitudes on two sides and sine value of equivalent phase difference of the voltages on two sides and inversely proportional to the equivalent reactance. Regulation of phase shift of the phase-shifting transformer can realize the power regulation objective and expand the interconnected power systems' stable operation areas within the power limit.

8.2.3 Simulation Technologies of VFTs

The application of modern new power technologies is usually based on numerous in-depth simulation studies. A VFT simulation model established through universal power system analysis software is important for better study and application of the new technology. Based on an in-depth analysis of basic structure and static, dynamic, and electromagnetic transient characteristics of a VFT, this book reaches a power flow steady-state equation, an electromechanical transient equation, and an electromagnetic transient equation for a VFT. These equations describe the basic relations of the main parameters of a VFT such as rotor speed, transmission power, driving torque, voltage, and current of the main circuit, ignition angle of the control drive circuit and basic principles of interconnection, power transmission, and flexible operation control of a VFT, thus theoretically proving that VFTs can make interconnected power systems more stable.

In order to perform studies on the application of VFTs in power systems, the power flow model, electromechanical transient model, electromagnetic transient model, and short-circuit current calculation model for VFTs have been studied and established in this book with help of program functions such as PSASP, EMTP, and PSCAD/EMTDC and realized in corresponding programs. These simulation models can realize seamless connection with other power system component models with the help of existing power analysis software, effectively simulate the operating characteristics of VFTs in different power system conditions and faults because of their great adaptability, and provide effective simulation methods for the studies, design, optimization, and regulation of VFTs in power systems.

8.2.4 Control Technologies of VFTs

As power transmission equipment for asynchronous grids, VFTs are typical multi-input multi-objective control objects, so are flexible and have efficient control over VFTs with help of the kinematic mechanics, the electromagnetic conversion law, and the grid operation law, which is important for the safe and economical operation of grids. The design of control systems is the key to realize the functions of VFTs.

To meet the operational control need of VFTs, we propose a three-level control system including component-level control, device-level control, and system-level control for VFTs, establish a main control block diagram based on methods such as PID and determine the corresponding parameters in this book. According to the simulation results, these models have great robustness and can adapt to different power system conditions and operation modes, and effectively realize main functions such as torque control, speed control, voltage phase regulation, transmission power control, start and switching, reactive voltage control, low-frequency oscillation suppression, power system frequency regulation, and power flow optimization for VFTs to realize the main control objectives of VFTs.

8.2.5 System Characteristics of VFTs

As for the connection of a VFT to a power system, it is necessary to study the VFT's functions, its actual role in the power system, and its possible impacts on related equipment of the power system. Thus, it is necessary to analyze the operating characteristics of a VFT in different conditions systematically.

To meet the grid development need in China, we establish a fully digital simulation system for power systems with different scales (including VFTs) with help of the developed VFT simulation model and control system, the recommended VFT model parameters and software such as PSASP, EMTP, and PSCAD/EMTDC and perform digital simulations regarding power flow, stability, and electromagnetic transient under different conditions and for different operations and faults in this book. The simulation result verifies effectiveness of the developed VFT simulation model and control system and proves that the good control and regulation performance of VFTs can realize functions such as interconnection, power transmission, flexible control, and characteristic improvement for asynchronous grids. A VFT has a power step response time of about 400 ms mainly because of its large rotational inertia. The response time can meet the need for power control and regulation and make power systems more stable. Furthermore, VFTs need no additional preventive actions because they generate no new harmonics for power systems or subsynchronous oscillation risk for neighboring generator units.

8.2.6 Low-Frequency Oscillation Suppression by VFTs

Low-frequency oscillation is now an important influential factor for the safe operation of interconnected power systems. After a VFT is connected to a power system, its damping of low-frequency oscillation of the power system is a great factor that is an important reason for the application of VFT technology. Furthermore, the way to control low-frequency oscillation of the sections of a whole power system with help of local signals is an important issue.

To solve the low-frequency oscillation between large interconnected power systems, an adaptive low-frequency oscillation damping control method based on VFTs and the Prony method is proposed in this book. In the case of any structural change of a power system, this method can identify low-frequency oscillation mode and damping characteristics of the power system automatically, optimize and regulate the damping controller parameters and effectively suppress low-frequency oscillation of the synchronous interconnected power system, providing a new control means for the low-frequency oscillation of power systems. This method is highly adaptive because it has passed the simulation verification of typical four-generator power systems and practical complicated power systems.

8.2.7 Technical and Economic Characteristics of VFTs

Technical and economic advantages are important factors for the promotion of new equipment. An in-depth analysis on the technical and economical characteristics of VFTs is an important content for the promotion of VFTs. VFTs have functions similar to those of phase-shifting transformers when used in synchronous grids or to DC back-to-back transmission systems when used in asynchronous grids. In this book, the advantages of VFTs, phase-shifting transformers, and DC back-to-back devices are compared from their principles, technical characteristics, economical characteristics, and so on.

According to document studies and overall analysis, compared with a phase shifting transformer, a VFT has advantages such as extensive stepless regulation of phase-shifting angles, no harmonics, good damping characteristics, application to the interconnection of asynchronous grids, and so on; compared with a DC back-to-back transmission system, a VFT has advantages such as a simple structure, fewer components, low loss, high reliability, less land occupation, simple maintenance, a small need for reactive power configuration, high system adaptability, easy expansion, stepless regulation, no dead zones of operation, and no need for additional voltage support, making it obviously technically and economically advantageous when the transmission power is less than hundreds of thousands of kilowatts, or in power transmission to weak grids or passive power systems.

8.3 In-Depth Studies of VFTs

VFTs can realize bilateral power transmission and stepless regulation for asynchronous grids, improve the control and regulation capabilities, improve safety and stability of strong power systems, interrupt the influence of faults on interconnected grids effectively, suppress the low-frequency oscillation of power systems, realize basic functions such as frequency regulation, power flow optimization, oscillation damping, power support, powering on islands for power systems, and have marked technical and economic advantages, showing a prospect for the interconnection of large grids. Especially in China, VFTs can be used to realize the marginal interconnection of asynchronous grids and the electromagnetic looped network uncoupling of synchronous grids, improve the connection reliability of wind power, and so on. The theoretical equations, simulation model, control system, and simulation tools of VFTs have been studied systematically in this book. The further studies focus on the following aspects.

8.3.1 Structure Design of VFTs

Due to professional limit, no in-depth studies about the physical structure, magnetic field distribution, and winding design of VFTs have been covered in this book. However, these are the key to realizing the functions of VFTs. In future studies, it is therefore necessary to strengthen the application of methods such as the magnetic field theory and the finite element method, study the structure design and material technologies of VFTs, and propose more accurate equations and more efficient structure design proposals based on consideration of factors including resistance loss, flux leakage, and so on.

8.3.2 Control Study of VFTs

With many control functions and many mutually included control objectives, a VFT has a complicated control system. In this book, the basic control over VFTs is mainly realized through conventional PID control technology. At present, there are many power system control theories, which are important references for the control study of VFTs. In future studies, it is necessary to use new control theories actively according to the characteristics of VFTs and the practical application need, promote adaptability, and robustness of the control systems of VFTs and design the protection systems of VFTs.

8.3.3 Simulation Tools of VFTs

VFT is a kind of new equipment and a new technology as well, so the basic equations and basic simulation tools are developed and studied in this book. In future studies, it will be necessary to strengthen the study of the simulation tools of VFTs: improve and prefect VFT modules through the existing software functions, and develop corresponding simulation models in accordance with the needs of different simulation software to provide simulation technologies and platforms for related in-depth studies.

8.3.4 Application of VFTs in Projects

The pilot work in VFTs is critical for technology improvement. In future studies, it will be necessary to study the pilot work in projects according to the development needs of grids in China and other countries in order to improve the application technologies of VFTs and to provide a new solution for the interconnection of large grids that solves some practical problems, and make the power system become more adaptive and friendly.

Appendix

Application of VFTs in Projects

A.1 Overview

As a new technology for grid interconnection, VFTs have been used in the Langlois Substation in Canada, and the Laredo and the Linden Substations, both in the USA. For a comprehensive understanding of the design, construction, testing, and application of VFTs, this Appendix includes a study on the VFT documents publicly published on magazines, papers, and conferences, and a summary of the structure design, engineering demonstration, construction, and application of VFTs in these documents is provided for reference.

A.2 Main Structure and Systematic Control of a VFT

A VFT system includes a rotary transformer, a DC motor, a DC rectifier, a step-down transformer, a capacitor bank, and a circuit breaker. Figure A.1 shows the wiring diagram of a VFT system designed by GE for the Langlois Substation in Québec, Canada.

A VFT mainly includes a rotary transformer, a DC drive motor, and a collector ring. This is a structure simpler than that of a DC back-to-back transmission system. Figure A.2 shows the rotating part of a VFT. Figure A.3 is the sectional view of a VFT. Figure A.4 shows the outside view of a VFT.

As the core part of a VFT, a rotary transformer includes a rotor winding, a stator winding, and a core. Figure A.5 shows the relation between the rotor and stator winding of a VFT. Figure A.6 is an exterior view of the stator of a VFT.

As the electrical part of a VFT connecting its rotor winding and an external power system, a collector ring includes a three-phase collector ring, a brush, and a bushing. Figures A.7–A.9 show the related pictures.

For good performance, each VFT needs to be tested before delivery. For easy operation, the tests are usually done based on a transverse layout. Figures A.10 and A.11 are the circuit diagram and test scene, respectively, of a VFT delivery test.

A VFT is a large piece of power equipment and requires complicated installation. Figure A.12 shows a scene of VFT stator installation.

The nerve center of a VFT is the control and protection system and Figure A.13 shows a schematic diagram of control and protection in a VFT.

Variable Frequency Transformers for Large Scale Power Systems Interconnection: Theory and Applications,
First Edition. Gesong Chen, Xiaoxin Zhou, and Rui Chen.

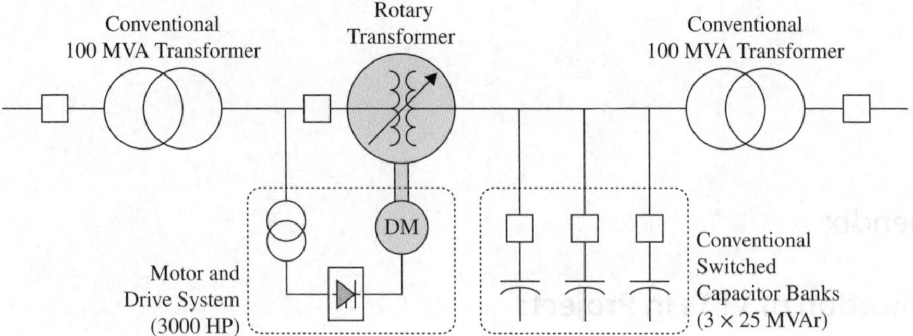

Figure A.1 Line diagram of a VFT system [1].

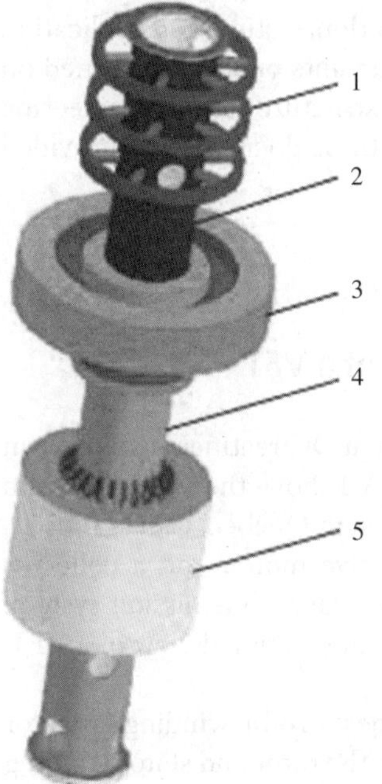

Figure A.2 Rotating part of a VFT [3]. 1 = collector ring; 2 = collector ring (upper) shaft; 3 = drive motor rotor (armature); 4 = rotor shaft (shaft); and 5 = rotor core and winding.

A.3   The World's First VFT Station: Langlois Substation

The first VFT in the world, the Langlois Substation in Québec, Canada, was put into operation in October 2003 to interconnect with the grid in New York. Figure A.14 shows its wiring diagram [1].

The VFT included a 105 MVA/17 kV rotary transformer, a 3000 HP DC motor, a variable-speed drive system, three switched capacitor banks with a respective capacity of 25 MVA, and two 120/17 kV conventional booster transformers. In phase II of the project, another VFT of the same capacity and other related equipment will

Figure A.3 Sectional view of a VFT [1].

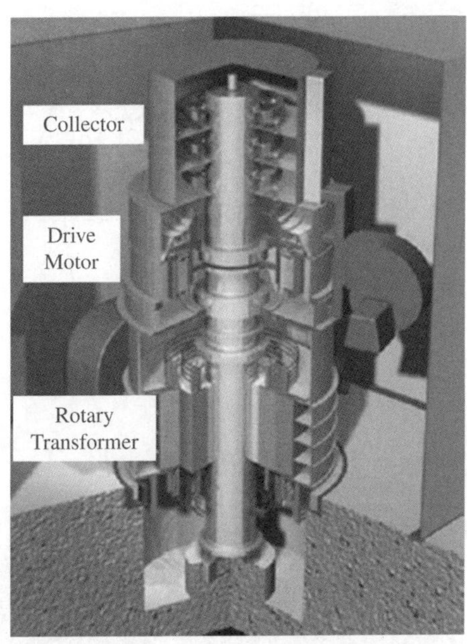

Figure A.4 Outside view of a VFT [1].

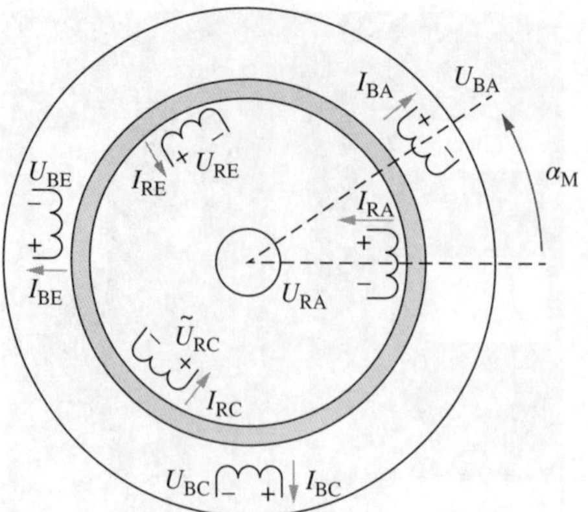

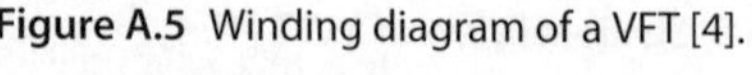

Figure A.5 Winding diagram of a VFT [4].

Figure A.6 Outside view of a VFT stator [4].

be provided. Figure A.15 is the long-term layout of the project and Figure A.16 is the scene of the project.

According to the overall and systematic tests and studies done by GE and the local employer for the first VFT for interconnecting the grids in New York and Québec, the good characteristics and performance of VFTs were verified again. Figure A.17 gives the results of the power step response test for the VFT.

In Figure A.17, curve 1 shows step commands of the VFT; curve 2 shows the actual transmission power of the VFT; and curve 3 shows the relative angle between stator

Figure A.7 Assembly drawing of the collector ring of a VFT [3].

Figure A.8 Connection diagram of the brush and collector ring of a VFT [5].

Figure A.9 Outside view of the collector ring of a VFT [5].

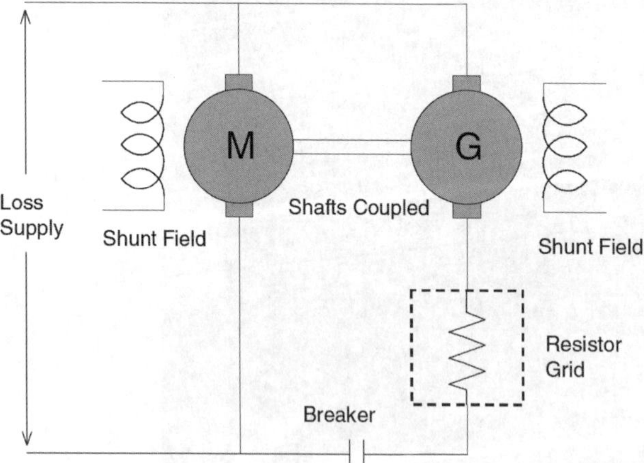

Figure A.10 Circuit diagram in the delivery test of a VFT [2].

Figure A.11 Delivery test scene of a VFT [2].

Figure A.12 Stator installation of a VFT [6].

and rotor of the VFT. The core equipment of the VFT, that is, a rotary motor, has great inertia and the water turbine units in the interconnected power systems also have great inertia, so the step response time of the VFT is approximately 0.4 s, which can meet the general control need but is slower than the response time of a HVDC.

Figure A.18 shows the course that power of the VFT of the Langlois Substation changes from 0 to 100 MW, −100 MW, and 0 in sequence. According to the figure, transmission power of the VFT can be regulated continuously and cross zero smoothly

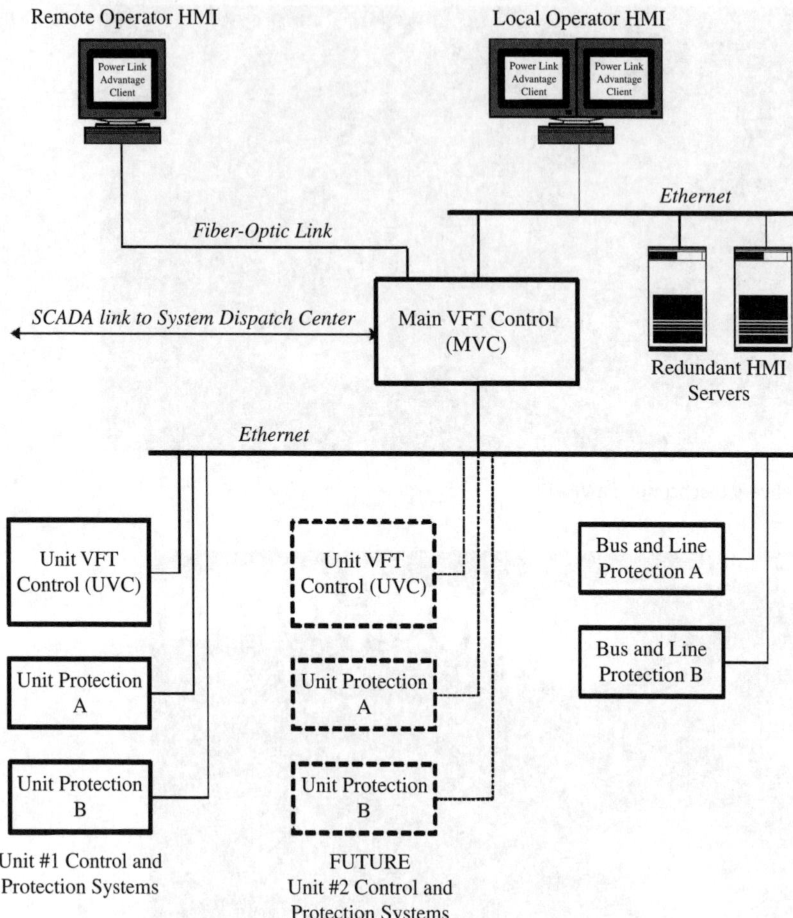

Figure A.13 Schematic diagram of the control and protection system in a VFT [1].

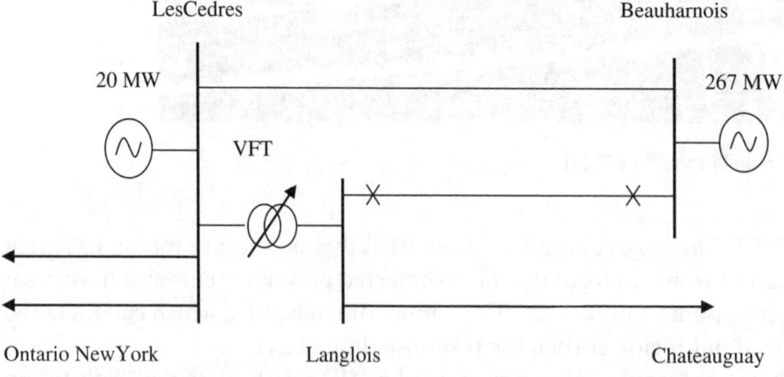

Figure A.14 The world's first VFT: The Langlois Substation in Québec, Canada.

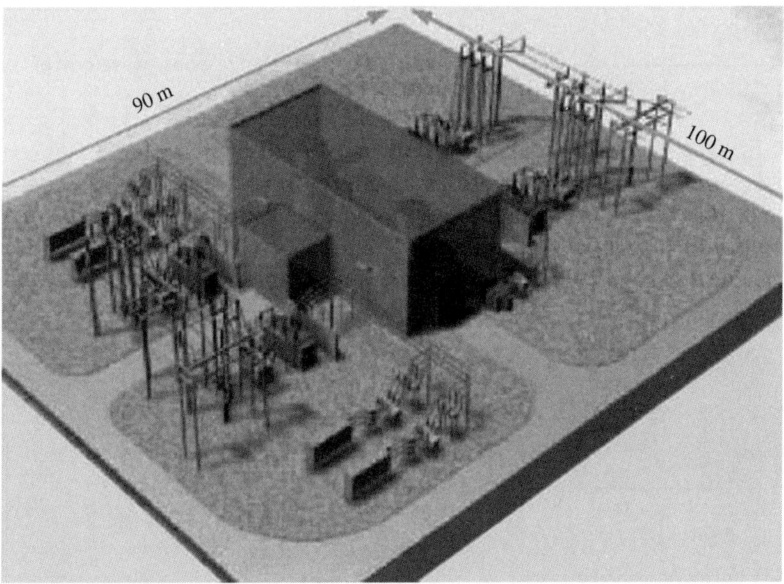

Figure A.15 Layout of the substation with the world's first VFT [1].

Figure A.16 Scene of the substation with the world's first VFT [7].

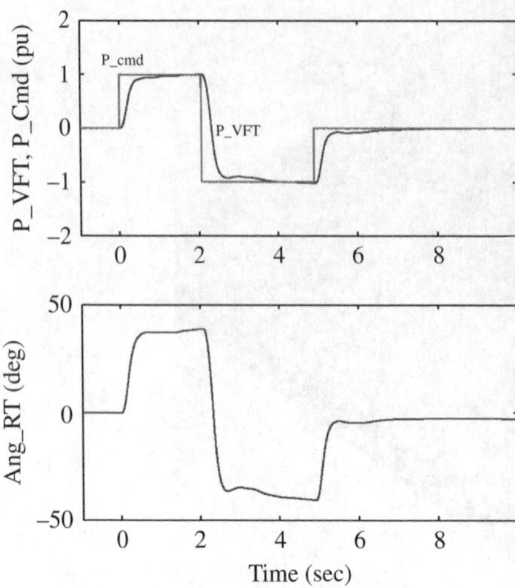

Figure A.17 Power step response waveform of the VFT [1].

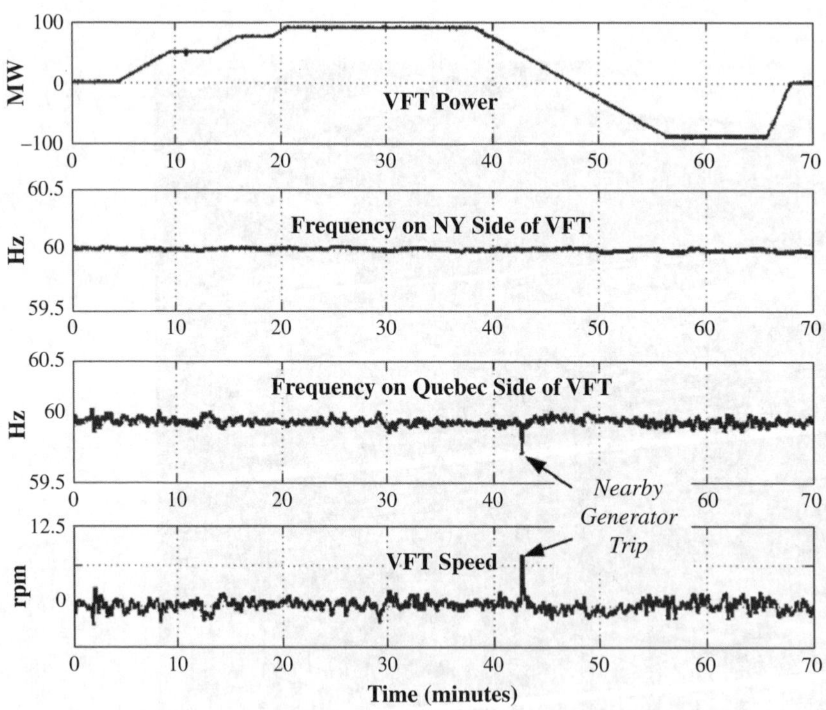

Figure A.18 Power transmission waveform of the VFT in the case of frequency changes on the two sides [1].

while the transmission power of a HVDC should be at least 10% of rated capacity of the HVDC and has a dead zone of operation between −0.1 and 0.1 pu. In addition, a unit of the grid in Québec tripped at the 42nd minute of the test and caused frequency distortion for the power side on the side, but transmission power of the VFT was not influenced greatly, however, it kept stepless regulation, which suggested that the VFT could regulate characteristics and withstand faults well.

On disturbance, the VFT could serve as a governor. See Figure A.19 for the characteristics. The figure indicates the reglation characteristic of the power systems (including the VFT) in the case of highly insufficient power of the power systems after a unit of the weak power system on one side tripped: After the unit with a transmission power of 45 or 35 MW was disconnected before the fault, the frequency of the power system on the side dropped greatly, so the VFT enabled its frequency regulation function to drop its transmission power to 10 MW and stabilize the frequency of the power system on the fault side at 59.2 Hz. According to the figure, the VFT could still power on the isolated grid with a fault of a large power loss to keep its frequency stable when the frequency of the isolated grid dropped to the dead zone.

Successful development and long-time operation tests of the world's first VFT further proved the adaptability of VFTs. VFTs have gradually gained interest from users and power system researchers. Some power enterprises and researchers have begun overall studies of the characteristics of VFTs.

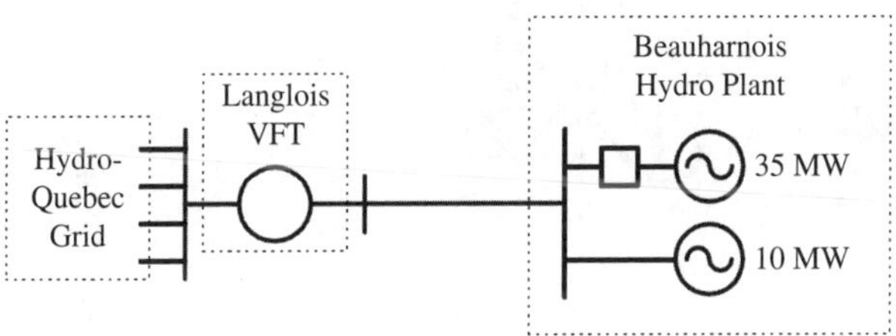

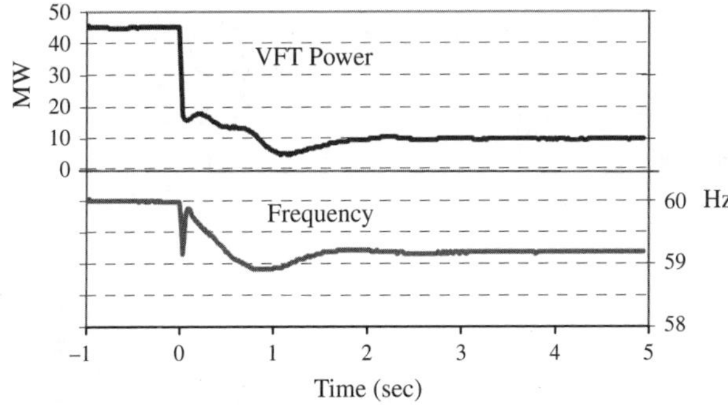

Figure A.19 Governor characteristics of the VFT [1].

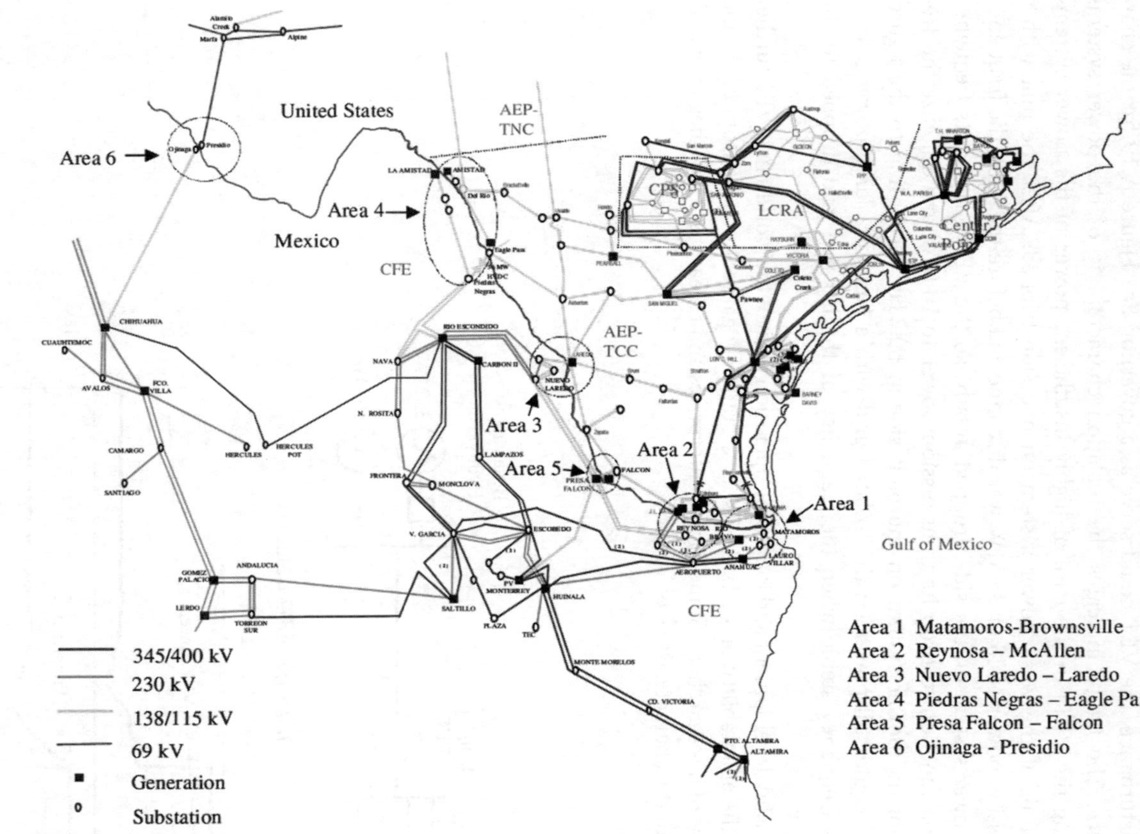

Figure A.20 Wiring diagram of the power systems of CFE and AEP [8].

A.4   The World's Second VFT Station: Laredo Substation

In consideration of the good performance of the world's first VFT in its practical tests and operation, VFTs have been used more in North America. According to the study and evaluation for the VFT in the Laredo Substation, VFTs are the most economical and adaptable interconnection mode for a region with a slow increase in power demand because they can solve poor power system stability due to the load increase from 2007 to 2010 well. VFTs are economical for the grids with a quick power load increase but a slow power source increase. Brownsville has such a power load increase.

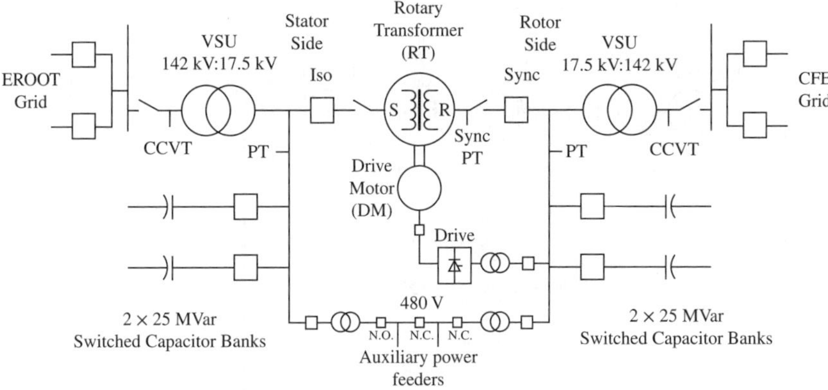

Figure A.21 Main wiring diagram of the VFT of the Laredo Substation [9].

Figure A.22 Outside view of the Laredo Substation [4].

The world's second VFT was financed by AEP, manufactured by GE, and installed in the Laredo Substation in southwest Texas to interconnect the grids of Texas and Mexico. The project was put into operation in 2007. The project included a 100 MW/17 kV rotary transformer, a 3750 HP DC motor, a variable-speed drive system, four switched capacitor banks with a respective capacity of 25 Mvar, and two 142/17.5 kV conventional generator booster transformers. Figures A.20–A.22 are, respectively, the wiring diagram of the power systems of CFE and AEP, the main wiring diagram of the VFT of the Laredo Substation, and the outside view of the Substation.

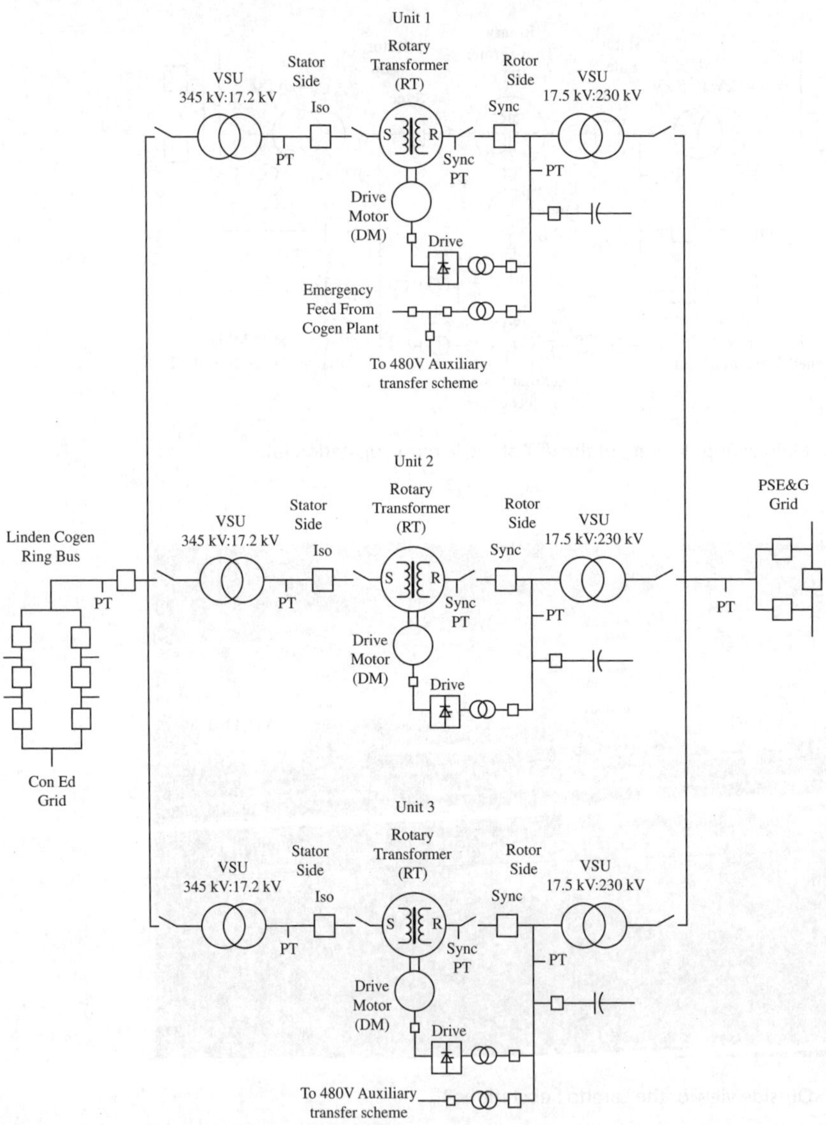

Figure A.23 Main wiring diagram of the Linden substation [10].

A.5   The World's Third VFT Station: Linden Substation

Based on numerous earlier studies and comparisons, three 100 MVA VFTs were installed between the PJM grid and the NYISO grid and put into operation in early 2010. The project had a compact structure and a small land occupation, which adapts to the problem of land resource shortage of upgrading of old grids. Figure A.23 shows the main wiring diagram of the Linden Substation; Figure A.24 shows the wiring diagram between the PJM grid and the NYISO grid; and Figure A.25 shows the outside view of the compact Linden Substation.

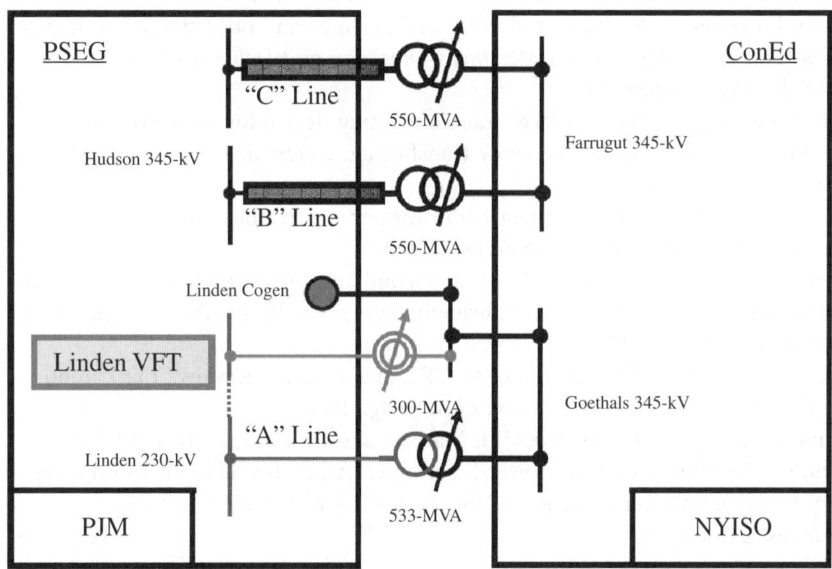

Figure A.24 Wiring diagram between the PJM grid and the NYISO grid [4].

Figure A.25 Outside view of the compact Linden Substation [2].

References

1 R.J. Piwko, E.V. Larsen, C.A. Wegner, Variable-frequency transformer – A new alternative for asynchronous power transfer, *Inaugural IEEE PES 2005 Conference and Exposition in Africa, Durban, South Africa*, 2005.

2 P. Truman, N. Stranges. A direct current torque motor for application on a variable frequency transformer. *Power Engineering Society General Meeting*, 2007.

3 A. Merkhouf, S. Upadhyay, P. Doyon. Variable frequency transformer: An overview. *Power Engineering Society General Meeting*, 2006.

4 P.E. Marken, J.J. Marczewski, D'Aquila R, et al. VFT-a smart transmission technology that is compatible with the existing and future grid. *Power Systems. PES'09*, 2009.

5 P. Marken, J. Roedel, D. Nadeau, et al. VFT maintenance and operating performance. *Power and Energy Society General Meeting: Conversion and Delivery of Electrical Energy in the 21st Century*, 2008.

6 H. Dayu. A new type of "face to face" interconnecting device for asynchronous power grids based on variable frequency transformer. *Electrical Equipment*, 2007, 8(7): 116–118 [in Chinese].

7 E. Paul. P.E. Marken. Variable frequency transformer – A simple and reliable transmission technology. *EPRI HVDC Conference*, 2009.

8 P. Hassink, V. Beauregard, R. O'Keefe, et al. Second and future applications of stability enhancement in ERCOT with asynchronous interconnections. *Power Engineering Society General Meeting*, 2007.

9 E.R. Pratico, C. Wegner, E.V. Larsen, et al. VFT operational overview: The Laredo Project. *Power Engineering Society General Meeting*, 2007.

10 E.R. Pratico, C. Wegner, P.E. Marken, J.J. Marczewski, First multi-channel VFT application –The Linden Project, Power Electronic/FACTS Installations to improve power system dynamic performance, at the *2010 IEEE PES T&D Conference, New Orleans*, 2010.

Index